AF480050

Environmental
BIOTECHNOLOGY

Environmental
BIOTECHNOLOGY

- Edited By -

Professor (Dr.) Arvind Kumar

2020
Daya Publishing House®

A Division of

Astral International (P) Ltd
New Delhi 110 002

Published by   :   **Daya Publishing House®**
A Division of
Astral International Pvt. Ltd.
– ISO 9001:2015 Certified Company –
4736/23, Ansari Road, Darya Ganj
New Delhi-110 002
Ph. 011-43549197, 23278134
E-mail: info@astralint.com
Website: www.astralint.com

Digitally Printed at   :   **Replika Press Pvt. Ltd.**

Dedicated

to

My Father

Late Shri Brahmadeo Sinha

A Noted Freedom Fighter, A Gandhian,
A Philosopher, A Humanitarian,
An Educationist &
An Environmentalist

PREFACE

Biotechnology encompasses all such biological processes, which have industrial or commercial application. Though, it appears to be a new discipline of bio-sciences but is as ancient as can be tracted back to the dawn of human civilization when the caveman learnt poisoning his arrows with YEW (Taxol used for treating cancer). During the course of development human necessities and curiosities progressively pave the way for industrial growth and development where biological processes are taken into cosideration for economic growth of country as well as protection and improvement of human health putting the promise research on the back burner. In fact, there may hardly be any biological process that does not have commercial utility, but one should have wisdom to differentiate between facts and fantasies. Some of the aspects of biotechnology that may be brought to fruition with existing technology and bioresource having down-to-earth are identified in the context of our nation.

The increasing awareness of biotechnological processes towards proper management and conservation of environment has inspired general mass to adopt eco-friendly approaches. It is thus become imperative to evolve strategies for rational exploitation of natural resources for human welfare without disturbing existing eco-balance of the environment.

Therefore, keeping in veiw of the importance of environment and biotechnology, I have ventured to bring out such a review volume with a hope to improve eco-friendly biotechnological strategies for the sake of socio-economic and environmental stability.

In this book, all 64 chapters are written by authors who are acknowledged experts in their respective fields, with the intention to provide sufficient depth of the subjects to satisfy the needs at a level, which will be comprehensive and interesting. The book shall be useful to the students, teachers, scientists and researchers in the field of Environmental biotechnocrats and other people with similar interest.

My special thanks and appreciation go to the scientists whose contributions have enriched this volume. I wish to express my sincere gratitude to Professor Dr. Indu Dhan, Hon'ble Vice Chancellor, S.K. University, Dumka who has been a source of constant inspiration. I am also grateful to Professor M.C. Dash, Vice Chancellor, Sambalpur University; Professor H.R. Singh, Pro-Vice Chancellor, University of Allahabad; Professor M.P. Singh, Pro-Vice Chancellor, V.B. University, Hazaribag; Professor N.C. Datta, Calcutta University; Professor Ajit Varma, J.N.U., New Delhi; Professor S.N. Kaul, NEERI; Professor

Tanmoy Bhattacharya, Midnapore University; Professor S.K. Konar, Kalyani University; Professor P.C. Mishra, Sambalpur University; Professor Sudhendu Mandal, Viswa Bharati University; Professor P.P. Sood, Saurashtra University, Rajkot; Professor B.D. Joshi, G.K. University, Hardwar; Professor A.K. Mittal, B.H.U., Varanasi; Professor U.S. Bagde, Mumbai University; Professor D.B. Tembhare, Nagpur University; Professor B.N. Pandey, Magadh University; Professor J. Ojha, Bhagalpur University; Professor S.P. Roy, Bhagalpur University; Professor M. Raziuddin, V.B. University; Professor M.V. Subba Rao, Andhra University; Professor P.S. Murthy, Bangalore University; Professor P. Natrajan, Kerala University; Professor S.P. Hosmani, Mysore University; Professor R. Ramani Bai, Chennai University and Dr. M.P. Sinha, Ranchi University who have been a source of continuous encouragement and inspiration to me.

I am especially thankful to Dr. Sikandar Prasad Yadav, Principal, S.P. College, Dumka for his encouragement. I am indebted to respected Shri Tribhuvan Podar of University Department of Zoology, Bhagalpur University, Bhagalpur for encouragement. I also acknowledge the incentives provided by one of my research scholars, Dr. Chandan Bohra, Head of Zoology, B.S.K. College, Barharwa for helping me actively in bringing out this book.

I also express my deep sense of gratitude to my parents whose blessings have always prompted me to pursue academic activities deeply. I am also thankful to my sweet wife, Professor Kumari Bimla and my two lovely sons, Kumar Pallav Shivashankaran and Kumar Prasun Ramakrishnan whose natural smiles extended to me relief all through this tiresome endeavour.

Last but not the least, I am also thankful to Mr. Anil Mittal, Proprietor of Daya Publishing House, Delhi for publishing this book. Finally, I will always remain a debtor to all my well-wishers for their blessings, without which this book would not have come into existence.

Dumka,

Prof. Arvind Kumar

LIST OF CONTRIBUTORS

A. Arun	P.G. Unit of Microbiology, Thiagarajar College, Madurai (Tamil Nadu), India
A. Bharani	Dept. of Environmental Sciences, Tamil Nadu Agricultural University, Coimbatore - 641 003
A. Hemasundaram	Department of Chemistry, S.V. Univesity, Tirupati-517502
A. Jeyasankar	Entomology Research Institute, Loyola College, Chennai-600 034, India
A. Sahul Hameedu	P.G. Unit of Microbiology, V.H.N.S.N. College, Virudhunagar, Tamil Nadu
A.J.A. Ranjit Singh	Dept. of Biology, Sri Paramkalyani College, Alwarkurichi, India-627 412
A.K. Patra	Regional Research and Technology Transfer Stations (OUAT) Chiplima, Sambalpur - 768025, Orissa
A.N. Kannappan	Department of Physics, Annamalai University, Annamalai Nagar-608 002
A.S. Shanware	Rajiv Gandhi Vikas Biotechnology Centre, L.I.T. Premises, Nagpur University, Nagpur-440 033 (M.S.) India
Anil Handa	Department of Mycology and Plant Pathology, Dr. Y.S. Parmar University of Horitculture and Foresty, Nauni, Solan - 173230 (H.P.)
Anitha Subash	Department of Biochemistry, Avinashilingam Deemed University, Coimbatore - 641043
Anjana	Post Graduate Department of Botany, (Microbiology Lab), Karnataka University, Dharwad-580003
Arun Handa	Department of Silviculture and Agroforestry, University of Horticulture and Forestry, Nauni, Solan (H.P.)-173230

Arvind Kumar	Environmental Biology Research Unit, Post Graduate Department of Zoology, S.K. University, Dumka-814 101
Avtar Singh	Deparment of Agronomy and Agromet, Punjab Agricultural University, Ludhiana-141 004 (Pb.)
B. Jayanti	Biochemistry Laboratory, Department of Bioscience, Barkatullah, University Bhopal, India
B. Kotaiah	Professor & Head, Dept. of Civil Engg., S.V.U. College of Engg., Tirupathi-517502
B. Sannappa	Department of Sericulture, University of Agricultural Sciences, GKVK, Bangalore-560 065
B.N. Pande	Dept. of Environmental Science, Dr. B.A. Marathwada University, Aurangabad-431004 (M.S.)
Bhupesh Gupta	Department of Mycology and Plant Pathology, Dr. Y.S. Parmar University of Horitculture and Foresty, Nauni, Solan - 173230 (H.P.)
C. Doreswamy	Sericulture College, Chintamani-563125
C. Jeyapaul	Dept. of Biology, Sri Paramkalyani College, Alwarkurichi, India-627 412
C. Maruthanayagam	A.V.C. College (Autonomous), Department of Zoology Mannampandal, Mayiladuthurai-609 305
C. Padmalatha	Dept. of Zoology, Rani Anna Govt. College for Women, Tirunelveli, India-627 008
C. Rathika	PG Unit of Microbiology & Zoology, Thiagarajar College, Theppakulam (PO), Madurai-09
C.N. Manoj	Department of Chemistry, Gandhigram Rural Institute-Deemed University, Gandhigram-624 302, Dindigul District, Tamil Nadu, India
C.V. Saraswat	Department of Silviculture and Agroforestry, Dr. Y.S. Parmar University of Horticulture and Forestry, P.O. Nauni (Solan)-173230 (H.P.)
Chandan Bohra	Ecological Research Laboratory, Department of Zoology, B.S.K. College, Barharwa-816 101 (Jharkhand)
Chandan See	Centre of Mining Environment, Indian School of Mines, Dhanbad-826004 (Jharkhand)
D. Augustine Selvaseelan	Dept. of Environmental Sciences, Tamil Nadu Agricultural University, Coimbatore - 641 003
D. Belsare	Biochemistry Laboratory, Department of Bioscience, Barkatullah, University Bhopal, India

D.L. Bharamal — Department of Zoology, Shri Pancham Khemaraj Maha Vidyalaya Sawantwadi - 416 510, Maharashtra, India

D.N. Singh — Regional Research and Technology Transfer Stations (OUAT) Chiplima, Sambalpur - 768025, Orissa

D.S. Cheema — Department of Vegetable, Punjab Agricultural University, Ludhiana-141004 (Punjab)

Dharmesh Gupta — Department of Mycology and Plant Pathology, Dr. Y.S. Parmar University of Horticulture and Foresty, Nauni Solan - 173230 (H.P.) India

Dr. S.K. Maiti — Centre of Mining Environment, Indian School of Mines, Dhanbad-826004 (Jharkhand)

G. Gandhi — Department of Human Genetics, Guru Nanak Dev University, Amritsar-143005

G. Sharmila — A.V.C. College (Autonomous), Department of Zoology Mannampandal, Mayiladuthurai-609 305

G. Suresh — Department of Chemistry, S.V. Univesity, Tirupati-517502

G.B. Shinde — P.G. Dept. of Biochemistry, L.I.T. Premises, Nagpur University, Nagpur-440 003 (M.S.) India

G.H. Pandya — Senior Scientist, National Environmental Engineering Research Institute, Nehru Marg, Nagpur-440 020

G.K. Thilaka — P.G. Unit of Microbiology, Thiagarajar College, Madurai (Tamil Nadu), India

Geeta B. Patil — Post Graduate Department of Botany (Microbiology Lab.) Karnatak University, Dharwad-580003

Gokulananda Patel — Department of Business Administration, Sambalpur University, Jyoti Vihar, Burla, Sambalpur-768019, Orissa

H.C. Lakshman — Post Graduate Department of Botany, (Microbiology Laboratory) Karnataka University, Dharwad - 580003 (India)

Hemandeep Kaur — Department of Vegetable, Punjab Agricultural University, Ludhiana-141004 (Punjab)

I.V. Ramana Reddy — Assoc. Professor, Dept. of Civil Engg., S.V.U. College of Engg., Tirupathi-517502

J. Arjun — Department of Zoology, Lumding College Lumding-782447

J.M.L. Gulati — Regional Research and Technology Transfer Stations (OUAT) Chiplima, Sambalpur - 768025, Orissa

J.S. Kang — Deparment of Agronomy and Agromet, Punjab Agricultural University, Ludhiana-141 004 (Pb.)

Jasjit Singh — Department of Agronomy, Punjab Agricultural University, Ludhiana-141004 (Punjab)

K. Elumalai — Entomology Research Institute, Loyola College, Chennai-600 034, India

K. Sasikumar — Department of Botany, Annamalai University, Annamalainagar - 608 002, Tamil Nadu, India

K. Thillai — P.G. Unit of Microbiology, Thiagarajar College, Madurai, Tamil Nadu, India

K.S. Pant — Department of Silviculture and Agroforestry, Dr. Y.S. Parmar University of Horticulture and Forestry, P.O. Nauni (Solan)-173230 (H.P.)

K.T.K. Anandapandian — PG Unit of Microbiology & Zoology, Thiagarajar College, Theppakulam (PO), Madurai-09

L.A. Yeragi — Department of Botany, Institute of Science, Mumbai-400032, India

L.N Bhardwaj — Department of Mycology and Plant Pathology, Dr. Y.S. Parmar University of Horticulture and Forestry, Nauni, Solan-173230 (H.P.)

M. Das — Department of Zoology, Gauwahati University, Gauwahati-781014

M. Elayarajan — Department of Soil Science and Agricultural Chemistry, Tamil Nadu Agricultural University, Coimbatore-3

M. Sakthivel — P.G. Department of Zoology, Kamaraj College, Thoothukudi-628 003.

M. Srinivasan — CAS in Marine Biology, Annamalai University, Parangipettai-608502 (Tamil Nadu)

M.G. Sethuraman — Department of Chemistry, Gandhigram Rural Institute-Deemed University, Gandhigram-624 302, Dindigul District, Tamil Nadu, India

M.M. Mishra — Regional Research and Technology Transfer Stations (OUAT) Chiplima, Sambalpur - 768025, Orissa

M.S. Tiwana — Deparment of Agronomy and Agromet, Punjab Agricultural University, Ludhiana-141 004 (Pb.)

Mrs. Jaya Vikas Kurhekar — Department of Microbiology, Dr. Patangrao Kadam College, Sangli–416416, Maharashtra, India

N. Bimola Devi — Post Graduate Studies Centre, HRDRI, Canchipur-3

N. Kumar — Department of Human Genetics, Guru Nanak Dev University, Amritsar-143005

N. Raja — Entomology Research Institute, Loyola College, Chennai-600 034, India

N. Sangbanbi Devi Seram — Department of Botany, Imphal College, Imphal-1

N.K. Bohra — Arid Forest Research Institute, Jodhpur (Rajasthan)

N.S. Nagarajan — Department of Chemistry, Gandhigram Rural Institute-Deemed University, Gandhigram-624 302, Dindigul District, Tamil Nadu, India

N.V.S. Naidu — Department of Chemistry, S.V. Univesity, Tirupati-517502

P. Balaji — Post Graduate Department of Microbiology, V.H.N.S.N. College, Virudhunagar-626001

P. Jothimani — Department of Environmental Sciences, Tamil Nadu Agricultural University, Coimbatore-3

P. Prema — Post Graduate Department of Microbiology, V.H.N.S.N. College, Virudhunagar-626001

P.D. Thakur — Department of Mycology and Plant Pathology, Dr. Y.S. Parmar University of Horticulture and Foresty, Nauni Solan - 173230 (H.P.) India

P.R. Kumar — Indian Agricultural Research Institute, Regional Station, Katrain (Kullu Valley), Himachal Pradesh-175129

P.S. Dheenan — Post Graduate Department of Microbiology, V.H.N.S.N. College, Virudhunagar-626001

Pankar Panwar — Department of Silviculture and Agroforestry, University of Horticulture and Forestry, Nauni, Solan (H.P.)-173230

Prasad — Aquatic Biotechnology and Fish Pathology Lab., M.J.P. Rohilkhand University, Barelly-243006

Prof. (Dr.) M.H. Fulekar — Life Sciences, University of Mumbai, Kalina, Vidyanagari, Santacruz (E), Mumbai-98

Prof. U.S. Bagde — Life Sciences, University of Mumbai, Kalina, Vidyanagari, Santacruz (E), Mumbai-98

R. Govindan — Department of Sericulture, University of Agricultural Sciences, GKVK, Bangalore-560 065

R. Kaur — Department of Human Genetics, Guru Nanak Dev University, Amritsar-143005

R. Manivanan	Mathematical Modelling Centre, Central Water & Power Research Station, Pune-411024 (India)
R. Panneerselvam	**Department of Botany, Annamalai University, Annamalainagar - 608 002, Tamil Nadu, India**
R. Parvatham	Department of Biochemistry, Avinashilingam Deemed University, Coimbatore - 641043
R. Priya	Department of Chemistry, Gandhigram Rural Institute-Deemed University, Gandhigram-624 302, Dindigul District, Tamil Nadu, India
R. Rajaram	CAS in Marine Biology, Annamalai University, Parangipettai-608502 (Tamilnadu)
R. Rajesh Kumar	P.G. Unit of Microbiology, V.H.N.S.N. College, Virudhunagar, Tamilnadu
R.C. Rana	Department of Forest Products and Utilization, Dr. Y.S. Parmar University of Horticulture and Forestry, Nauni, Solan - 173 230 (H.P.)
R.K. Arora	Division of Entomology, Faculty of Agriculture, Sher-e-Kashmir University of Agricultural Sciences & Technology-J, Udheywalla, Jammu-180002
R.R. Kamdi	Rajiv Gandhi Vikas Biotechnology Centre, L.I.T. Premises, Nagpur University, Nagpur-440 033 (M.S.) India
R.T. Chaudhari	School of Environmental and Earth Sciences, North Maharashtra University, Jalgaon, 425001 (M.S.) India
R.V. Patil	Prof. & Head, Deptt. of Chemistry, Bharati Vidyapeeth's, Dr. Patangrao Kadam Mahavidyalaya, Sangli (M.S.)
Radha Munagala	Department of Biochemistry, Avinashilingam Deemed University, Coimbatore - 641043
Rajendra Gartia	Department of Statistics, Sambalpur University, Jyoti Vihar, Burla, Sambalpur-768019, Orissa
Rajendra Kumar	Senior Scientist, Division of Genetics, IARI, New Delhi-110012
Ramakrishna Naika	Department of Sericulture, University of Agricultural Sciences, GKVK, Bangalore-560 065
Ramakrishna Naika	Sericulture College, Chintamani-563125
Rekha Goyal	Reproductive Physiology Section, Department of Zoology, University of Rajasthan, Jaipur-302004

S. Chidambaram Pillai	Lecturer (S.G.) in Botany, P.G. Department of Botany, V.O. Chdambaram College, Tuticorin-628008
S. Dey	R.S.I.C. N.E.H.U. Shillong
S. Ignacimuthu	Entomology Research Institute, Loyola College, Chennai-600 034 India
S. Kerur	Post Graduate Department of Botany, (Microbiology Lab), Karnataka University, Dharwad-580003
S. Mahimai Raja	Dept. of Environmental Sciences, Tamil Nadu Agricultural University, Coimbatore - 641 003
S. Muthalagi	P.G. Department of Zoology, Kamaraj College, Thoothukudi-628 003
S. Muthukumaravel	PG Unit of Microbilogy, V.H.N S.N. College, Virdhunagar, Virdhunagar dt.
S. Sreedhar Reddy	Lecturer, Dept. of Civil Engg., Vellore Institute of Technology, Vellore-632014
S.A. Salgare	Department of Botany, Institute of Science, Mumbai-400032, India
S.C. Chowfla	Department of Mycology and Plant Pathology, Dr. Y.S. Parmar University of Horticulture and Foresty, Nauni Solan - 173230 (H.P.) India
S.D. Khambe	Head, Dept. of Microbiology, Miraj Mahavidyalay, Miraj (M.S.)
S.G. Ahmad	Associate Professor and Head, Dept. of Agri. Structures & Envirn. Engineering, College of Agricultural Engineering, J.N.K.V.V. Jabalpur-482004 (M.P.)
S.G. Nanaware	Department of Zoology, Shivaji University, Kolhapur-416004 Maharashtra, India
S.K. Aher	New Arts, Commerce and Science College Parner, Dist. Ahmednagar-414302 (M.S.)
S.K. Kaushal	College of Basic Sciences, CSKHPKV, Palampur-176062
S.R. Bamane	Reader in Chemistry, Deptt. of Chemistry, Bharati Vidyapeeth's, Dr. Patangrao Kadam Mahavidyalaya, Sangli (M.S.)
S.R. Singh	Post Graduate Studies Centre, HRDRI, Canchipur-3
S.R. Thorat	School of Environmental and Earth Sciences, North Maharashtra University, Jalgaon, 425001 (M.S.) India

S.S. Pande — Rajiv Gandhi Vikas Biotechnology Centre, L.I.T. Premises, Nagpur University, Nagpur 440 033 (M.S.) India

S.U. Meshram — Rajiv Gandhi Vikas Biotechnology Centre, L.I.T. Premises, Nagpur University, Nagpur 440 033 (M.S.) India

S.V. Thite — Shri Chhatrapati Shivaji College, Shrigonda, Dist. Ahmednagar (M.S.)

Seema Bhadauria — Microbiological Research Laboratory, Department of Botany, R.B.S. College, Agra - 282002 (U.P.)

Shashikala — Biochemistry Laboratory, Department of Bioscience, Barkatullah University Bhopal, India

Siddagangaiah — Agricultural Marketing and Co-operation, University of Agricultural Sciences, GKVK, Bangalore-560 065

Suresh C. Joshi — Reproductive Physiology Section, Department of Zoology, University of Rajasthan, Jaipur-302004

Usha Rana — College of Basic Sciences, CSKHPKV, Palampur-176062

Usha Sharma — Department of Mycology and Plant Pathology, Dr. Y.S. Parmar University of Horticulture and Forestry, Nauni, Solan-173230 (H.P.)

V. Arumugam — Department of Physics, Annamalai University, Annamalai Nagar-608 002

V. Mariappan — Department of Biology, Gandhigram Rural Institute, Deemed University, Gandhigram-626302 (Tamilnadu)

V.K. Kondawar — Senior Scientist, National Environmental Engineering Research Institute, Nehru Marg, Nagpur-440 020

Versha Rani — Microbiological Research Laboratory, Department of Botany, R.B.S. College, Agra - 282002 (U.P.)

Venkatesh Babu, J. — P.G. Unit of Microbiology, Srimad Andavar College, Trichy, Tamilnadu

Vinay Verma — Department of Animal Science, M.J.P. Rohilkhand University, Barelly-243006

W. Kaur — Department of Human Genetics, Guru Nanak Dev University, Amritsar-143005

Yogendra — Department of Animal Science, M.J.P. Rohilkhand University, Barelly-243006

CONTENTS

(xxii)

ENVIRONMENTAL BIOTECHNOLOGY

Edited By :   Professor (Dr.) Arvind Kumar

Published By :   DAYA PUBLISHING HOUSE

1

GREEN BIOTECHNOLOGY IN POLLUTION ABATEMENT : AN OVERVIEW

• *Arvind Kumar and Chandan Bohra**

Environmental Biology Research Unit, Post Graduate Department of Zoology,
S.K. University, Dumka-814 101
* Ecological Research Laboratory, Department of Zoology, B.S.K. College,
Barharwa-816 101 (Jharkhand)

Abstract

Population explosion, industrialization and urbanization are the factors which led to the problem of environmental pollution. Use of chemical fertilizers and pesticides have further degraded our natural resources by their adverse effects. Phytoremediation has been emerged as a eco-friendly technique for the sutainable development. This technology is called as "green technology". During phytoremediation, plants exclude the metals, extract and store them in their body or convert them into volatile form that can be released into air. Chelation, compartmentalization and biotrans formation are the mechanisms for heavy metal detoxification in plants. However, phytoremediation technique can be improved by using soil amendements, plant-microbe interaction and genetic engineering etc.

Introduction

Rapid industrialization and urbanization in the last century due to enormous technological advancements have led to the problem of environmental pollution. Indiscriminate and reckless use of chemical fertilizers and pesticides have affected the ecosystem considerably. Prolonged and over usage of chemicals on soil results in decline of soil fertility due to salt accumulation, water logging, deterioration of environment, human health hazards and poor sustainability of agricultural lands (Kumar and Bohra, 2002). The outcome of consumption and production activities is the generation of large amounts of waste and release of pollutants. The rapid degradation of the soil has poised a threat to the sustainability of the agricultural system. Over last few decades, the impact of science and technology on society and ecosystem has intensified the deterioration of biological wealth as well as ecosystem leading to loss of biodiversity and natural resources. Agriculture of forties, which was eco-friendly, has now become fully chemicalized with new farming technologies and its commercialization. Recent scientific studies have been oriented to find the cost-effective techniques for pollution abatement. In this response, the potential of biological organisms, *viz.*, microbes, algae, aquatic macrophytes and other plants, for detoxification of environmental pollutants has been well recognized (Atlas, 1981; Sakaguchi *et al.*, 1981; Varshney, 1983; Leahy and Colwell, 1990 and Kumar and Bohra, 2002). Use of micro-organisms for the removal of hazardous toxic pollutants from soil and ground water is called as bioremediation (Brown, 1995 and Adriano *et al.*, 1998). The process of cleaning toxic pollutants from soil and ground water as well as heavy metal contamination by the use of plants is called as phytoremediation (Cunningham and Ow, 1996; Arthur *et al.*, 1998 and Salt *et al.*, 1998). This technology is also called as "Green Technology".

The use of microorganisms for the removal of hazardous toxic pollutants from soil and ground water has been tried since long and have proved to be successful. Similarly, plants have also been known for their process of cleaning heavy metal contamination. During phytoremediation, plants exclude the metals, extract and store them in their body or convert them into volatile form that can be released into air. According to Salt *et al.* (1998) and Shrivastava and Shrivastava (1998), the use of plants for its own support system offers a cheap, renewable and promising method for sustainable development. *Thlaspi caerulesens* and Poplar are the most potential hyper accumulator plant for phytoremediation of toxic heavy metals. Several crop species has been also identified as phytoremedials, among which Indian mustard in the most promising. Terrestrial and aquatic plants offer a low cost, sustainable and renewable technique to remove the herbicide and pesticide pollution from soil and ground water for environmental support system. The idea of phytoremediation is not new one. Hundreds of years ago, Europeans first noticed that certain plants grow in abundance near natural deposits of zinc and nickel. It has been known that some plants readily absorb materials that can be quite dangerous to humans. A series of fascinating scientific discoveries combined with an interdisciplinary research has developed this idea into a cost effective and eco-friendly technology called phytoremediation. Thus, phytoremediation is the use of plants to make soil contaminants non-toxic and is often termed as bio-remediation, botanical-remediation, and green remediation etc. The idea of using rare plants which hyperaccumulate metals to selectively remove excessive soil metals was introduced by Chaney (1983), and gained public exposure in 1990.

Phytoremediation may be defined as a system in which certain plant working together with soil organisms, can transform contaminants into harmless and valuable forms. It is an affordable technology and is most useful when pollutants are within the root zone of plants (top three to six feet). It may be the only economically feasible technology for the sites where pollutants spread over a large area. It is an *in situ* approach. The process is relatively inexpensive because it uses the same equipment and supplies used in agriculture. For phytoremediation, it is important to select not only a plant which is capable of degrading the pollutant but also one that will grow well in the specific environment. The pollutants are phytotoxic and affect the normal physiological and genetic constituent of plants but certain plants are tolerant and accumulate toxic chemicals to overcome these problems through the process of phytochelation and metalothioneins in their cells. Phytochelation is a process that makes the plant capable of binding the metals in the form of peptides. Some plants are rich in antioxidants such as ascorbic acid, phenolic compounds, beta carotene etc and help them to scavenge the toxic compounds.

Process of Phytoremediation

Phytoremediation takes advantage of plants nutrient utilization processes to take in water and nutrients through roots, transpire water through leaves and acts as a transformation system to metabolize organic compounds such as oils and pesticides or they may absorb and bioaccumulate toxic trace elements including the heavy metals cadmium, lead and selenium. Based on these facts, phytoremediation is divided into following area:

Phytoextraction: In this method plants accumulate heavy metals in their roots and shoots which are later on, burnt up (Salt *et al.*, 1984 and Gleba *et al.*, 1999). Plants accumulating heavy metals in high concentration (> 1000 mg/g) at a rapid rate with high biomass production are called as hyperaccumulators. Italian serpentine plant (*Alyssum bertolonii*) and *Thalaspi caerulescens* are the examples some hyper-accumulator plants which accumulate heavy metals like Zn and Cd (Brown *et al.*, 1994 and 1995; Tolra *et al.*, 1996 and Robinson *et al.*, 1998). Zn can also be phytoextracted by oat, barley and Indian mustard (Ebbs and Kochian, 1998).

Rhizofiltration: Rhizo means root. In this method, plant roots absorb and consolidate heavy metals from effluents (Salt *et al.*, 1998) by adsorption, accumulation, translocation and precipitation. It is a cost-effective process and the plants performing this method are very fast growers with regards to root development. According to Ebbs and Kochian (1998), *Brassica juncea* has been reported to be a fast accumulator of Cd, Pb, Sr and Ni etc. It allows *in situ* treatment, minimizing disturbance to the ecosystem.

Phytostabilization: This process reduces the mobility of the contaminant and prevents migration to the ground water or air and it reduces bioavailability for entry into food chain. Some plants secrete malic and citric acid which in turn convert metals, like aluminium into its unavailable form. Hybrid poplar grasses are widely used for phytostabilization.

Phytovolatilization: It is the uptake and transpiration of a contaminant by a plant, with the release of the contaminant to the atmosphere from the plant. Lewis *et al.* (1966) was first to show that plant species are able to volatilize selenium. Evans *et al.* (1968) identified volatile selenium compound, released from selenium accumulator *Astragalus racemosus*, as dimethyl diselenide.

Phytoremediation of Heavy Metals from Soil: Plants respond to heavy metals toxicity in a variety of different ways. The general mechanisms for heavy metal detoxification in plants and other organisms include chelation, compartmentalization and biotransformation.

Chelation: It is a mechanism for heavy metal detoxification in plants by a ligand. Rauser (1999) reviewed the role of several ligands in plants. Two major classes of heavy metal chelating peptides exists in plants. These are phytochelatins and metallothioneins. The phytochelatins (PCs) are a group of metal binding polypeptides. PCs accumulate excess heavy metals in the vacuoles of plants and are considered to be indicators of metal stress. Phytochelatin metal complexes are actively transported into plant vocuoles by a group of organic solute transporters that are directly energized by ATP (Salt and Rauser, 1995).

Compartmentalization: Within cells Cd and phytochelatins accumulate in the vacuole and this accumulation appears to be driven by a Cd/H^+ antiport and an ATP-dependent PC transporter (Salt and Wagner, 1993; Salt and Rauser, 1995).

Biotransformation: Some, toxic metals can be reduced in plants by chemical reduction of element. It is likely that terrestrial plants biotransform arsenic.

Metal accumulating plants are invariably restricted to metalliferous soils found in different regions. *Thlaspi caerulescens* is a metallophyte which is able to accumulate Zn upto

4% in leaf dry matter (Tolra *et al.*, 1996). It is very useful in phytoremediation of metal contaminated soil. However, *Brassica juncea, B. napus* and *B. oleracea* exhibited moderately enhanced Zn accumulation (Ebbs *et al.*, 1997). *Salix* sp., *Betula pendula, Mint piperita, M. arvensis, Agrostis capillaris* and *Festuca rubra* are also reported to be useful for phytoremediation of Zn (Duncan *et al.*, 1995; Ebbs *et al.*, 1997; Punshen and Dickinson, 1997 and Zheljazkov *et al.*, 1999).

Ambrosia trifida, Mosla scabra, Commelina communis and *Lemna minor* etc are reported to be useful in the phytoremediation of copper. *T. caerulescens, A. trifida* and *Salix viminalis* etc are useful for Cd phytoremediation (Gregev *et al.*, 1997; Kang *et al.*, 1998 and Nedelkoska and Doran, 2000).

B. juncea is very useful in the phytoremediation of lead (Begonia *et al.*, 1998) Indian mustard seems to be the best known terrestrial species for selenium phytoremediation (Pilon Smits *et al.*, 1999). For arsenic and mercury contaminated soil, bioremediation by microorganisms are very useful.

Phytoremediation of Heavy Metals from Water: According to Brooks *et al.* (1998) aquatic phytoremediation system involve:

(i)   Floating water hyacinth (*Eichhornia crassipes*), and

(ii) Submersion of the rhizosphere of terrestrial plants in order to remove metal contaminants (rhizofiltration).

According to Nzengung *et al.* (1999), there are two phytoprocesses for the remediation of polluted water:

(a)   Uptake and phytodegradation of metals in branches and leaves of trees, and

(b)   Rhizodegradation.

Ali *et al.* (1999) investigated that plants and algae growing in water bodies accumulate appreciable amount of metals and some plant species such as *Polygonum amphibium* and *Potamogeton crispum* that accumulate heavy metals could be significant biomonitoring agents and useful for phytoremediation technology to restore water quality. Roots of *Salix* is the most efficient among all. According to Zhu *et al.* (1999), water hyacinth which is an aquatic floating plant is an important accumulator of Cd and Cr. It also helps in the phytoextraction of Se and Cu. Qian *et al.* (1999) identified many species for phytoremediation such as umbrella plant (*Cyperus alternifolius*) for Mn and Cr; water zinnia (*Wedelia trilobata*) and parrot's feather (*Myriophyllum brasiliense*) for Cd and Ni; smartweed (*Polygonum hydropiperoides*) for Cu and Pb; water lettuce (*Pistia stratiotes*) for Cu, Hg, As and Se and smartweed for Cr and Pb. From a phytoremediation point of view, smartweeds is probably the best plant species for removal of trace element from wastewater due to its faster growth and higher plant density. Pilon Smits *et al.* (1999) identified parrot's feather, *Juncus riphioides* and *Typha latifolia* etc as wetland species for Se phytoremediation.

Phytoremediation of Soil Polluted with Herbicides and Pesticides: Pradhan *et al.* (1998) identified three plant species *i.e.* Lucerne, switch grass (*Panicum virgatum*) and little bluestem grass (*Schizachyrium scoparium*) for phytoremediation of soils contaminated with polynuclear aromatic hydrocarbons (PAHs). Contamination by pentachlorophenol (PCP) can be removed by crested wheat grass (Miller *et al.*, 1998) and perennial ryegrass (*Lolium perenne*) (Ferro *et al.*, 1999). Felsot *et al.* (1997) has suggested land farming as a disposal method for pesticide-contaminated soil which could be made more efficient by amendments with different organic nutrients.

Strategies to Improve Phytoremediation

(a) Using Soil Amendments: To clean contaminated soil and water, use of metal accumulating plants is the most recent eco-friendly and cost effective technology. According to Blaylock *et al.* (1997) and Terry *et al.* (1998), the recent discovery of certain chelating agents and soil conditioners has greatly facilitated metal uptake by soil-grown plants. Soil conditioners may be used to provide necessary factors for successful phytoremediation. Animal manures, biosolids, composts, bulking agents and plants are some soil conditioners used in phytoremediation. Huang and Cunningham (1996) and Blaylock *et al.* (1997) suggested that Indian mustard has the capability to accumulate high concentrations of lead when soils are ammended with synthetic chelates, like EDTA. It facilitates Pb transport into xylem and increases its translocation to shoots. Bennett *et al.* (1998) has suggested that the efficiency of phytoextraction can be improved through agronomic practices such as fertilization. Ebbs *et al.* (1997) suggested when soil was ammended with Gro-Power (a commercial soil amendement that enhnaces soil structure and fertility), removal of Zinc by plant shoots was doubled. Huang *et al.* (1998) reported that Uranium phytoextraction can be increased by adding organic acids such as acetic acid, citric acid and malic acid etc.

(b) Plant-microbe Interaction: Siciliano and Germida (1998) suggested that plant bacteria association are important to degrade mixtures of mono and di-chlorinated benzoic acids. Plant soil microbe interaction plays an important role in the degradation of pentachlorophenol (PCP) by crested wheat grass (Miller *et al.*, 1998).

(c) Genetic Engineering: Seguin *et al.* (1998) suggested that advances in genetic engineering also opened the ways to enhance the bio-degradable and reclamation abilities of naturally occuring microorganisms and higher plants as well as to incorporate these useful genes in agricultural crops. Rugh *et al.* (1998) over expressed *Escherichia coli* gsh II gene encoding glutathione syathetase in Indian mustard to develop transgenic plants. This plant has now increased capacity to accumulate and tolerate heavy metals. Transgenic plants accumulated significantly more Cd than the wild type and showed enhanced tolerance to Cd at both seedling and mature plant stages.

Advantages and Limitations of Phytoremediation: Although phytoremediation has not been used extensively, yet, it has many advantages:

(i) It is low cost compared to present "mechanical" methods of soil remediation.

(ii) It is faster than natural attenuation.

(iii) It is passive and solar driven.

(iv) It can greatly reduce the amount of contaminated materials going to soil.

Still, phytoremediation is new and not fully developed. Followings are some of the limitations of phytoremediation:

(i) It is generally slower than most other treatment methods and is climate dependent.

(ii) It usually requires nutrient additions.

(iii) Some metals and other contaminant concentrations can be toxic to some plants.

(iv) More than one phytoremediation method may be required for mixed contaminant sites.

(v) To utilize agricultural machinery for planting and harvesting, the site must be large enough.

Phytoremediations has been used to clean up metals pesticides, solvents and landfill lechates. Although phytoremediation as a clean up technique is not yet widely applied, momentum for its use is expected to build, particularly in application niches where other technologies are less suitable or do not exist. This practice is increasingly used to remediate sites contaminated with heavy metals and toxic organic compounds. In the present era of information technology and biotechnology, both can be weaved together so that this "green" technology could be used to grow crops on contaminated soils and clean up the heavy metal contaminated sites for sustainable future. It is the right time to take the decision how to increase our agricultural productivity and at the same time try and replenish our resources for the days ahead. Unless the latest tools of science and technology are applied for sustainable development of agriculture, hunger will persist and the green revolution will fade. Therefore, research effort is must to see the applicability and feasibility of green biotechnology in form of phytoremediation techniques in Indian scenario to prevent environmental hazards and for reclamation of soil and water resources.

References

Adriano, D.C., Bollag, J.M., Frankenberger, W.T. and Sims, I.C. 1998. In : Bioremediation of contaminated soils (Eds. Adriano, D.C., Bollag, J.M. and Frankenberger, W.T.). p. 385.

Ali, M.B., Tripathi, R.D., Rai, U.N. and Singh, S.P. 1999. Chemosphere, 39 : 2171-2182.

Arthur, E.L., Coats, J.R., Kearney, P.C. and Roberts, T. 1998. In : Pesticide remediation in soils and water (ed. P.C. Kearney). Wiley International, New York, U.S.A. p. 381.

Atlas, R.N. 1981. Microbial degradation of Petroleum hydrocarbons : An environmental perspective. Microbial. Rev. 45 : 180-209.

Begonia, G.B., Davis, C.D., Begonia, M.F.T. and Gray, C.N. 1998. Bul. Env. Cont. Toxicol. 61 : 38-43.

Bennett, F.A., Tyler, E.K., Brooks, R.R., Gregg, P.E.H. and Stewart, R.T. 1998. In: Plants that hyper accumulate heavy metals. pp. 249-259.

Blaylock, M.J., Salt, D.E., Dushenkov, S., Zakharova, O., and Gushhman, C. 1997. Enhanced accumulation of Pb in Indian mustard by soil applied chelating agents. Environ. Sci. Technol. 31 : 860-865.

Brooks, R.R., Robinson, B.H. and Brooks, R.R. 1998. Plants that hyper accumulate heavy metals. pp. 203-226.

Brown, S.L., Chaney, R.L., Angle, J.S. and Baker, A.J.M. 1994. J. Env. Quality. 23 : 1151-1157.

Bronw, S.L., Chaney, R.L., Angle, J.S. and Baker, A.J.M. 1995. Soil. Sci. Soc. Am. J. 59 : 125-133.

Brown, K.S. 1995. Bio-Science, 45 : 579-582.

Chaney, R.L. 1983. Plant uptake of inorganic wastes. In : Land treatment of hazardous wastes (Eds. Parr, J. E., Marsh, P.B. and Kia, J.M.). Park Ridge, I. L. Noyes Data Corp. pp. 50-76.

Cunningham, S.D. and Ow, D.W. 1996. Promises and prospects of phytoremediation. Plant Physiol. 110 : 715-719.

Duncan, H.J. McGregor, S.D., Pulford, I.D., Wheeler, C.T., Brink, W.J. Bosman, R. and Arendt, F. 1995. In : Contaminated soil '95 (eds. Brink, W.J. and Bosman, R.) pp. 1187-1188.

Ebbs, S.D., Lascat, M.M., Brady, D.J., Cornish, R., Gorden, R. and Kochian, L.V. 1997. J. Env. Qual. 26 : 1424-1430.

Ebbs, S.D. and Kochian, L.V. 1998. Em. Sci. Tech. 32 : 802 - 806.

Evans, C.S., Asher, C.J. and Johnson, C.M. 1968. Isolation of dimethyl diselenide and other volatile selenium compounds from *Astragellus revemosus*. Aust. J. Biol. Sci. 21 : 13-20.

Felsot, A.S., Dzantor, E.K., Kruger, E.L., Anderson, T.A. and Coats, J.R. 1997. In : Phytoremediation of Soil and Water contaminants (Eds. E.L. Cruger, and T.A. Andarson) pp. 77-91.

Ferro, A.M., Rock, S.A., Kannedi, J.J.H. and Turner, D.L. 1999. Ent. J. Phytoremed. 1 : 289-306.

Gleba, D., Borsjuk, N.V., Borsjuk, L.G., Kneer, R., Poulev, A., Skarzhimskaya, M., Dushenkov, S., Lomgendera, S., Gleba, Y.Y. and Raskin, I. 1999. Proc. Nat. Acad. Sçi. U.S.A. 96 : 5973-5977.

Greger, M., Landberg, T. and Prost, R. 1997. Proc. Third International Conference on Bio-geo-chemistry and Trace Elements, Paris. France. pp. 505-511.

Huang, J.W.W. and Cunningham, S.D. 1996. Lead phytoextraction : species variation in lead uptake and translocation. New Phytol. 134 : 75-84.

Huang, W., Blaylock, M.J., Kapulnik, Y. and Ensley, B.D. 1998. Env. Sci. Tech. 32 : 2004-2008.

Kang, B.H., Shim, S.I., Lee, S.G., Kim, K.H. and Chung, I.M. 1998. Kor. J. Weed. Sci. 18 : 262-267.

Kumar, A. and Bohra, C. 2002a. A sorry state of coal fields in Jharkhand Pradesh with special reference to water pollution and its control for sustainable agriculture: A critical review. In : Environmental Pollution and Agriculture. (Ed. A. Kumar). A.P.H. Publishing Corporation, New Delhi. pp. 1-14.

Kumar, A. and Bohra, C. 2002b. Impact of environmental stress on the growth behaviour of water hyacinth, *Eichhornia crassipes* with speical reference to removal of pollutants. In : Ecology and Ethology of Aquatic biota. (Ed. A. Kumar), Daya Publiching House, Delhi. pp. 345-353.

Leahy, J.G. and Colwell, R.R. 1990. Microbial degradation of hydrocarbons in the environments. Microbiol. Rev. 54 : 305-315.

Lewis, B.G., Johnson, C.M. and Delwiche, C.C. 1966. Release of volatile selenium compounds by plants. Collection procedures and preliminary observations. J. Agril. Food. Chem. 14 : 638-640.

Miller, E.K., Boydston, J.A. and Dyer, W.E. 1998. ASAE Paper No. PNW 98-104. p. 7.

Nedelkoska, T.V. and Doran, P.M. 2000. Biotech. Bioeng. 67 : 607-615.

Nzengung, V.A., Wang, C.H., Harvey, G. and Wang, C.H. 1999. Env. Sci. Tech. 33 : 1470-1478.

Pilon Smits, E.A.H., Desouza, M.P., Hong, G., Amini, A., Bravo, R.C., Payabyab, S.T. and Terry, N. 1999. J. Env. Qual. 28 : 1011-1018.

Pradhan, S.P., Conrad, J.R., Paterck, J.R. and Sirvastava, V.J. 1998. J. Soil. Contm. 7 : 467-480.

Punshon, T. and Dickinson, M.N. 1997. New Phytologist. 137 : 303-314.

Qian, J.H., Zayed, A., Zhu, Y.L., Yu, M. and Terry, N. 1999. J. Env. Qual. 28 : 1448-1455.

Raugh, C.L., Senecoff, J.F., Meagher, R.B. and Merkle, S.A. 1998. Nat. Biotech. 16 : 925-928.

Rauser, W.E. 1999. Structure and function of metal chelators produced by plants : the case for organic acids, amino acids, phytin and metallothioneins, Cell. Biochem, Biophys. 31 : 19-48.

Robinson, B.H., Leblanc, M., Petit, D., Brooks, R.R., Kirkman, J.H. and Greggm, P.E.H. 1998. Pl. and Soil. 203 : 47-56.

Sakaguchi, T., Nakajima, A. and Horikoshi, T. 1981. Studies on the accumulation of heavy metals elements in biological systems. XVIII. Accumulation of molybdenum by green microalgae. Euro. J. Appl. Microbiol. Biotech. 12 : 84-89.

Salt, D.E. and Wagner, G.J. 1993. Cadmium transport across tonoplast of vesicles from oat roots. Evidence for a Cd^{2+}/H^+ antiport activity. J. Biol. Chem. 268 : 12297-12302.

Salt, D.E. and Rauser, W.E. 1995. MgATP-dependent transport of phytochelatins across tonoplast of oat roots. Plant Physiol. 107 : 1293-1301.

Salt, D.E., Blaylock, M., Kumar, N.P.B.A., Dushenkov, V., Ensley, B.D., Chet, I. and Raskin, I. 1998. Phytoremediation : A novel strategy for the removal of toxic metals from the environment using plants. Biotech. 13 : 468-474.

Seguin, A., Lapointe, G., Charest, P.J., Bruce, A. and Palfreyman, J.W. 1998. Forest Prod. Biotech. pp. 287-303.

Shrivastava, A.K. and Shrivastava, P. 1998. Proc. Nat. Acad. Sci. India. Sec. B. Biol. Sci 68: 199-215.

Siciliano. S.D. and Germida, J.J. 1998. Env. Toxicol. Chem. 17 : 728-733.

Terry, R.E. and Wllace, A. 1998. In: Hand book Soil Conditioners. (Ed. A. Wallace). pp. 551-573.

Tolra, R.P., Poschenrieder, and Barcelo, J. 1996. Nutrition. 19 : 1531-1540.

Varshney, C.K. 1983. Water pollution and management. Wiley Eastern Limited. U.K.

Zhelijazkov, V.D., Jeliazkova, E.A., Craker, L.E., Yanko, B., Georgieva, T., Kolev, T., Kovatcheva, S., Stanev, A., Margina, N., Caffini, J., Bernath, L., Craker, A.J. and Giberti, G. 1999. Acta. Hort. 500 : 111-117.

Zhu, Y.L., Zayed, A.M., Qian, J.H., Desouaza, M. and Terry, N. 1999. J. Env. Qual. 28 : 339-344.

ENVIRONMENTAL BIOTECHNOLOGY

Edited By :   **Professor (Dr.) Arvind Kumar**

Published By :   **DAYA PUBLISHING HOUSE**

2

BIOTECHNOLOGY OF CROP PLANT ENVIRONMENT

• *S.G. Ahmad*

Associate Professor and Head, Department of Agriculture Structures and Environmental
Engineering, College of Agricultural Engineering, J.N.K.V.V. Jabalpur-482 004 (M.P.)

Abstract

In agriculture from field preparation to disposal of final produce, agro-ecological environment and agro-ecosystem play mist significant role. In the particular case efforts have been made to established relationship with agro-ecological crop plant ecosystem. The case will help in environmental study of different agro-climatic zone of the country for understanding the biotechnology of crop plant environment.

Most parts of Indian remains hot and dry from April to July. The daily temperature in certain areas may go upto 40°C and the relative humidity as low as 35% (Kanl and Sukumaran 1984). As a result of these conditions it is not possible to grow different types of crop plants in different agro-climatical zone all the year round and maintain quality. By controlling temperature, relative humidity, moisture content, air movement etc one can control and create favourable crop plant environment.

Introduction

Agriculture has the responsibility of meeting the basic food and fiber requirements of the society. As the society enlarges and becomes more sophisticated, the requirements also increase and becomes more varied, therefore, agricultural practices in a society must keep on improving if the society has to continue its upward growth. The agro-ecological diversity in the country makes it necessary to study the ambient environment of the place and analyse its for crop plant environment. Agriculture in India has made excellent progress after independence. Better varieties has been bred, improved agronomic and engineering practices have been developed, but crop plant environment more need control and study of check measures. It has been proved that prehistoric man learned the cropping season and developed methods of growing for his needs. As agriculture developed man expanded his knowledge in this area. The crop plant environment is a living science, and carries of biological processes, is quite different than that of study inert materials. This is a complex phenomenon and needs considerable study of parameters, causing the change.

Jabalpur sub-zone is situated almost in the centre of the country (at 23°09" North-latitude, 79°58", East longitude) and an altitude of 411 m, above mean sea level. It is on the south slope of vast Kymore Plateau.

Among the twelve agro-climatic zones of the state, Jabalpur lies in Kymore Plateau and Satpura Hill Zone–Rewa and Satna, Part of Panna, Seoni and Katni tehsil of Jabalpur are the district in this zone with an area of 4.99 mha which is 11.27% of the total area of the state. It receives annual rainfall of 1000-1500 mm. Jabalpur and Rewa represents slightly different ecological characteristics featured of the zone (Dr. Gyendra Singh–Farm Mechanization in M.P. CIAE, Bhopal, T. B. No. CIAE/2000/82 5.7).

The prevailing agro-climate of the region affect crop production to a greater extent in comparison to any single variable or input. Therefore, farm production at place can be improved by understanding the agro-climate of the area and by making adjustments with it. This is possible only by understanding prevailing argo-climate and crop's of the region, their inter-co-relation and biotechnology of crop plant environment.

Characterization of agro-climate environment of any location, considering all the variables of the physical environment which directly and/or indirectly have bearing's on crop production are useful to planners, for better management of farming operation and conserving the natural resources. It will further help in deriving all other information on crop production which is of agronomic relevance.

However, factor of uncertainty should always be considered because of unusual deviations that may occur in weather conditions at any particular location.

Our knowledge about plant physiology suggests that for a plant of given genetic make up the factors that affect the plant growth are.

1. Light
2. Temperature
3. Air composition, and
4. Nature of root medium.

The majority of the engineering and agronomic practices modify and try to control the nature of root medium. There is practically no way to substantially modify or control light, temperature and air composition parameters in open field cultivation, irrigation and wind breaks provide some degree of influence. Crop production in open fields still remains to be contingent upon good weather conditions with the increasing human population, rising level of sophistication, increasing competition for resources and unpredictable climatic changes. The traditional open field cultivation needs to be reassessed.

Materials and Methods

Plant growth is determined by genetic make up of the plant and environment. The genetic make up established the ultimate limits beyond which the plant can not grow and develop. The environment determines how much of the potential growth inherent in a cultivar can be realized. Major improvements in the genetic potential can be expected as plants are redesigned to take advantage of new information about how the environment affects plant growth and new techniques for maintaining optimum growing conditions.

It is probably safe to say that all of the potential growth of cultivars now available is not realized in most production programes; all of the environmental factors affecting growth and development are not maintained at optimum levels at all stages of growth; neither the plant no the environment are managed with the precision necessary for producing maximum growth.

Environmental Factors Affecting Plant Growth

Temperature, light gases, water and essential mineral elements. Some of these factors are influenced or changed by the greenhouse structure; others are not changed by the greenhouse structure.

1. Changed by Greenhouse Structure Temperature – Soil and air light – visible and infrared (heat)

   Gases – atmospheric CO_2, water vapour (relative humidity) and pollutants.

2. Not changed by Greenhouse Structure.

Water – Soil moisture. Soil mixtures (physical and chemical properties) including water, soil oxygen, temperatures, mineral elements, pH and soluble salts.

The effect of environmental factors on plant growth is exceedingly complex – they interact in an infinite number of combination: for example – an optimum level of CO_2 can not be established without taking into consideration light, moisture and fertility levels.

Optimum Levels for Environmental Factors

1. Seasonal changes in solar radiation (light intensity, plant and air temperatures)
2. Plant species (crop).
3. Age or stage of development of the plant.

Basic Physiological Processes in Plants

1. Photosynthesis, respiration, protein synthesis food manufacture and utilization.
2. Transpiration and water absorption cell enlargement, turgidity, tissue succulence.
3. Mineral element absorption essential for growth, influence carbohydrate utilization.
4. Flower initiation – vegetative *vs.* reproductive growth.

Temperature

For plants adequately are supplied with water and mineral elements as supplied with water and photosynthetic rates are determined mainly by radiant energy, carbon dioxide and temperatures. A statement about the optimum temperature for photosynthesis can not be made without considering the radiant flux and the carbon dioxide concentration.

Optimum day-night temperatures (air)

- Related to light and CO_2 levels.
- Seasonal changes related to light.
- Plant age or size, higher when plants are young, reduced as plants mature.
- Low temperatures are used to harden plants before transplanting outdoors.

Temperature Affect

- Crop timing.
- Flower initiation and development (bulb crops, hydrangeas).
- Soil temperatures water and mineral absorption, leaf temperatures,
- Water loss and moisture stress within the plant.

Light

Photosynthesis, the most important component of crop metabolism, is the process by which light energy is used to reduce carbon dioxide to sugars. Process dependence on different wavelengths. Photosynthetic rate in white light increases with increasing illuminance light unit. It is limited by some other environmental factor such as temperature or carbon dioxide.

Intensity

- Determines yield and quality through its effect on the supply of carbohydrates.
- Adjusted to light requirements of crop species.
- Light intensity affected by greenhouse cover, angle of roof, orientation, greenhouse roof shape, structural components.
- Plant spacing is designed to obtain adequate light levels for all leaves on plant.
- Influences post-harvest life of cut flowers.

Duration

- Controls flower initiation and development.
- Short or long-days are regulated to maintain vegetative growth to achieve adequate plant size before flowers form.
- A temperature relationship exists with day length, may be critical for initiation or uniform flower development.
- Qualitative short or long day plants, daylength requirement absolute.
- Quantitative short or long day plants, daylength speeds up or slows down flowering but does not prevent it.

Quality

- Influenced by greenhouse cover.
- Can alter form and development.
- A major factor when light from electric lamps issued.

Atmospheric Gases

Air movement among the plants is relatively slow and the carbon dioxide utilized in the photosynthesis may not be replaced at the leaf surface at a rate sufficient to maintain the process.

Carbon dioxide

- Basic raw material for photosynthetic process normal level 300 ppm.
- May decrease to 100-150 ppm in greenhouse air when temperature differential between indoor-outdoor approaches below 0°C.
- CO_2 levels of 1000-2000 ppm increase crop growth.
- Moving air also helps to replenish CO_2 concentrations at leaf surface.

Oxygen

- Atmospheric levels generally not limiting.
- O_2 in soil, however, may become deficient in poorly drained soils.

- Can be consumed in tight plastic combustion; ethylene results as a product of incomplete combustion; causes sleepiness of carnations, prevents bud initiation.

Water Vapour or Relative Humidity

For a crop which is well watered, air humidity may not have any effect. But, if water is in short supply, higher air humidity can help in reducing rate of transpiration and hence the irrigation requirement, Air humidity upto 70% are generally safe from the point of view. of pathological effects.

Effect

- Influences transpiration (water loss) and incidence of infection by plant pathogens.
- Increased by humidification misting.
- Decreased by ventilation and heating air.

Air Pollutants

- Toxic at very low concentrations.
- A problem in suburban areas.
- Can be produced by greenhouse boilers, unit heaters, or CO_2 generators if combustion is not complete or if fuel contains sulphur.

Soil Conditions

Water

- Obtained from soil.
- Must be absorbed in sufficient quantities to replace amounts lost by transpiration.
- Yield and quality determined by water supply, extent of moisture is stress, influences plant texture and substance.
- The key question concerning the management of soil moisture is the frequency of irrigation.
- How much water to apply is determined by capacity of soil to hold water-saturate soil plus 10% excess to prevent buildup salts.

Results and Discussion

The whole cropping system in Jabalpur sub-zone is mostly rainfed and lies in rice wheat crop zone. The practice of keeping the land fallow during Kharif is known as Haveli System. It is very common system and results in under utilization of land and resources, which consequently influences the levels of productivity, type of crop and time of sowing needs.

The soil in the region is medium black with medium N.P.K. status. The depth varies from 2 to 3 m and water holding capacity is about 150 mm. The productivity of most or crops in the region is low as compared to all India average.

Kharif season starts with the commencement of the South-West monsoon, which decides the prospects of farming. The duration of crop growing period depends upon date of onset and date of withdrawal of South-West monsoon, Kharif crops of about 90-100 days are grown.

After harvest of Kharif crops Rabi crops are taken with the left-over soil moisture available in the seeding zone. Early withdrawal of South-West monsoon limits the chances of taking second crop.

Thus, most of the unirrigated area in the region is left fallow in the Rabi season, particularly when the South-West monsoon recedes earlier than normal dates.

(i) The climatic environment of the Jabalpur sub-zone can be divided into three seasons.

(ii) The hot season from the middle -March to the middle of June.

(iii) The monsoon season from the middle of June to the end of September (Kharif season).

The cold season from November to mid-March (Rabi season). October is a transition month from the monsoon to winter conditions. The monsoon months receive about 90 per cent of annual rainfall and July and August are the rainiest months.

Each component of weather has been studied in detail. The data base used in this study consists of weekly average of different weather elements for the 10 years period. It was obtained from Meteorological Observatory of India of Indian Meteorological Department at Adhartal, Farm J.N. Krishi Vishwa Vidyalaya, Jabalpur. The weather elements considered were Maximum Temperature, Rainfall, Morning Relative Humidity, Evening Relative Humidity, Wind Velocity and Sunshine Hours.

The graphs were plotted for each weather elements for the past 10-year period *i.e.* from 1987-1996. They showed nearly similar trend for all the years. These graphs clearly showed the extreme values of weather elements for the 10-year period. The average of weekly data for the 10 years were calculated for further analysis.

References

Anonymous. 1987-1996. Annual Reports of Meteorological Observatory of Indian Meteorological Department at Adhartal Farm J.N.K.V.V. Jabalpur.

Bohra, C.P., Ojha, T.P. 1985. Technical Report of CIAE, "Proceedings of the Summer Institute on Green House Design and Environmental Control", Bhopal, May 1985.

Helickson, A., Mylo, Walker N. John. 1983. "Ventilation of Agricultural Engineering Structures, Michigan, U.S.A.

Maheshwari, R.C., Chandra, P., Bohra, C.P. 1983. "Scope of Greenhouse Research in India" Engineering Today (ISAE) . Vol. 7 No. , Sept.–Oct. 1983, Bhopal 8-12.

Sharma, A.K. Shukla, R.K. (1982), "Design of Non-thermal Glass-house in Tropical Areas". Project College of Agricultural Engineering, Jabalpur.

3

VERMICULTURE BIOTECHNOLOGY, EXPERT SYSTEM FOR HOTEL WASTE MANAGEMENT

• *S.R. Bamane*, R.V. Patil** and S.D. Khambe****

*Reader in Chemistry, **Prof. & Head, Deptt. of Chemistry,
Bharati Vidyapeeth's, Dr. Patangrao Kadam Mahavidyalaya, Sangli (M.S.)
*** Head, Deptt. of Microbiology, Miraj Mahavidyalay, Miraj (M.S.)

Abstract

Prototype expert system has been developed for selection of a suitable hotel waste management system. Hotel industry increased vigorously during last two decades in most of the urban areas in India. The survey of some hotels (Grade I, II, III) as well as some of the 5-star hotels in metropolitan cities showed that various kinds of solid wastes get generated and thrown out in nearby area without caring for any kind of treatment. These waste can be grouped as dry waste and wet waste. Amongst the various methodologies for treatment of the solid waste from hotels the reuse and recycling methodology of dry waste whereas vermicomposting method for treatment of wet biodegradable solid waste found to be most suitable, feasible and economical method because;

1. It is pollution free and purely natural process.
2. The entire quantity of solid waste will get converted into good quality organic manure.
3. The process is very simple and huge greater flexibility in scale of operation.
4. Do not require highly technical personnel.
5. As the process is aerobic the microbial decomposition oxidises the organic matter to a fairly stable state resistant to odour production.
6. The vermicompost is a valuable product when used as a soil conditioner. It is a source of micro- and macro- elements in addition to many important biochemicals which are required for the growth of plant.
7. This compost can be used for any type of crop, plants and trees.
8. The process requires not much external energy.
9. The compost can be enriched with additional bacteria and microelements.

Key Words : Hotel waste, Organic Fertilizer, Biodegradable waste, Vermicomposting, Micro-organisms.

Introduction

Expert system is one the important technologies to emerge in the last decade for solution of problem which require great deal of experience and judgement. Environmental Science is also one of such field. In recent years, there have many attempts at developing expert system in the field of environmental science. These expert systems are related to waste water management (BATA, WATERX expert system for anaerobic digester), air pollution control and monitoring process (expert system for gas purification process, for power plant) but only few attempts have been made for this field by developing a prototype expert system. The expert system is able to advise hotel management personnel and aid them in making suitable judgements.

Garbage is an unavoidable consequence of prosperous high technology. Hotel *i.e.* Hospitality Industry as it is called has increased multifold during last two decades in most of the urban centres in India. In big and medium cities, due to various socio-economic factors, there is sudden increase in number of big, small and roadside eateries around every nook and corner of the city. This had contributed to a large quantum of solid waste generated in big cities. It is a problem not only in India but throughout the world. The world garbage reminds us of an overflowing garbage bin, with rag pickers rummaging through the garbage and animals looking for something to eat. The daily garbage production is increased and it has becoming big problem to the municipalities. The municipal machinery is becoming helpless to manage the waste disposal.

The survey of some hotels (Grade I, II, III) as well as some of the 5-star hotels in metropolitan cities showed that various kinds of solid wastes get generated and thrown out in nearby area. These can be grouped as follows:

Dry waste : Paper, plastic bags, paper napkins, bread wrappers, plastic and glass bottles, oil cans, glass jars, juice and cold drink bottles, wrappings of fruit, straw, paper dishes, metal bottles, rubber, leather, plastics, bed sheet cloths, napkins etc.

Wet Waste : Kitchen wastes, peels of vegetables, rotten fruits, fruit seeds, fruit skin pealing and remnants, waste vegetable, roots and stalks, wasted flowers, rotten food, milk, used tea wet powder, vegetarian/non-vegetarian food wasted in the dishes, bones, egg shells, garden waste etc.

The present practice is to throw the entire waste as it is in a nearby dust bins. The ill maintained and not so regularly cleaned dust bins pose a pathetic scene to watch. It is always invariably surrounded by stray dogs, donkeys, stray cattle and most unfortunately rag pickers. The organic waste so strewn around gets fast degraded and starts stinking badly. The bad stink further prohibits the house wives and attendants to deposit the fresh waste into bins proper and thus the waste starts getting spread around setting the chain of filth and unhygienic conditions to a further degree.

Available Technologies for Treatment of Solid Waste from Hotels

1. Reuse and recycling
2. Sanitary land filling
3. Burning in boiler
4. Anaerobic digestion and energy recovery
5. Composting
6. Vermicomposting
7. Manure is value-added product

Eco-friendly material as it eliminates the odour, Fly problem.

Motivate the individuals to practice segregation of waste at sources encouraging means such as vemircoposting the hotel owner to handle their own waste and dispose it of through environmental friendly.

The process of vermicomposting is depicted in the following flow chart representation.

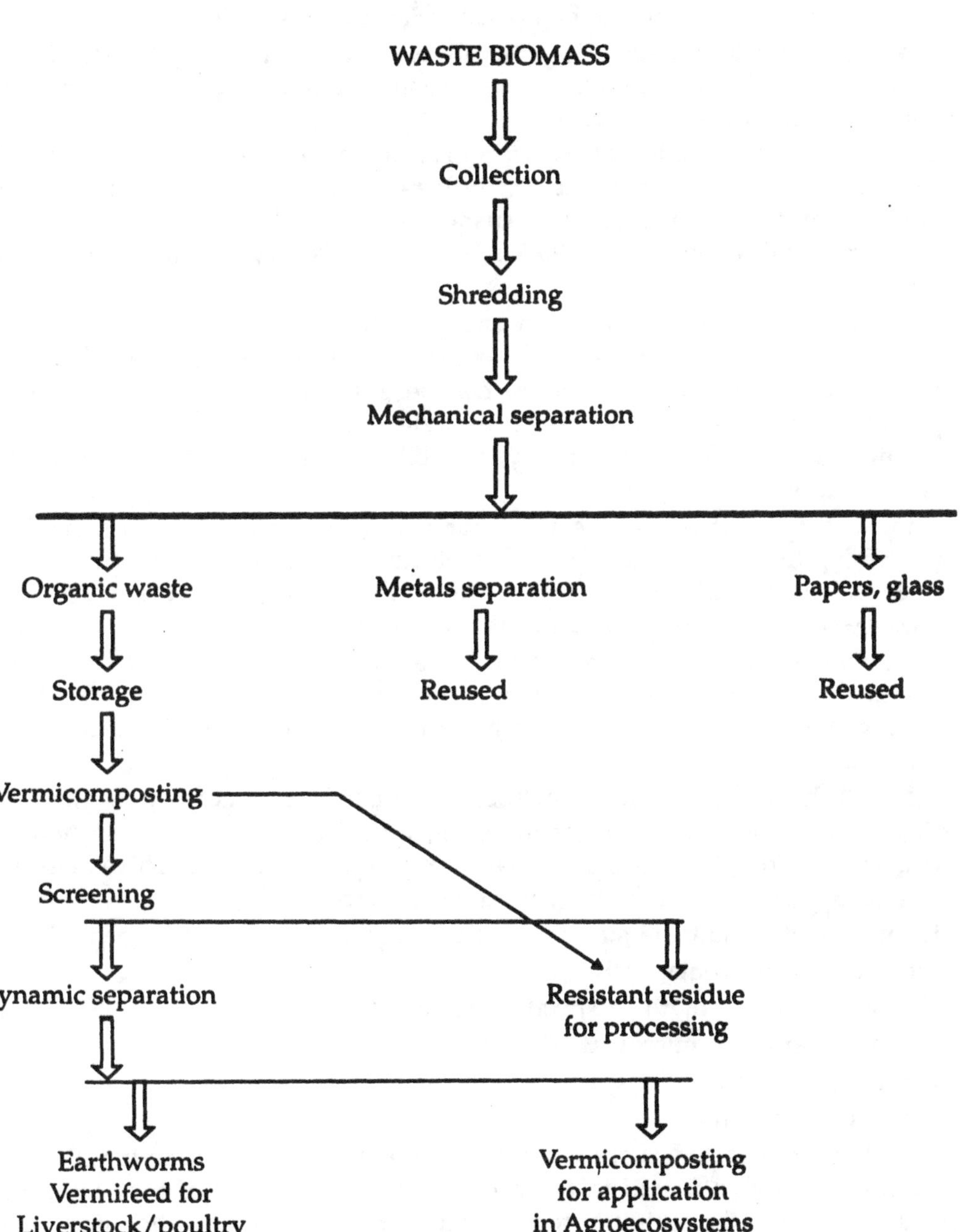

Vermicomposting

1. Reuse and recycling for dry solid waste is the method wherein the most of the material can be converted into other useful material or used as it is after proper refinement. Metal cans, plastic materials including bottles, cans, metallic caps, paper wastes, plastic bags, construction and debris waste, ragged cloths like bed sheets, curtains, napkins towels etc can be recycled to convert into fresh materials.

2. Sanitary land filling method is generally used for inorganic materials like soil, sand, canker, minerals etc. The major hotel waste is organic in nature. It is light weight and very difficult to transport.

3. To get energy by burning of the waste is not feasible. Paper can be burn with other fuels in boiler, but transportation will become headache. It will also cause air pollution problems by producing suspended particulate matter. The calorific value of the material is found to be 800 Kcal/Kg which is very less than other conventional fuels.

4. The waste is cellulose, lipid (fat) and protein in nature and contain much easily degradable material. For methane gas recovery, it has to use in Bio-Gas Plant along-with cattle dung. As the material can be easily get degraded anaerobically, the process is feasible. But the problem is with high cost involvement, high technical requirement. If failed, it is not easy to stabilise the system and the disposal of outlet is again a problem.

5. Composting method can be used for this waste but as the material is purely putrefying in nature, it may pose a smell nuisance. It is also necessary to add some other easily degradable material during composting (like cattle dung).

6. For acceleration of composting reaction it is necessary to use special species of micro-organisms by which the time for mineralization can be reduced to considerable extent. The process can be further accelerated by using the natural digesters *i.e.* earthworms. Special species of earthworms are identified which can be utilised for this process.

Detailed study is carried out on microbial and earthworm activity during hotel waste composting, which is a point of major importance in designing, planning, evaluation and operating a vermin-compost process as a waste management device. This is also needed as the process of composting has biological limitation factors like,

1. Requirement of suitable strains of micro-organisms and earthworms.
2. Efficiency of microbial activity.
3. Cultivation and growth of specific species of earthworms.
4. Size and nature of microbial ad earthworm population.
5. Environmental conditions.
6. The substrate nature.

If any of the above factors is present in less or much more than optimum, the functioning of the entire process is inhibited in the proportion to the extent of the factor. Thus, it is very necessary to study the optimum requirement for successful operation and maintenance of the process. In the context of this, present investigation is conducted on the topic of role of micro-organisms in the treatment of hotel waste through aerobic vermicomposting.

Material and Methods : A Case Study

The daily waste from the individual hotel is initially segregated at source. Separate two bins were provided. In first bin all dry non-degradable garbage like plastic, paper, bottles, tins and metallic wastes were stored and kept aside which will be collected by rag pickers. In another bin wet biodegradable garbage like vegetable waste, food waste, fruit peels, egg shells, bones etc were collected.

The vermicompost is prepared from the compost which contains a mixed microbial culture of decomposing micro-organisms. The decomposed material is further subjected for digestion by earthworms. A special variety of earthworm Isenia *foetida* was used for this purpose. The culture thus prepared also contains eggs of earthworms. An earthern pot of 1 ft diameter which is generally used for domestic flower plants was used which was filled with specially prepared vermicompost to a height of about 4 inches at bottom. Then the waste generated daily was put into the same in a properly cut and meshed form. A small porous cover was put on the top to avoid fly breeding and to maintain aerobic conditions. The watering of the material is done only if the material looks dry. The watering was made so as to maintain moisture content to about 60 per cent. The normal function of the system is confirmed by smell of the contents. A foul smell indicates overloading which may warrant use of one or more pots, undercharging of excess water or addition of more bio-culture. Once the pot is filled, it is covered with thin layer of loose earth and is daily watered for about ten to fifteen days. The process takes about 20 to 25 days to degrade the waste completely depending on waste quality and quantity available daily.

The chemical[1,2,3] and microbiological[4] analysis was made after complete degradation of the material. The isolation and identification of micro-organisms involved in the process is carried out with help of Burgey's manual of determinative bacteriology and other methods[5,6,7].

Result and Discussion

The micro-organisms isolated and identified from the vermicompost prepared are shown in Table 3.1. Which shows that the normal heterotrophic organisms are responsible for the degradation of the hotel waste. No any pathogenic micro-organisms detected during study. The earthworms are helping to speed up the process.

Table 3.1 : Organisms Isolated and Identified from Vermicompost Prepared from Hotel Waste

BACTERIA	*Aerobacter spp., Agrobacter spp., Acetobacter aceti, Bacillus spp., Cellulomonas flavigera, Citrobacter fruindii, Pseudomonas spp., Zoogloea ramigera.*
FUNGI	*Aspergillus niger, Aspergillus flavus, Mucur pusillus, Penicillium notatum, Rhizopus nigiricans.*
ACTINOMYCETES	Micromonospora purpura, Nocardia farcinica Streptomyces bobili, Streptosporangium spp., Thermomonospora curvata.
Plate Count :	Bacteria $- 4 \times 10^8$
	Fungi $- 43 \times 10^6$
	Actinomycetes $- 10 \times 10^4$

Note : The common organisms observed in 6 samples are reported.

The chemical analysis of the vermicompost prepared from hotel waste is shown in Table 3.2. This shows that the material is pure organic and contains the nutrients required for growth of plants. The compost can be used as soil conditioner. In order to ensure safe application of vermicompost the standards are shown in Table 3.3 may be ensured.

Conclusion

This paper discusses the applicabilitly of an expert or knowledge based approach to hotel waste management system. A prototype expert system has been developed for this purpose.

The prototype expert system developed for hotel waste management can serve as a tool for selecting the best alternative for the purpose.

The method is found to be excellent for the hotels who have problem of disposal of daily waste. It also gives following benefits:

Table 3.2 : Chemical Analysis of Vermicompost from Hotel Waste

Sr.	Parameters		Value*
1.	pH (Saturated)		7.50 - 7.6
2.	Conductivity (Saturated) m, mhos/cm		0.40 - 0.50
3.	Nitrogen %		2.00 - 2.50
4.	Phosphorus as P205 %		2.00 - 2.50
5.	Potassium as K2O %		3.20 - 3.40
6.	Organic Carbon %		46.0 - 50.0
7.	Total Ash %		31.0 - 34.0
8.	Total Volatile Solids %		66.0 - 69.0
9.	C/N		23.0 - 20.0
10.	Chlorides %		0.30 - 0.32
11.	Iron	ppm	600 - 700
12.	Manganese	ppm	50 - 60
13.	Zinc	ppm	8 - 10
14.	Copper	ppm	15 - 18
15.	Calcium %		1.4 - 2.0
16.	Magnesium %		0.15 - 0.28

*Values on dry weight basis. The average values of six samples.

Table 3.3 : Standards for Compost
In order to ensure safe application of vercompost the following standards may be ensured:

Parameters	Maximum acceptable Concentration (in ppm.)
Arsenic	20
Cadmium	20
Chromium	300
Copper	500
Lead	500
Mercury	10
Nickel	100
Zinc	2500

1. Reduction of solid waste at source, which will help in lowering down the load of city solid waste disposal system.

2. Various fungi, bacteria and actinomycetes are responsible for bioconversion of hotel waste. These organisms can be used effectively as inoculum to accelerate composting process.

3. The time required to complete stabilisation and formation of most stable compost prepared from spentwash and pressmud with 3:1 proportion was observed to be 30 days when optimum conditions were maintained.

4. The microbiological process involved in the degradation of the hotel waste includes initial fungal predominance followed by thermiphilic microbial reaction and then mesophilic reaction.

5. The health problem will be reduced due to segregation of perishable solid waste.
6. Stinking sites and the dirty blots on the city will be out.
7. An excellent quality organic manure will be made which will be useful as soil conditioner.
8. A better yield of agricultural products will be a plain bonus.

References

AOAC. 1995. Methods of analysis. Association of Official Agricultural Chemists. 7th Edn. Washington D.C., pp. 113-231.

APHA. 1985. Standard methods for examination of water and wasterwater. 15th Edn. American Public Health Association, pp. 1-1193.

Bear, F.E. 1964. Chemistry of the soil. 2nd Edn. Oxford and IBH Publ. Co. Pvt. Ltd., pp. 1-515.

Buchnan, R.E. and Gibbons, N.E. 1975. Bergy's manual of determinative bactriology. 8th Edn. The Williams and Wilkins Co., Baltimore, pp. 1-1267.

Gilman, J.C. 1975. A manual of soil fungi. Oxford and IBH Publ. Co., New Delhi, pp. 1-450.

Government of India. Report on SMW in Class I cities in India. 1999.

Kononova, M.M. 1966. Soil organic matter. Oxford Pergamon, Press, pp. 139-150.

Manohar. P. Ramesh. Reddy., Ind. J. Env. & Heatlh. Jully. 2000.

Manual on Solid Waste Management 2000. Published by AIILSG. Mumbai.

The Gazette of India., Ministry of Environment & Forest Notification, New Delhi.

U.S.E.P.A. EPA. Guide for infectious waste management. EPA/530 SW 86014. Mav 86 P.V.

Zhu, X.X. and Simron, A.R. 1966. Expert System for water Treatment plant Operation Journal of Environment Engineering. 122(9): 822.

ENVIRONMENTAL BIOTECHNOLOGY

Edited By :   Professor (Dr.) Arvind Kumar

Published By :   DAYA PUBLISHING HOUSE

4

EVALUATION OF BIOCLEANING EFFICIENCY OF FEW CYANOBACTERIAL STRAINS IN FISH FARM EFFLUENTS

• *S. Chidambaram Pillai*

Lecturer (S.G.) in Botany, P.G. Department of Botany,
V.O. Chidambaram College, Tuticorin

Abstract

Fresh water cyanobacterial strains of *Anabaena variabilis*, *Phormidium valderianum* and *Nostoc calcicola* were screened in fish farm effluent in laboratory bioassays to compare their biocleaning efficiency. A. variabilis and P. valderianum responded well compared with *Nostoc calcicola* and the results were interpreted and probable reasons were discussed.

Key Words : Fish farm effluent treatment, Biocleaning, Cyanobacterial strains, Growth parameters.

Introduction

Rejection of used waters in the natural water bodies raises increasing concern on water pollution. Unfortunately aquaculture has assumed the dimension from farming to industry which generate and discharge assimilable pollutant waste rich in nitrates, phosphates and ammonia which magnify the eutrophication problems. As micro-algae especially cyanobacteria have been considered ideal waste removers, an attempt has been made to use few strains to compare their biocleaning efficiency of fish farm effluents. The effect of this effluent on the growth of these algae has also been undertaken.

Materials and Methods

Freshwater fish farm effluent was collected from the Manju fish farm, Karambai, Kallidaikurichi, Tirunelveli District, Tamil Nadu. Cyanobacterial strains of *Anabaena variabilis*, *Phormidium valderianum* and *Nostoc calcicola* were used from the culture collections of Department of Botany. V.O. Chidambaram College, Tuticorin and have been maintained in the BG11 medium under continuous illumination of light at 1400lux and 27±2°C. The effluent was filtered through cotton to remove unwanted debris etc and used as such without dilution.

Experiments were conducted in duplicates. To study the efficiency of these cyanobacteria in biocleaning the effluent, the treatments were undertaken for a total duration of 20 days under laboratory conditions. One ml of uniform suspension of *Anabaena*, *Phormidium* and

Table 4.1 : Characteristics of Fish Farm Effluent (mg/l) except pH

Parameters	Day (0)
pH	7.4
DO	6.2
CO_3^-	0.0
HCO_3^-	3.6
TS	175
BOD	4.2
COD	32
Sulphide	2.8
Nitrate	3.8
Nitrite	BDL
Phosphate	4.60
Ammonia	13.74

BDL – Below Detectable Limits.

Nostoc species was inoculated as initial inoculum in all the flasks. The effluent samples were periodically (every 10th day) analysed for various physico-chemical parameters using APHA 1980. Growth of these cyanobacterial strains was measured in terms of chlorphyll-a *carotenoids* and protein as the biomass component (Subramanian *et. al.*, 1994).

Results and Discussion

Characteristics of fish farm effluent has been depicted in Table 4.1. Table 4.2 represents the chemical characteristics of the effluent under different treatments using the cyanobacterial strains and effluent influenced change in few biochemical constituents of these strains are depicted in Table 4.3.

Table 4.2 : Chemical Characteristics of Fish Farm Effluent before and after Cyanobacterial Treatments (mg/l) except pH

Parameters	Day	*Anabaena* sp. 10th day	*Anabaena* sp. 20th day	*Phormidum* sp 10th day	*Phormidum* sp 20th day	*Nostoc* sp. 10th day	*Nostoc* sp. 20th day
pH	7.4	9.95	9.82	9.60	9.58	8.52	9.20
DO	6.2	7.7	7.6	8.7	8.4	7.2	7.8
CO_3^-	0	48	46	32	32	44	42
HCO_3^-	36	–	–	–	–	–	–
TS	175	145	140	132	129	158	151
BOD	4.2	2.8	2.2	2.0	1.6	2.6	2.1
COD	32	28	22	24	20	22	20
Sulphide	2.8	1.6	1.2	2.6	2.4	2.8	2.2
Nitrate	3.80	3.52	2.98	1.98	1.18	3.24	3.04
Nitrite	BDL	0.112	BDL	0.207	BDL	0.132	BDL
Phosphate	4.6	2.4	1.8	3.2	BDL	3.8	2.6
Ammonia	13.74	4.74	2.82	2.36	1.48	5.68	4.64

BDL – Below Detectable Limits.

Table 4.3 : Chlorophyll-a, Carotenoids and Protein Content of Three Cyanobacterial Strains before and after Effluent Treatment

Organisms	Initial inoculum 'o' day	Control 20th day	Treatment Days 10th day	Treatment Days 20th day
CHLOROPHYLL.A (mg ml⁻¹)				
Anabaena sp.	0.94 ± 0.18	2.28 ± 0.16	2.44 ± 0.17	2.45 ± 0.30
Phormidium sp.	0.72 ± 0.16	1.78 ± 0.14	2.04 ± 0.15	2.28 ± 0.16
Nostoc sp.	1.68 ± 0.24	2.57 ± 0.30	1.51 ± 0.23	0.76 ± 0.16
CAROTENOIDS (mg ml⁻¹)				
Anabaena sp.	0.0413 ± 0.020	0.0463 ± 0.020	0.0634 ± 0.020	0.0682 ± 0.020
Phormidium sp.	0.0148 ± 0.013	0.0185 ± 0.015	0.0226 ± 0.016	0.0244 ± 0.017
Nostoc sp.	0.020 ± 0.028	0.026 ± 0.027	0.0178 ± 0.075	0.0128 ± 0.021
PROTEIN (µg ml⁻¹)				
Anabaena sp.	166.07 ± 0.040	170.66 ± 0.81	133.66 ± 0.72	100.98 ± 0.62
Phormidium sp.	126.44 ± 0.700	136.53 ± 0.73	145.26 ± 0.75	139.40 ± 0.73
Nostoc sp.	86.12 ± 1.780	92.55 ± 1.85	77.69 ± 1.69	60.37 ± 1.49

Inoculated cyanobacteria are known to grow fairly efficiently in fish farm effluents and their growth was measured in terms of chlorophyll, carotenoids and protein changes.

In the fresh effluent the pH value was slightly alkaline and in the cyanobacteria screened effluent a marked increase in pH. The increased pH is related to the depletion of inorganic carbon during photosynthesis by these algae and this typical capacity of cyanobacteria to bring about changes in pH is evident in this study also. Increased pH often favours the growth of cyanobacteria (Tam and Wong, 1989; Manoharan and Subramanian 1993).

Cyanobacteria screened effluent showed a pronounced increase in dissolved oxygen also. This increment is due to algal photosynthetic activity and also the solubility of oxygen from the atmosphere under laboratory condition in which they are incubated. High dissolved oxygen in batch cultures enhance the nutrient removal which subsequently reduce the oxygen demands of the effluents (Kankal *et al.*, 1987).

The fresh effluent showed only bicarbonate ions. These algae by metabolising the bicarbonate ions, increase the pH value. HCO_3 utilization by algae' is an adaptive phenomenon (Senthil and Goyal, 1996). The gradual withdrawal of CO_2 from HCO_3 resulted in the liberation of OH ions which has made the effluent water more alkaline.

Cyanobacterial treatments in fish farm effluent showed considerable reduction in total solids. This reduction observed might be due to the uptake of dissolved solids by these organisms. Similar observation of solids removal by *Oscillatoria* and *Halobacterium* sp. combination in effluent waters had already been documented (Uma and Subramanian, 1990).

Cyanobacteria treated fish farm effluent showed reduction in both BOD and COD levels. BOD reduction in wastewaters using algal cultures had been reported (Pouliot *et al.*, 1989; Tripathi *et. al.*, 1991). Similar observation of COD reduction by *Oscillatoria chlorina* screened effluent had also been documented earlier (Rana and Palria, 1987). It is clear from this study that these algae not only retrieve inorganic nutrients, but also helps in reducing BOD and COD level.

Sulphide is generally toxic with lower pH values. This toxicity can be overcome with increased pH by the algae during their active growth in the effluent (Gilles Laliberte *et. al.*, 1990). In this study *Nostoc* and *Phormidium* species screened effluent showed no remarkable removal in the sulphide levels when compared with *Anadaena*, sp. treated effluent which may be due to poor utilization. The utilization of sulphide during photosynthesis by *Oscillatoria* species obtained from the sulphide enriched waters can not also be ruled out here (Cohen *et. al.*, 1975).

Nitrate serves as nitorgen source for the growth of algae. High level of nitorgenous compounds in wastewaters can be effectively removed by algae in general (Tam and Wong, 1989), cyanobacteria in particular (Proulx and dela Noue, 1988). Even after a prolonged exposure in the effluent, *Anabaena* and *Nostoc* species showed poor utilization of nitrate nitorgen which might be due to the fact that these algae during their growth might differentiate heterocysts, the site of nitrogen fixation, which contribute poor removal of nitrate in the effluent. Similar observation was made by Haselkon (1978). Also the ammonical nitrogen utilization instead of nitrate nitrogen by these algae leaving substantial nitrate level in the effluent. On the other hand, *Phormidium* screened effluent showed apparent assimilation of all the nitrogen sources from the effluent. Varma and Wilcomb (1964) contend that nitrite nitrogen promotes better algal growth than ammonia nitrogen in wastewaters.

Nitrite at low concentration serves as nitrogen source for the growth of cyanobacteria. In the fresh effluent the nitrite level was well below the detectable limits. But after cyanobacterial treatment, within 10 days its level has marked increase in all the case which may be due to the reduction of nitrate into nitrite which in turn is removed completely with in another 10 days. Similar initial level increase and subsequent depletions of nitrite in paper mill effluent treatment using the cyanobacterium *Oscillatoria pseudogeminata* has been made earlier (Manoharan and Subramanian, 1992).

While *Anabaena* and *Nostoc* sp. showed poor utilization of phosphates, *Phormidium* screened effluent showed its level below detectable limits. The capcaity of cyanobacteria to remove large amount of phosphorous from wastewater had been documented earlier (Chan *et al.*, 1979, Tam and Wong, 1989). Filamentous cyanobacteria are not only able to retain these pollutants for a longer period, but can be suitably harvestd easily (Kumar and Sharma, 1975). The removal of phosphate may be attributed with their uptake during the active growth and this observation of inverse relationship between increased algal growth and decreased phosphate concentration in wastewaters Charles Neos and Varma (1966) can not also be ruled out.

The effluent after treatment with these three strains showed considerable reduction of ammonia within 20 days. This reduction of ammonia level may be due to effective uptake by these micro-algae to maintain their metabolism which is in accordance with the observation made by (Shelef *et al.*, 1978). As ammonia did not seem to interfere with these algal growth, they penetrate the cell membrane by diffusion which could explain its removal from the effluent.

Influence of Effluent on the Biochemistry of Three Cyanobacterial Strains

Due to varied physiological adaptation of these algae in the effluent, there is a significant change on the biochemical composition of the organisms. The growth of these organisms in terms of chlorophyll-a, carotenoids and proteins showed considerable variations.

During the treatment period, the two strains *Anabaena* and *Phormidium* sp. showed remarkable increase in chlorophyll-a and carotenoid pigments then the *Nostoc* sp. which showed gradual reduction in the above pigment composition indicating their poor adaptation in the effluent.

As far as the total protein level is concerned, the fish farm effluent had brought down the total protein content of *Anabaena* and *Nostoc* species during the treatment period. Contrary to this in *Phormidium* sp., a initial level increase and subsequent reduction in protein level was noticed. The decrease in protein proportion can be expected when nitrogen and phosphorous have been taken up by these algae during their active growth phase (Talbot and de la None, 1993). Similar observation of protein level reduction in *Oscillatoria* sp. was made while treating a paper mill effluent (Manoharan and Subramanian, 1992). Also the environmental stress induced modification in the protein level while they grow in the effluent can not also be ruled out (Kimpel and Key, 1985; Bhagwat and Apte, 1989).

Cyanobacteria are not a metabolically sluggish group of organisms as they play a major role in removing various pollutants as their nutrients from the effluents. Based on the discussions of the above results, it can be concluded that of the three strains employed in the above study, *Anabaena variabilis* and *Phormidium valderianum* were found the better choice candidates which were minimally influenced by the adverse condition of the effluent, but helped in removing/reducing the fish farm pollutants maximally.

References

APHA 1980. Standard methods for Examination water and wastewater. 15th Ed. American Public Health Association, New York.

Bhagwat, A.A. and Apte, S.K. 1989. Comparative analysis of protein induced by heat shock, salinity and osmotic stress in the nitrogen fixing cyanobacterium *Anabaena* sp. Strain L.31 Journal of Bacteriology. 171 : 5187-5189.

Chan, K.Y., Wong, K.H. and Wong, P.K. 1979. Nitrogen and Phosphorus removal from sewage effluent with high salinity by *Chlorella salina*. Environ. Pollu, 18 : 139-146.

Charles, Neos and Varma, M.M. 1966. The removal of phosphate by algae. Water and sewage works pp. 456-459.

Cohen, Y., Padan, E. and Shilo, M. 1975. Facultative anoxygenic photolosyn thesis in the cyanobacterium *Oscillatoria limnetica*. Journal of Bacteriology 123 : 855-861.

Gilles Laliberte, Daniel Proulx, Niels De pauw and Joel de la Noue 1990. Algal Technology in wastewater Treatment pp. 283-297.

Haselkorn, R., 1978. Heterocysts, Ann. Rev. Plant Physiol, 29 : 319-344.

Kankal, N.C., Nema, P., Gokhe, B.H. and Mehata, C.G. 1987. Dissolved oxygen and detention time as process parameters in sewage treatment by aerated lagoon. IACPW Tech. Annual. 14 : 61-68.

Kimpel, J.A. and Key, J.L. 1985. Heat shock in plants. Trends Biochem. Sci. 85 : 353-357.

Kumar, H.D. and Sharma, V. 1975. Experimental studies on the uptake of mineral nutrients by algae isolated from polluted habitats. Biochem Physiol Flanzen (BPP) 166 : 221-232.

Manoharan, C. and Subramanian, G. 1992. Interaction between paper mill effluent and the cyanobacterium *Oscillatoria pseudogeminata var, unigranulata*. Pollution Research 11(2) : 73-84.

Manoharan, C. and Subramanian, G. 1993. Feasibility studies on using cyanobacteria in effluent treatment. Indian Journal of Environmental Health, 35(2) : 88-96.

Proulx and de la Noue 1988. Removal of macro-nutrients from wastewaters by immobilized microalgae. In Bioreactor immobilized enzymes and cell fundamental and applications. Ed.M.Moo-Young. Elsevier Applied Science. pp. 301-310.

Pouliot, Y., Buelna, G., Racine, C. and de la Noue 1989. Culture of cyanobacteria for tertiary wastewater treatment and biomass production. Biological wastes 29 : 81-91.

Rana, B.C. and Palria, S. 1987. Pollution abatement of certain wastewaters using biological and chemical methods. Indian Journal of Ecology 14(1) : 1-6.

Senthil, C. and Goyal, S.K. 1996. Inorganic carbon assimilation in algae and its response to elevated levels of carbondioxide. Phykos 25(1&2) : 1-12.

Shelef, G., Moraine, R., and Oron, G. 1978. Photosynthetic biomass production from sewage. In Ergebnisse der Limnologie, Heft II : Micro-algae for food and feed. Edited by C.J. Soeder and R. Binsack. Stuttgart, Germany. pp. 3-14.

Subramanian, G., Uma, T., Thajuddin, N., Sekar, S., Sundararaman, M., and Sophia Rajini. 1994. Laboratory manual, STORMCUP, NFMC, Bharathidasan University, Trichy.

Talbot, P. and J. de la Noue. 1993. Tertiary treatment of wastewater with *Phormidium bohneri* (Schmidle) under various light and temperature conditios. Water Research 27(1) : 153-159.

Tam, N.F.Y. and Wong, Y.S. 1989. Wastewater nutrient removal by *Chlorella pyrenoidosa* and *Scenedesmus* sp. Environ. Pollut. 58 : 19-84.

Tripathi, B.D. and Shukla, S.C. 1991. Biological treatment of wastewater by selected aquatic plants. Environmental Pollution 69, 69-78.

Uma, L. and Subramanian, G. 1990. Effect of cyanobacteria in effluent treatment. (Proceedings of National symposium in cyanobacterial Nitrogen fixation). Indian Agricultural Research Institute, New Delhi, pp. 437-444.

Varma, Man M. and Wilcomb, Maxwell, J. 1964. Effect of Light Intensity on Photosynthesis. Water and Sewage Works 110, 191.

ENVIRONMENTAL BIOTECHNOLOGY

Edited By :   Professor (Dr.) Arvind Kumar

Published By :   DAYA PUBLISHING HOUSE

5

ESTIMATING GENETIC DIVERGENCE AMONG DIFFERENT PROVENANCES OF WILD POMEGRANATE (*PUNICA GRANATUM* L.)

• *K.S. Pant, Pankaj Panwar & C.V. Saraswat*

Department of Silviculture and Agroforestry,
Dr. Y.S. Parmar University of Horticulture and Forestry, P.O. Nauni (Solan) (H.P.)

Abstract

Genetic divergence among fifteen provenances of *Punica granatum* was studied in Himachal Pradesh – India, using Mahalanobis D^2 statistics. The study reveals that maximum D^2 value of 416.09 occurred between S_3 (Narag) and S_{15} (Nagwain) provenances. It was observed that number of fruits contributed maximum 33.33 per cent divergence followed by viability 18.10 per cent. On clustering provenances on basis of D^2 statistics five different clusters were recognised showing differences among them. The intra-cluster differences was highest (7.72) in cluster II, whereas, inter-cluster genetic difference was maximum (20.04) between clusters III and IV.

Key Words : Punica granatum, Genetic divergence, Mahalanobis D^2 statistics.

Introduction

The tree improvement programme starts with the study of available variation in the entire range of species distribution and delineation of provenances capable of providing the best adapted trees, Successful tree improvement programmes are those in which proper provenances are used and there is sufficient genetic divergence between the population, Owing to the multiple use of this species, as its root and bark contains tannin and alkaloids, its use in treatment of tapeworms, diarrhoea, epilepsy etc. Therefore, large scale plantations of wild pomegranate are coming up which are mostly raised through locally available seed sources which may not be of desired quality. To improve the species using improvement programmes, estimation of genetic divergence is imperative. It was in this reference that the present investigation was carried out in University of Forestry, Solan, India.

Material and Methods

A pilot survey of population of *Punica granatum* L. in Himachal Pradesh was undertaken to identify the sites where this species occurs in wild stands. The sampling procedure included delineation of the whole area under the species into number of sites depending upon the altitude, aspect and variation on morphological characters. In this way, fifteen provenances were selected in five districts of Himachal Pradesh. The name and geographical location of the selected sites is given in Table 5.1. The selected provenances were confined to 32°32'18"N latitude and 76°06'11"N to 77°52'18"E longitude. Five naturally occurring trees of wild pomegranate which were mature, having approximately same age, and crown size were selected from each site, as per the method adopted by Pozdnajakov, 1969 and Dumitriu-Tataranu, 1970. Eight fruits from each selected tree were collected depending upon their direction (two each from South, North, East and West) and were mixed together. Average weight of fruit per tree, fruit length, number of fruit per tree, rind thickness, total sugar, total soluble solids, acidity, viability of seed (Bonner, 1974) and per cent germination of seed was determined immediately after harvesting the fruits. The seeds were sown in the nursery in the month of June. No fertilizer was applied and beds were kept moist and weed free. 200 seeds of each tree were sown in 2.5m × 1.2m × 0.25m raised bed, in 15 cm spaced lines. Seed to seed distance was kept 10cm and seeds were placed at a depth of 1.5cm.

Table 5.1 : Geographical Locations of Different Provenances of Wild Pomegranate (*Punica granatum* L.) in Himachal Pradesh

Code number	Provenances	District	Altitude (m)
S1	Oachghat	Solan	1200
S2	Kandaghat		1150
S3	Darlaghat		1250
S4	Daro Deoria	Sirmour	1000
S5	Narag		1250
S6	Bagthan		1300
S7	Basantpur	Shimla	950
S8	Ghanati		1050
S9	Shoghi		950
S10	Sundernagar	Mandi	900
S11	Rewalsar		800
S12	Sarkaghat		900
S13	Kullu	Kullu	1300
S14	Banjar		1250
S15	Nagwain		1300

Source: Toposheets of Sirmous, Solan, Mondi, Kullu and Shimla districts of Himachal Pradesh.

Table 5.2 : D² Values in Matrix Form for Different Wild Pomegranate (*Punica granatum* L.) Provenances (Pooled Basis)

	1	2	3	4	5	6	7	8	9	10	11	12	13	14	15
1	0	54.11	47.96	74.93	154.91	103.57	102.79	92.85	77.51	109.87	109.42	85.50	172.81	110.51	131.02
2			70.01	60.46	126.60	62.86	61.47	56.29	38.45	100.82	90.25	113.13	175.93	95.27	193.70
3				50.40	163.09	91.78	112.02	101.66	99.29	126.45	100.81	104.86	191.77	124.51	181.92
4					62.07	87.24	63.99	112.69	107.34	176.31	125.85	146.71	265.89	144.67	253.50
5						164.61	65.98	212.63	167.43	297.14	245.58	245.54	387.20	237.49	416.09
6							149.59	38.34	60.54	81.12	53.75	65.12	107.14	39.73	136.14
7								127.55	104.47	216.98	177.88	214.74	297.57	190.80	332.75
8									63.00	91.55	65.82	97.00	119.50	62.84	135.10
9										44.66	50.18	62.31	119.11	59.89	174.44
10											24.97	22.86	48.44	28.36	91.57
11												31.46	62.02	19.11	100.63
12													49.86	19.74	67.96
13														46.11	51.01
14															79.95
15															0

Table 5.3 : D² Values Arranged in Increasing Order for Magnitude for Different Wild Pomegranate
(*Punica granatum* L.) Provenances (Pooled Basis)

1	2	3	4	5	6	7	8	9`	10	11	12	13	14	15
47.96(3)	38.45(9)	47.96(1)	50.40(3)	62.07(4)	38.34(8)	61.47(2)	38.34(6)	38.45(2)	22.86(12)	19.11(14)	19.74(14)	46.11(14)	19.11(11)	51.01(13)
54.11(2)	54.11(1)	50.40(4)	60.46(2)	65.98(7)	39.73(14)	63.99(4)	56.29(2)	44.66(10)	24.97(11)	24.97(10)	22.86(10)	48.44(1)	19.74(12)	17.96(12)
74.93(4)	56.29(8)	70.01(2)	62.07(5)	126.60(2)	53.75(11)	65.98(5)	62.84(14)	50.18(11)	28.36(14)	31.46(12)	31.46(12)	49.86(11)	28.36(10)	79.95(14)
77.41(9)	60.46(4)	91.78(6)	63.99(7)	154.91(1)	60.54(9)	102.79(1)	63.00(9)	59.89(14)	44.66(9)	50.18(9)	49.86(13)	51.01(15)	39.73(6)	91.57(10)
85.50(12)	61.47(7)	99.29(9)	74.93(1)	163.09(3)	62.86(9)	104.47(9)	65.82(11)	60.54(6)	48.44(13)	53.75(6)	62.31(9)	61.02(11)	46.11(13)	100.63(11)
92.85(8)	62.86(6)	100.81(11)	87.24(6)	164.61(6)	65.12(12)	112.02(3)	91.55(10)	62.31(12)	81.12(6)	62.02(13)	65.12(6)	107.14(6)	59.36(9)	131.02(1)
102.79(7)	70.01(3)	101.66(8)	107.34(9)	167.43(9)	81.12(10)	127.55(8)	92.85(1)	63.00(8)	91.55(8)	65.82(8)	67.96(15)	119.11(9)	62.84(8)	135.10(8)
103.57(6)	90.25(11)	104.86(12)	112.69(8)	212.63(8)	87.24(4)	149.59(6)	97.00(12)	77.51(1)	91.57(15)	90.25(2)	85.50(1)	119.50(8)	79.95(15)	136.14(6)
109.42(11)	95.27(14)	112.02(7)	125.85(11)	237.49(14)	91.78(3)	177.88(11)	101.66(3)	99.29(3)	100.82(2)	100.81(3)	97.00(8)	172.81(1)	95.25(2)	174.44(9)
109.87(10)	100.82(10)	124.51(14)	144.67(14)	245.48(11)	103.57(1)	190.80(14)	112.69(4)	104.47(7)	109.87(1)	100.63(15)	104.86(3)	175.93(2)	110.51(1)	181.92(3)
110.51(14)	113.13(12)	126.45(10)	146.71(12)	245.54(12)	107.14(13)	214.74(12)	119.50(13)	107.34(4)	126.45(3)	109.42(1)	113.13(2)	191.77(3)	124.51(3)	197.76(2)
131.02(15)	126.60(5)	163.09(5)	176.31(10)	297.14(10)	136.14(15)	216.98(10)	127.55(7)	119.11(13)	176.31(14)	125.85(4)	146.71(4)	265.89(4)	144.67(4)	253.50(4)
154.91(5)	175.93(13)	181.92(15)	253.50(15)	387.20(13)	149.59(7)	297.57(13)	135.10(15)	167.43(5)	216.98(7)	177.88(7)	214.74(7)	297.57(7)	190.80(7)	332.75(7)
172.81(13)	193.70(15)	181.77(13)	265.89(13)	416.09(15)	164.61(5)	332.75(15)	212.63(5)	174.44(15)	297.14(5)	245.48(5)	245.54(5)	387.20(5)	237.49(5)	416.09(5)

Table 5.4 : Contribution of Each Character to Divergence Among the Different Wild Pomegranate
(*Punica granatum* L.) Provenances (Pooled Basis)

Characters	Rmd thickness	No. of fruits	Fruit length	Height upto first branch	No. of leaves	Acidity	Total sugar	Total soluble solids	Viability	Seedling Height	Diameter	Plant (%)	Percent germmation	Fruit weight
Percent Contribution	0.95	33.33	5.71	3.81	6.67	1.91	11.43	3.81	18.10	9.05	2.86	3.81	2.86	3.81

Observations on seedling parameter *viz.* seedling height, collar diameter, number of leaves per seedling and height up to first branch were taken after one and a half year of sowing *i.e.* in the month of December. Considering fifteen sites as treatment and individual trees of each sites as replication, genetic divergence was estimated by Mahalanobis D^2 statistics and the genotypes were grouped on the basis of maximum generalized distance using Tocher's method as described by Rao, 1952. The data for above characters were taken for two consecutive years and were pooled.

Results and Discussion

Pooled data analyzed for estimating generic divergence between 15 provenances of wild pomegranate ($S_1 - S_{15}$), data of which is presented in Tables 5.2 to 5.6, reveals that the provenances have marked genetic divergence among themselves.

The Mahalanobis D^2 value among provenances is given in Table 5.2 in matrix form and in increasing order in Table 5.3, The maximum D^2 value (416.09) was observed between S_5 and S_{15} provenances followed by 387.20 between S_5 and S_{13} provenances, 332.75 between S_7 and S_{17} provenances. The minimum D^2 value of 19.11 was observed between S_{11} and S_{14} provenances.

The per cent contribution of different characters towards divergence showed that number of fruits contributed to the maximum (33.33%) divergence followed by viability (18.10%), total sugar (11.43%) while the minimum (0.95%) contribution toward divergence was observed for rind thickness (Table 5.4). On the basis of D^2 analysis the genotypes were grouped into six cluster (Table 5.5). The maximum number (six) of genotypes were grouped in cluster-I (S_6, S_9, S_{10}, S_{11}, S_{12}, S_{14}),

Table 5.5 : Clustering of Genetic Divergence Among Different Wild Pomegranate (*Punica granatum* L.) Provenances (Pooled Basis)

Clusters	Population
Cluster I	S_{10}, S_{14}, S_{11}, S_{12}, S_9, and S_6
Cluster II	S_1, S_3, S_2 and S_4
Cluster III	S_{13} and S_{15}
Cluster IV	S_5
Cluster V	S_7
Cluster VI	S_8

Cluster-II had four genotypes (S_1, S_2, S_3 and S_4) while cluster III had two genotypes (S_{13} and S_{15}). The remaining clusters had one genotype representing cluster IV (S_5), Cluster V (S_7) and cluster VI (S_8).

Table 5.6 : Average Intra and Inter Cluster Distance (D values) Among Different Wild Pomegranate (*Punica granatum* L.) Provenances

	I	II	III	IV	V	VI
I	6.65	10.21	9.50	16.20	13.26	8.35
II		7.72	13.99	11.25	9.22	9.53
III			7.14	20.04	17.75	11.28
IV				0	8.12	14.58
V					0	11.29
VI						0

The average intra- and inter-cluster D values are presented in Table 5.6. The generalized intra-cluster (D) distance was maximum in cluster II (7.72) followed by cluster III (7.14) while the minimum distance was in cluster I (6.65). Among inter-clusters maximum genetic difference was recorded between clusters III and IV (20.04) followed by clusters III and V (17.75). The least inter-cluster difference was observed in cluster IV and V (8.12).

The assessment of genetic divergence between the provenances using Mahalanobis generalized distance (D^2) analysis and then grouping of these genotypes in different clusters showed that the selected genotypes were highly divergent among each other. Provenances falling in different clusters are also sharing, by and large, geographical continuity. However, considerable divergence is shown by some of the provenances. Since these genotypes were selected from different locality of Himachal Pradesh, they showed varied differences among each other which may be attributed to the edaphic and/or climatic conditions prevailing in these areas. Besides this, the genetic constitution of this species must be the cause of huge variation in these traits. Therefore, the variation observed may be due to the different genetic architecture as a result of breeding system, level of heterogeneity and adaptation to diverse environmental conditions. The pattern of group cancellations proved that geographical diversity need not necessarily be related to genetic diversity. This was in line with the results obtained by Shwe *et al.* (1972) and Ayyamperumal (1991) in Soybean; Suthamathi and Dorairaj (1994) in Napier grass. This means that geographic diversity, though important, may not be the only factor in determining genetic divergence. Genetic diversity is the outcome of several factors, including topographical diversification. Therefore, selection of genotypes for hybridization should be based on genetic diversity rather than geographical diversity (Singh, 1993).

The clustering of genotypes from different locality into one cluster indicated the genetic similarity among the germplasm coming from different geographic region, as observed by Jaylal (1994) in Soybean; Suthamathi and Dorairaj (1994) in *Pennisetum purpureum*. This may also be due to the fact that the unidirectional selection practised for a particular trait in several places produced similar phenotypes which were aggregated in one cluster irrespective of their distant geographic origin (Singh and Bains, 1968). On the other hand, many genotypes originating from one region were scattered over different clusters. Such genetic diversity among the genotypes of common geographic origin could be due to the factors like heterogeneity, genetic architecture of the populations, past history of selection, developmental traits and degree of general combining ability (Murthy and Arunachalam, 1966).

The study also provided information on the characters that contributed maximum to the total divergence among genotypes. The traits causing maximum towards genetic divergence was number of fruits followed by viability, total sugar while rind thickness contributed minimum divergence. The characters contributing maximum to D^2 value have given more emphasis for the purpose of fixing priority of parents in hybridization programme (Singh, 1993). Low variability for these traits in such a wide variety of genotypes may also suggest high degree of constancy and heritability of these traits.

References

Ayyamperumal, A. 1991. Genetic analysis in *Glycine max* (L.) (Merrill). Ph.D. Thesis, Tamil Nadu Agricultural University, Coimbatore, India.

Bonner, F.T. 1974. Seed testing. In : Seeds of woody plants in the United States. Forest Service, USDA, Washington, D.C. pp. 136-152.

Dumitriu - Tataranu, I. 1970. Studies with a view to the selection of some provenances and forms of larch occurring naturally in Rumania. For. Abst. 37, 792: 1976.

Jaylal, L. 1994. Genetic divergence in soybean for physiological and yield attributes under rainfed conditions. Indian J. Genet. 54(4): 418-424.

Murthy, B.R. and Arunachalam, V. 1966. The nature of divergence in relation to breeding systems in some crop plants. Indian J. Genet. 26 : 188-198.

Pozdnajakov, A.A. 1969. Morphological characters for the visual determination of age generations of *Abies siberica* For. Abst. 32,1243: 1971.

Rao, C.R. 1952. Advanced statistical methods in biometric research. John Willey and Sons. Inc., New York: 60 - 80

Shwe, U.H.; Murthy, B.R., Singh, H.B. and Rao, U.M.B. 1972. Genetic divergence in recent elite strains of Soybean and Ground nut in India. Indian J. Genet. 32: 285-298.

Singh, N.D. 1993. Analysis of genetic divergence in *Bambusa tulda* Roxb. for North-east India. Advances in Hort. and Forestry, 3: 187-191.

Singh, R.B. and Bains, K.S. 1968. Genetic divergence for gaining out turn and its components in upland cotton. Indian J. Genet. 28(3) : 262-268.

Suthamathi, P. and Dorairaj, M.S. 1994. D^2 analysis for fodder yield in Napier grass (*Pennisetum purpureum* (K) Schum). Indian J. Genet. 54(3) : 225-228.

ENVIRONMENTAL BIOTECHNOLOGY

Edited By :   Professor (Dr.) Arvind Kumar

Published By : DAYA PUBLISHING HOUSE

6

PRELIMINARY GENETIC DAMAGE STUDIES ON IMPACT OF USING UNDERGROUND WATER RESOURCES

• G. Gandhi*, R. Kaur, W. Kaur and N. Kumar

Department of Human Genetics, Guru Nanak Dev University, Amritsar

*(Corresponding author)

Abstract

Mahal village lies to the north-west of Amritsar about 0.5 km away from an open waste water drain which collects effluents from various industrial units and municipal wastes. The drain water has heavy metals in high concentrations. Settlements along the drain, like Mahal, use underground water for their anthropogenic activities. The groundwater of Mahal has a high bacterial load and seepage from the drain may also be contaminating it with various substances from industrial effluents as well as heavy metals. Hence a preliminary study was attempted to assess whether usage of this water resource could cause genetic damage in some village residents. Studies for scoring DNA damage using the Single Cell Gel Electrophoresis assay in peripheral blood lymphocytes and the Micro-nucleus test in urothelial cells were carried through. Tail lengths of the comets and number of cells with comets far exceeded the values in an age- and sex-matched control group. The micro-nucleus test also depicted the significantly elevated damage in the residents. The exposure of Mahal residents to genotoxicant(s) cannot be refuted since diverse tissues of the same individuals have exhibited significant damage to the genetic damage. A larger scale investigation and characterization of the genotoxicants will form a sequel of this work.

Key Words : Groundwater usage, Wastewater drain, Genetic damage.

Introduction

Only about 4.9% of the world's natural water is suitable for direct human use of which two-thirds is groundwater. Hence, ground water has always been an important and in some parts of the world even a major source of freshwater supply. However, water sources have been adversely affected both qualitatively and quantitatively by all kinds of human activities on land, in air and in water. Rather, ground water serves as a sink for wastes from accidental spills, run-offs of agricultural and domestic fertilizers, leaky sewers as well as oil tanks. Various groundwater contaminants are known to include substances which are toxic or carcinogenic (Botkin and Keller, 1995, DeFlora *et al.*, 1997). India too has overtly polluted water like other developing countries. The waste water of the city of Amritsar (31.6°E, 74.9°N) has been analyzed for its constituents. Mahajan *et al.* (1994) analysed wastewater from one of the city's drains (Tung Dhab) and reported it to contain heavy metals like zinc, cadmium, lead and copper with amounts varying from different disposal points and in different seasons. The authors suggested that the lower values of heavy metals in June than in December may be due to dilution brought about the greater usage of water and heavy rainfall in June. This Tung Dhab drain water was also analysed by Dhillon *et al.* (1997) for various parameters like pH, electrical conductivity, content of carbonates, bicarbonates, chlorides, calcium, magnesium, sodium, nutrients, potential and the concentration of heavy metals. They too reported the presence of lead, cadmium, nickel, chromium and observed that the quality of the drain water also differed from place to place depending on the discharge of industrial effluents and domestic wastes into it.

Grover and Kaur (1999) reported that the drain water increased the frequency of chromosome aberrations in the Allium cepa root tip cells.

The Tung Dhab drain forms an integral part of the sewerage network of Amritsar city. Its width is 4 ft (starting) and 45 ft (outfall) and it has a length of 20 km, with a catchment area of 80.63 square miles. The discharge of wastewater is at a rate of 400 cusec (outfall) and it has a starting capacity for 31 cusec. The Tung Dhab drain also carries defilement from the household, agriculture wash-off and industrial run-offs. Though the effluents of some industries are treated, however the physico-chemical characteristics of the drain water indicate that it is still a potent source of various constituents both of anthropogenic and non-anthropogenic nature. Along the drain, settlements have arisen with people depending on drain water for agricultural practices and on groundwater resources for their household water supply.

Mahal village is located 8 km to the west of Amritsar. It is a linear settlement extending from 0.5 km from the Tug Dhab drain to about 4.5 km away, covering an area of 30,000 sq ft with 1285 residence housings 6002 individuals (Singh, 2001, personal commun.). The population comprises Jats, Mazbhi Sikhs, Sharmas, Bishnois, and Nais. They are involved in agriculture or work as shopkeepers, electrician, shoemakers, richshaw pullers and government employees. There is no sewerage network in the village. The residents use handpump drawn water from 60-70 ft deep bores. The underground water of Mahal has also been reported to contain a high bacterial load (Dhami and Thukral, 2001). The effect of utilization of ground water (which has a high probability of being contaminated from village activities and seepage from Tung Drain) on the genetic make-up of the residents of Mahal village was hence investigated using both the Single Cell Gel Electrophoresis (SCGE) assay and the Micro-nucleus (MN) test.

The single cell gel electrophoresis or comet assay has been successfully applied to evaluate the genotoxicity of chemicals. It is also used for environmental biomonitoring and may also be considered a very important tool for (geno) toxicological evaluation (Tice *et al.*, 1992, Betti *et al.*, 1993).

Micro-nuclei are accentric chromosome or chromatid fragments or whole chromosomes which fail to incorporate into the main nucleus at cell division (Fenech and Morley, 1985). The MN test on exfoliated cells has been extensively used to investigate both clastogenic and aneugenic effects arising from exposure to drugs and environmental contaminants (Santelli *et al.*, 1994, Moore *et al.*, 1996).

Materials and Methods

The present study was performed on a small group of non-related village residents as a preliminary investigation to observe whether usage of underground water resources can cause genetic damage in different tissues by scoring different genetic end-points so that a larger scale investigation may be undertaken. For this an initial survey was conducted to select residents of Mahal who were staying in the village for a long period or were born there. The socio-economic status of the selected residents was assessed by personal visits to their homes. Individual records for the residents were kept on pre-designed questionnaires (Carrano and Natarajan, 1996) for anamnesic data and on period of stay, use of ground water, bore length, disease incidence, etc. Detailed reproductive and family histories (upto three generations) for each selected resident were also recorded (Bennett *et al.*, 1995). The selected residents (n=10) and an age-, sex- and socio-economic status-matched control group (n=12) were explained the nature of the study and their written informed consent was taken.

The SCGE Assay : This is a highly sensitive technique for detecting DNA damage, very small amount of blood (100μl) is required, it is easy to perform and the results are obtained in a short time (Vijayalaxmi *et al.*, 1992). The protocol given by Singh *et al.* (1988) was slightly modified to accommodate the use of agarose-coated slides in lieu of frosted ones (Ahuja and Saran, 1999) and the use of silver nitrate to stain the comets instead of ethidium bromide (Delincee, 1985). Finger-prick blood samples (peripheral blood lymphocytes, PBLs; 300μl) were collected in microfuge tubes containing heparin and phosphate buffered saline (PBS). In brief, plain glass slides were layered with 1% normal melting point agarose (NMPA) subsequently coated with blood sample (25μl of whole blood and 0.5% of low melting point (LMP) agarose) and a final coat of 0.5% (NMPA) to sandwich the sample. The cells were lysed, electrophoresed and stained. The slide preparations were coded and scored under 40X of a binocular microscope. DNA damage resulting in fragmentation was observed to form comets with the tails towards the anode. For each sample, 500 cells were scored for comets and the comet lengths were measured using a calibrated ocular micrometer.

The MN test : This test is applicable to any exfoliated cells which can be easily obtained and the exfoliated urothelial cells have an advantage as they can be collected using a non-invasive procedure. The urothelial cells in the urine encounter abnormal osmotic pressures, pH changes and exposures to toxic metabolite. Therefore for the present study, the MN test on urothelial cells was performed after the method of Lehucher-Michel *et al.* (1996). Urine samples (5-7ml) were collected in phosphate buffered saline and transported on ice to the laboratory, centrifuged for 20' at 2500rpm and the cell pellet was fixed in chilled Carnoy's fixative. The suspension was centrifuged at 2500rmp for 10' and the cell pellet was used to make a smear preparation on a clean glass slide. This was stained with 10% Giemsa and coded. A total of 500-1000 cells were scored per individual at 40X for the presence of micro-nuclei which were then confirmed at 100X.

Results and Discussion

In this preliminary study, a group of ten Mahal residents (7 males, 3 femals) and 12 healthy individuals (6 females, 6 males – neither residing near an open wastewater drain nor drinking or using underground water), in the range of 22-60 years were investigated both for DNA damage in their PBLs and for aneugenic and/or clastogenic damage in their urothelial cells (Tables 6.1 and 6.2). All the females are housewives while the males work as agricultural labourers, general merchants and security staff. There was no incidence of disease in the subjects and their reproductive histories also did not reveal any aberrance. There were 7 individuals who had been living in Mahal since their birth and so have been

Table 6.1 : The Comet Assay and the Micronucleus Test in the Some Residents of Mahal Village

Age (yrs)	Duration of stay (yrs)	Occu-pation	Resi-dence from drain (km)	G vs. T+G	SES	Depth of bore (ft.)	Alcohol intake	Veg (V)/ Non-Veg (NV)	Smok-ing history	No. of Children	No. of PBLs scored	No. of dama-ged cells (%)	Mean Migra-tion length $[X]$ (μm)[a] ± S.E.	No. of Urothelial cells scored	No. of MNd cells	Frequ-ency of MNd cells
22	22(B)	Agric. labourer	0.5	G	M	70	+	NV	–	Un	500	93(18.6)	11.5±0.13	494	2	0.405
25	25(B)	Painter	0.5	G	L	70	–	V	–	3(2D, 1S)	500	100(20.0)	12.0±0.14	756	3	0.39
32	32(B)	Agric. Labourer	0.7	G	M	80	+	V	–	2 (1D, 1S)	500	105(21.0)	12.6±0.11	617	6	0.97
44	10	Security staff	0.7	G	M	70	–	V	–	2 (1D, 1S)	500	98(19.6)	11.3±0.11	650	4	0.61
50	50(B)	Agric. labourer	0.5	G	L	60	+	NV	–	2S	500	110(22.0)	12.0±0.14	509	2	0.39
48	10	General Merchant	0.5	T+G	M	70	–	V	–	4 (3D, 1S)	500	104(20.8)	12.6±0.11	902	6	0.66
60	35	General Merchant	0.5	G	L	70	–	NV	–	4 (1D, 3S)	500	120(24.0)	14.6±0.22	952	6	0.63
Total											3500	733(20.9)	12.6* ± 0.43	4880	29	0.57* ± 0.07
24	24(B)	Housewife	0.7	G	M	70	–	V	–	Un	500	86(17.2)	11.5±0.13	811	3	0.369
26	26(B)	Housewife	0.5	G	M	80	–	V	–	1S	500	83(16.6)	11.5±0.13	810	8	0.987
35	35(B)	Housewife	0.7	G	L	60	–	V	–	2S	500	100(20.0)	12.6±0.11	752	7	0.930
Total											1500	269(17.9)	11.8* ± 0.29	2373	18	0.76* ± 0.16
Grand Total											5000	1002 (20.04)	12.2* ± 0.12	7253	47	0.66* ± 0.03

* – Significant when compared to total C group ($p \leq 0.05$ and $p \leq 0.01$; Student's 't' Test).

a – Calculated as an average of individual DNA migration lengths of comets in that group.

Un – Unmarried; D – Daughter; S – Son; Un–Unmarried; no miscarriages/abortions/decreased children reported.

G – Groundwater; T+G – Tap and groundwater use; SES – Socio-economic status; M – Middle, L – Low.

using the ground water resources. Alcohol intake (n=3) and food habits (n=3) non-vegetarians) could be confounding factors only among some males since these habits are absent in the female group.

Table 6.2 : The Comet Assay and the Micronucleus Test in Normal Healthy Individuals (Control Sample)

Age (yrs.)	Occupation	SES	Alcohol intake	Veg (V)/ Non-veg (NV)	Smoking history	No. of children	Total cells scored	Cells showing DNA migration No. (%)	Mean migration length[a] (X) (μm) ± S.E.M.
The SCGE Assay									
22(F)	Housewife	M	–	V	–	Un	500	45(9)	6.0±0.20
25(F)	Housewife	M	–	V	–	1	500	30(6)	6.6±0.24
35(F)	Housewife	M	–	V	–	2	500	30(6)	6.6±0.24
Total (n=3)							1500	105(7.0)	6.4±0.16
24(M)	Sanitary worker	M	–	V	–	Un	500	35(7.0)	6.6±0.24
48(M)	Mess worker	M	–	V	–	2	500	36(7.2)	6.6±0.20
54(M)	Canteen worker	L	+	V	+	3	500	55(11.0)	8.0±0.32
Total(n=3)							1500	126(8.4)	7.0±0.38
Grand Total (n=6)							3000	231(7.7)	6.7±0.12

The Micronucleus Assay

Age (yrs.)	Occupation	SES	Alcohol intake	Veg (V)/ Non-veg (NV)	Smoking history	No. of children	No. of cells scored	No. of Micro-nucleated cells	Frequency of Micro-nucleated cells (%)
23(F)	Housewife	M	–	V	–	3	848	–	–
35(F)	Housewife	L	–	V	–	3	615	–	–
27(F)	Housewife	M	–	NV	–	3	724	–	–
Total (n=3)							2187		
20(M)	House servant	L	–	V	–	Un	689	–	–
35(M)	Rickshaw puller	L	–	NV	+	2	692	–	–
45(M)	Health worker	M	–	V	–	3	1025	–	–
Total (n=3)							2406		
Grand Total (n=6)							4593	–	–

a – Calculated as an average of individual DNA migration lengths of comets in that group;

M – Male; F – Female;

D – Daughter; S – Son; Un – Unmarried; no miscarriages/abortions/deceased childred reported;

SES–Socio-economic status; M – middle; L - Low.

Comets were observed in the PBLs of all individuals but with more cells showing damage in the study (20.04%) than in the controls (n=6; 7.70%). The DNA migration lengths (range 11.3±0.11 μm to 14.6±0.22 μm) were also significantly elevated in the residents (12.62±0.12 μm versus 6.70±0.12 μm in controls; tcal = 13.8; ttab = 4.14, df = 14, p≤0.001). There was however no significant difference for comet lengths between male and female residents. All the residents also had micro-nucleated (MNd) cells (2-8 MN/cells scored) but there were none in the controls. The per cent frequency of MNd cells was elevated in female residents (0.76±0.16) as compared to that in males (0.57±0.07). Therefore, these results indicate that DNA as well as chromosomal damage is evident in different tissues of Mahal residents.

Contaminated water bodies with industrial effluents from dump sites, paper and pulp units and sewage have been documented to induce genetic damage (Werner *et al.*, 1994, Mersch and Beauvis, 1997, Perez-Alzola and Santos, 1997, Zeng *et al.*, 1999). The heavy metals in the drain are also known genotoxicants and or carcinogens (Vaglenov *et al.*, 1999, Blasiak and Kowalik, 2000, Valverde *et al.*, 2001). Since the percolating water can pick up a large number of heavy metals and reach the aquifer system so it can contaminate the ground water resources (Jain *et al.*, 2000). There is also generally higher proportion of dissolved constituents in ground water than in surface water because of greater interaction of ground water than in surface water with various materials in geologic strata. Similarly, the presence of heavy metals and other industrial effluents in the drain and their percolation into the Mahal underground water system has potential to cause genetic damage, both at the DNA and chromosomal levels, as observed in the present pilot study. The small size of the study group highlights and throws into prominence the nexus between ground water usage and the mutagenic potential of its components. However, water analysis can further elucidate the nature of these toxicants and lay the ground for an exhaustive investigation since there are very many other Mahal like settlements along the Tung Dhab drain.

References

Ahuja, Y.R. and Saran, R. 1999. Alkaline single cell gel electrophoresis assay I. Protocol. Cytol Genet 34(1): 57-62.

Blasiak, J. and Kowalik, J. 2000. A comparison of the *in vitro* genotoxicity of tri- and hexa-valent chromium. Mutat. Res. 469 (1): 135-145.

Betti, C., Davini, T., Giannsesi, L., Loprieno N. and Barale, R. 1994. Microgel electrophoresis assay (comet test) and SCE analysis in human lymphocytes from 100 normal subjects. Mutat. Res. 307(1): 323-333.

Bennett, R.L., Steinhaus, K.A., Uhrich, S.B., Resta, R.G., Lochnerdoyle, D., Markel, D.S., Vincent, V. and Hamnishi, J. 1995. Recommendations for standardized human pedigree nomenclature. Am. J. Hum. Genet. 56(3): 745-52.

Botkin, D. and Keller, E. 1995. Environmental Science : earth as a living planet. John Wiley and Sons, New York, pp. 373-420.

Carrano, A.V. and Natarajan, A.T. 1986. Minimal criteria for population monitoring using cytogenetic techniques. Hand out prepared for WHO and IARC sponsored international course on "Health monitoring of human populations exposed to mutagens and carcinogens" at Cancer Research Institute, Tata Memorial, Bombay, 3(14): 1-40.

De Flora, S., Camoirano, A., Bagnasco, M., Bennicelli, C., Corbett, M.G.E. and Kerger, B.D. 1997. Estimates of the chromium (VI) reducing capacity in human body compartments as a mechanism for attenuating its potential toxicity and carcinogenicity. Carcinogenesis 18(3) : 531-537.

Delincee, H. 1995. Silver staining of DNA in the comet assay. Comet Newsletter #3, 1995.

Dhillon, M.S., Sandhu, R.S. and Mahajan, R.K. 1997. Physico-chemical characteristics of sewage water of Tung Dhab drain. In J. Environ. Protec. 17(8): 613-616.

Dhami, K. and Thukral, A.K. 2001. Potability of the ground water of some areas adjoining seasonal drain at Amritsar. In : Environmentally sustainable trade and development (Eds. J.S. Ahluwalia and P. Chaturvedi) World Environment Foundation, New Delhi, pp. 369-373.

Fenech, M. and Morley, A.A. 1985. Measurement of micro-nuclei in lymphocytes. Mutat. Res. 147: 29-36.

Grover, I.S. and Kaur, S. 1999. Genotoxicity of wastewater samples from sewage and industrial effluents detected by Allium root anaphase aberration and micro-nucleus assay. Mutat. Res. 426(2): 183-191.

Jain, C.K., Bhatia, K.K. and Kumar, S.R. 2000. Ground water contamination in greater Guwahati, Assam. J. Environ. Protec. 20(9): 641-648.

Lehucher-Michel, M.P., Amara-Mokrane, Y.A., Devictor, B. and Catilina, P. 1996. Micro-nuclei kinetics of exfoliated urothelial cells. Mutat. Res. 354: 1-7.

Mahajan, A., Sandhu, R.S. and Mahajan, R.K. 1994. Simultaneous evalnation of heavy metals in wastewater of Amritsar city by Stripping Voltametric technique. Ind. J. Environ. Protec. 14(1): 818-820.

Mersch, J. and Beauvais, M.N. 1997. The micronucleus assay in the Zebra mussel, Dreissena polymorpha, to *in situ* monitor genotoxicity in fresh water environments. Mutat. Res. 393(1-2): 141-149.

Moore, L.E., Smith, A.H., Hopenhayn, R.C., Biggs, M.L. and Kalman, D.A. 1997. Micro-nuclei in exfoliated bladder cells among individuals chronically exposed to arsenic in drinking water. Cancer Epidemiol. Biomarkers Prev. 6(1): 31-36.

Perez-Alzola, L.P. and Santos, M.J. 1997. *In vitro* genotoxic evaluation of conventional bleached and bio-bleached soft wood pulp mill effluents. Mutat Res. 395(2-3): 107-112.

Santelli, M.G.M., Cerqueria, E.M., Oliveira, C.T. and Pereira, C.A. 1994. Biomonitoring of nurses handling antineoplastic drugs. Mutat. Res. 322(3): 203-208.

Singh, K. 2001. (Mr. Karam Singh, male health worker from the Department of Health and Family Welfare, Punjab Govt. provided the information on statistics and residents of Mahal village on personal basis).

Singh, N.P., McCoy, M.T., Tice, R.R. and Schnieder, E.L. 1988. A simple technique for quantification of low levels of DNA damage in individual cells. Exp. Cell. Research. 175(1): 184-191.

Tice, R.R., Strauss, G.H.S. and Peters, W.P. 1992. High dose combination of alkylating agents with autologous bone marrow support in patients with breat cancer. Preliminary assessment of DNA damage in individual peripheral blood lymphocytes using the single cell gel electrophoresis assay. Mut. Res. 271: 101-113.

Vaglenov, A., Nosko, M., Georgieva, R., Carbonell, E., Creus, A. and Marcos, R. 1999. Genotoxicity and radio resistance in electroplating workers exposed to chromium. Mutat. Res. 446(1): 23-24.

Vijaylaxami, Tice, R.R. and Strauss, G.H. 1992. Assessment of radiation-induced DNA damage in human blood lymphocytes using the single-cell gel electrophoresis technique. Mutat. Res. 271(3): 243-252.

Valverde, M., Trejo, C. and Rojas, E. 2001. Is the capacity of lead acetate and cadmium chloride to induce genotoxic damage due to direct DNA-metal interaction. Mutagenesis. 16: 265-270.

Warner, M.L., Moore, L.E., Smith, M.T., Kalman, D.A., Fanning, E. and Smith, A.H. 1994. Increased micro-nuclei in exfoliated bladder cells of individuals who chronically ingest aresenic contamination water in Nevada. Cancer Epidemio. Biomarkers. Prev. 3: 583-590.

Zeng, D., Li, Y. and Lin, Q. 1999. Pollution monitoring of three rivers passing through Fuzhou city, People's Republic of China. Mutat. Res. 426: 159-161.

ENVIRONMENTAL BIOTECHNOLOGY

Edited By :   Professor (Dr.) Arvind Kumar

Published By :   DAYA PUBLISHING HOUSE

7

ROLE OF MOLECULAR MARKERS IN IMPROVEMENT OF COLE CROPS : AN OVERVIEW

• *Rajendra Kumar** and *P.R. Kumar*
* Senior Scientist, Division of Genetics, IARI, New Delhi-110012
Indian Agricultural Research Institute, Regional Station,
Katrain (Kullu Valley), Himachal Pradesh

Abstract

Cole crops comprises a group of highly differentiated plants *e.g.* cauliflower, cabbage, Kale, sprouting broccoli, Knol khol and Brussels sprouts originated from a single wild *Brassica oleracea* var. *oleracea* (*sylvestris* L.) commonly known as wild cabbage. Cole crops are presently grown all over the world from the tropical to the arctic regions. In India cauliflower and cabbage are the two most important crops grown while kohlrabi, though grown in certain areas, yet to be as popular as the other two. Of the two, cauliflower is most important covering more area than cabbage.

The conventional plant breeding has made tremendous contributions to the breakthrough in the global cole crops' production. The majority of the economically important characteristics of crop plants being polygenically inherited, have been characterized and manipulated by the help of quantitative genetic analysis. Recently the powerful conventional genetic approaches are being supplemented by development in the field of molecular biology.

Molecular markers

Molecular markers are heritable differences in nucleotide sequences of DNA from two different individuals, which follow a simple Mendelian pattern of inheritance. These differences are detected employing two basic techniques: (a) Southern blot hybridization and (b) Polymerase chain reaction (PCR). Restriction Fragment Length Polymorphism (RFLP) is the first molecular marker system to be conceived and developed (Botstein, 1980). It is based on the southern blot technique designed by Southern (1975). In this, a class of enzymes called restriction endonucleases is used, which recognizes a specific nucleotide sequence called target site and cleave large DNA molecules into small pieces at these sites. The DNA fragments of varying sizes so obtained are separated by electrophoresis in agarose gels and transferred to nylon membranes. The specific fragments on the membrane are identified by hybridization of a probe DNA followed by autoradiography. Size of the hybridizing fragments depends on the restriction sites. Difference in the location of the restriction sites in the DNA from different individuals/varieties leads to the variation in size of the hybridizing fragments and thus results in RFLP. On the other hand, markers such as random amplified polymorphic DNA (RAPD), sequence tagged site (STS) and amplified fragment length polymorphism (AFLP) are generated by enzymatic amplification of DNA by extension of oligonucleotide primers and therefore, are based on PCR.

The molecular markers offer several advantages over the other markers. These include: (1) abundance (2) co-dominance (3) phenotypic neutrality, (4) absence of epistasis and (5) developmental stage, tissue and environment independent expression. Large number of molecular markers can be obtained by using DNA probes from a variety of sources in combination with any of the commercially available restriction enzymes and a variety of DNA primers. This enables study of several markers in a single population for construction of high density linkage maps of cole plant genomes. The above attributes make molecular markers ideal for characterization of quantitative traits.

Construction of complete linkage maps

For a directed use of molecular markers in gene mapping and markers aided selection, it is essential to construct a high density complete linkage map. A linkage map consists of linkage groups, the number of the group being equal to the number of haploid chromosomes of the species under study. Each linkage group, in turn consists of genetic markers in a definite order. The distance between any two markers is indicated in centimorgan (cM), which is a function of the extent of recombination between the markers. Construction of linkage map involves the following steps: (1) Identification of the divergent parents, (2) Generation of mapping population, (3) Identification of polymorphic DNA probes and enzyme combinations/polymorphic DNA primers, (4) Establishing linkage, (5) Identification of linkage groups and (6) Assigning linkage groups to chromosomes.

Use of divergent parents helps in obtaining maximum number of markers due to high level of polymorphism at DNA level. Simultaneously, mapping and tagging of genes for which the parents differ, are facilitated. Once the marker is identified, its segregation is analysed in a mapping population. Four different mapping populations *viz.*, F_2, BC_1, doubled haploids and recombinant inbred lines (RILs) commonly used in genome mapping are derived from the F_1 hybrid of the two selected parents. The first three populations take the least time for their generation, whereas RILs take the maximum time (5-7 generation of selfing). Doubled haploids are obtained from *in vitro* culture of pollens/anthers taken from the hybrid. For the purpose of mapping genes underlying the quantitative variation, however, doubled haploids and RILs are better than F_2, since they can be replicated over locations and seasons.

Employing two point linkage test all markers are compared pair-wise. Markers showing linkage with each other are included in a group. The principle of maximum likelihood is applied to test the validity of linkage. A LOD score of 3 is taken as the minimum threshold. Once the linkage groups are identified, they are assigned to the chromosomes by the use of (1) *In situ* hybridization, (2) Known genetic or cytogenetic markers, (3) Chromosome stocks etc. A linkage map is considered complete and saturated if the number of the linkage groups correspond to the number of the haploid chromosomes and addition of more markers does not result in any significant increase in map length. The linkage map of rice is the most saturated among the crop plants (Harushima *et al.*, 1998). Several maps have already been developed for the cole crops involving crosses between different crops (Slocum *et al.*, 1990, Landry *et al.*, 1992, Kianian and Quiros, 1992, Quiros *et al.*, 1994; Camargo *et al.*,

1997, Kearsey *et al.*, 1996, Ramsay *et al.*, 1997 and Hu *et al.*, 1998). In the latest work Hu *et al.*, 1998 have assigned most of the linkage groups of aforesaid maps (developed by Kianian and Quiros, 1992, Landry *et al.*, 1992 and Camargo *et al.*, 1997). The base map developed in this study consisted of 167 RFLP loci in nine linkage groups plus eight markers in four linkage pairs covering 1738 cM.

Mapping of Genes

Molecular markers have been extensively used to map and tag many genes of agricultural importance influencing both qualitative and quantitative traits. The qualitative traits are monogenically inherited and show discrete variation. On the other hand, the quantitative traits are polygenically controlled show continuous variation. The methods used to map genes for these two kinds of traits are different. These are briefly outlined further.

(a) Monogenic traits

Genes for a few economic characters *viz.*, clubroot disease resistance (Landry *et al.*, 1992, Figdore *et al.*, 1993), nuclear recessive gene for male sterility (Barbeyron *et al.*, 1998) have already been mapped in the cole crops. Mapping of genes for monogenic traits in crop plants has involved three approaches: (i) use of a complete linkage map, (ii) use of near isogenic lines, and (iii) bulked segregant analysis of F_2 population. The first approach essentially requires availability of a complete genetic linkage map. Such maps need to be constructed based on segregation and linkage analysis of a large number of molecular markers using a mapping population (F_2, backcrosses, doubled haploids or recombinant inbred lines), which is generated from the F_1 hybrid of two divergent parent varieties/ lines. For the purpose of mapping a gene for disease resistance, markers are selected from all the chromosomes such that the maximum distance between any two adjacent markers does not exceed 20 centimorgan (cM). Co-segregation of any of these markers with trait phenotype in the mapping population reveals linkage. This facilitates immediate localization of the gene on the linkage map, since marker positions are already known. Although laborious, it is the most directed method of gene mapping.

The other two approaches do not require a linkage map. For instance, in case of the use of near-isogenic lines (which are developed by repeated backcrossing of the F_1 with the recipient parent and therefore, differ for a single gene), DNA markers distinguishing the two lines (say resistant *vs.* susceptible) are first identified. Co-segregation of these markers with the traits is analysed in a mapping population, often derived from the cross of the two isogenic lines (Martin *et al.*, 1991). Development of near-isogenic lines, however, takes many years. This method of gene mapping therefore, is best suited to crops, in which intensive backcross breeding has already been practised to generate isogenic lines.

Use of F_2 population in conjunction with a technique called bulked-segregant analysis, is the most rapid way of mapping genes (Michelmore *et al.*, 1991). In this approach, DNA from ten homozygous resistant F_2 plants are bulked to constitute a resistant bulk. Similarly, a susceptible bulk is made out of DNA from the homozygous susceptible F_2 plants. The DNA markers, which distinguish these two bulks are considered to have linkage with the target gene. Although, molecular linkage map is not a pre-requisite for its success, use of markers with known map position imparts greater efficiency to gene mapping in this approach.

(b) Polygenic traits

In cole crops QTLs for flowering time, petiole length (Camargo and Osborn, 1996), Clubroot resistance (Grandclement *et al.*, 1996, Landry *et al.*, 1992) and Black rot resistance (Camargo *et al.*, 1995) have already been reported. The method applied for characterization of monogenic traits are not appropriate for studying polygenic traits because of the difficulty in categorizing the individual plants of a mapping population into distinct phenotypic classes due to interaction of many genes and a significant role of environment in trait expression. Therefore, different statistical methods have been devised to deal with such traits.

The basic principle of determining whether a QTL is linked to a marker is to partition the mapping population into different genotypic classes based on genotypes at the marker locus and then use correlative statistics to determine whether the individuals of one genotype differ significantly with individuals of the other genotype with respect to the trait being measured. If the phenotypes differ significantly, it is interpreted that a gene(s) affecting the trait is linked to the marker locus used to classify the population. Three main methods of QTL mapping are (1) Single point/marker analysis, (2) Interval mapping and (3) Composite interval mapping.

Application in cole crops' improvement

Traditionally, desirable segregants are selected based on expression of the phenotype in an appropriate selection environment. Creation of such environments for large segregating populations routinely over generations is often difficult. In addition, identification of segregants in early generations requires greater efforts and time, when the trait is controlled by recessive genes. Moreover, it is impossible to pyramid genes for resistance for many races of pathogen based on resistance scoring instead of, if presence of only one gene offers some degree of resistance to all the races. Under these circumstances, markers can play a very significant role. Tightly linked markers located on either side of the target gene allow most efficient selection. Although, this method is yet to be integrated with conventional crop improvement schemes, it's potential use has already been demonstrated in rice by Dr. G.S. Khush and his colleagues at IRRI, Phillippines. Individual F_2 segregants carrying four genes. Xa4, Xa5, Xa13 and Xa21, which offer resistance to four different races of leaf blight pathogen *Xanthomonas oryzae*, could be efficiently identified using RFLP and PCR marker. With the development of saturated linkage maps and identification of PCR based flanking markers, marker assisted selection is going to be routinely practised in crop improvement schemes.

Employing molecular markers and appropriate statistical methods it is possible to characterize the polygenic traits in great detail. It is now possible to determine more precisely the number of environment interaction, presence of orthologous (QTLs) and genetic basis of heterosis and transgressive segregation. Specific areas of application of molecular markers in cole crop breeding includes :(1) Characterisation of germplasm resources, (2) Estimation of genetic diversity, (3) Identification and protection of crop varieties, (4) Marker assisted selection of QTLs for developing improved inbreds, varieties and hybrids, (5) Pyramiding of genes for biotic and abiotic stress tolerance, (6) Assessment and maintenance of genetic purity of A, B, R lines as well as inbreds, (7) Organelle DNA in testing genetic purity, and (8) Testing the genetic purity of commercial hybrids' seed lot.

Future Prospects

During the last decade, there has been rapid progress in this field, Several molecular marker systems have been developed. Saturated linkage maps have been constructed. Many traits of economic importance have been analyzed in great detail. Fine mapping of the QTLs has already been initiated in few cases. It is now possible to routinely practice marker assisted selection for qualitative traits in crops like rice, tomato, wheat, vegetables etc. In near future, MAS should be reality at least for the major QTLs. Ultimately it should be possible to isolate, clone and characterize the QTLs at molecular level.

References

Barbeyron, G., Boury, S., Thomas, G. *et al.* 1998. *Acta Horticulture*, 459: 149-156.

Botstein, D., White, R.L., Skolnick, M., *et al.* 1980. *Am. J. Human. Genet.*, 32: 314-331.

Bouhon, E.J.R., Keith, D.J., Parkin, I.A.P., *et al.* 1996. *Theor. Appl. Genet.*, 93: 833-839.

Camargo, L.E.A., Savides, L., Jung, G., *et al.* 1997. *J. of Hered.*, 88: 57-59.

Camargo, L.E.A., Williams, P.H. and Osborn, T.C. 1995. *Phytopathology*, 85: 1296-1300.

Figdore, S.S., Ferreira, M.E., Slocum, M.K., *et al.* 1993. *Euphytica*, 69: 33-34.

Grendclement, C., Laurent, F. and Thomas, G. 1996. *Theor. Appl. Genet.*, 93: 86-90.

Harushima, Y., Yano, M. *et al.* 1998. *Genetics*, 148: 479-494.

Hu, J., Sadowsky, J., Osborn, T.C. *et al.* 1998. *Genome*, 41: 226-235.

Kearsey, M.J., Ramsay, L.D., Jennings, D.E. *et al.* 1996. *Theor. Appl. Genet.* 92 363-367.

Kianian, S.F. and Quiros, C.F. 1992. *Theor. Appl. Genet.*, 84: 544-554.

Landry, B.S., Hubert, N., Crete, R. *et al.* 1992. *Genome*, 35: 409-419.

Mohapatra, T., Sharma, R.P. and Chopra, V.L. 1992. *Curr. Sci.*, 62: 482-484.

Mohopatra, T. 1997. Proceedings of the National seminar on Dec. 29-31, 1997, at C.M.S. College, Kottayam, Kerala.

Quiros, C.F., Hu, J. and Truco, M.J. 1994. *In :* Phillips, R.L. and Vasil, I.K. (eds.) *Advances in cellular and molecular Biology of Plants*. Vol. 1, *DNA based markers in plants*. Kluwer Academic Publ. Dordrecht, pp. 199-122.

Rafalsky, J.A. and Tingey, S.V. 1993. *Trends Genet.*, 9: 275-279.

Ramsay, L.D., Jennings, D.E., Bohuon, E.J. *et al.* 1997. *Genome*, 39 : 558-567.

Slocum, M.K., Figdore, S.S., Kennard, W.C. *et al.* 1990. *Theor. Appl. Genet.*, 80: 57-64.

Southern, E.M. 1975. *J. Mol. Biol.*, 98: 503-517.

ENVIRONMENTAL BIOTECHNOLOGY

Edited By :   **Professor (Dr.) Arvind Kumar**

Published By :   **DAYA PUBLISHING HOUSE**

8

RECYCLING OF RAW AND TREATED TEXTILE EFFLUENT BY FOREST TREES (*ACACIA FERRUGINEA* AND *DELONIX REGIA*)

• *R. Manivanan*

Mathematical Modelling Centre,
Central Water & Power Research Stations, Pune (India)

Abstract

The paper deals with recycling of textile effluent by irrigation to forest tree species such as *Acacia ferruginea* and *Delonix regia*. Height values are measured with reference to control, raw and treated effluent irrigation. *A. ferruginea* found to be low stem height where as *D. regia* is hight when irrigated with treated effluent. With reference to raw effluent irrigation, *D. regia* stem height is more than *A. ferruginea*. In the case of well water irrigation. *A. ferruginea* seen to be higher than *D. regia*. Hence, *D. regia* is observed to be more tolerant tree species with respect to raw and treated effluent irrigation. The statistical analysis like minimum, maximum, average, mode, median, correlation, covarience, standard deviation, standard error, frequency, skewness and kurtosis are calculated. All the above parameters are positive except kurtosis. More frequency is observed in treated effluent irrigation. Correlation found to be less than one in all the irrigation type.

Key Words : *Acacia ferruginea, Above ground height, Bicarbonates, Calcium, Carbonates, Delonix regia, Irrigation, Magnesium, Minerals, pH, Raw effluent and Treated effluent.*

Introduction

Rapid industrialisation and urbanization lead to the generation of large volumes of wastewater, the treatment and disposal of which can cause considerable deterioation of water quality. About 13,153 million gallons of wastewater is released every day from various urban and rural centres in India (Vimal & Talashikar, 1985). Short rotation forests consisting of *Salix* species were found to be suitable, vegetation filters with regard to the removal of nutrients from municipal sewage, sludge, leakage water and bioash. The municipal waste is purified by using natural circulation and production of biomass for energy purposes, (Perttu, 1993). Industrial effluents are considered as pollutants due to the presence of organic and inorganic compounds and suspended solids. Such industrial effluents destroy the fragile ecosystem. One of the most important industrial effluent is textile factory wastewater. These factories are located mainly in Gujarat, Maharashtra, Haryana, Karnataka and Tamil Nadu. Erode is a place known for several industries such as tanneries, oil mills and dyeing and textile factories. The textile mills consume large amount of water for various unit processes such as sizing, desizing, bleaching, mercerizing, dyeing and printing and discharge large amount of wastewater containing different types of toxic pollutants in to fresh water ecosystem. It is estimated that 12.65 litres of water is required for processing one metre of cloth which results in about 20 to 30 ppm of biological oxygen demand (Agarwal *et al.*, 1994). The wastewater from municipal sewage treatment plants and most industrial plants were discharged into streams, lakes or the ocean. Nutrients from the effulent accelerate eutrophication with consequent deterioration of water quality. These wastewater generally hindered the trees survival and growth. In a case study of teak plantations, irrigated with sewage observed that growth is significantly higher than the growth in ordinary water (Dakshindas *et al.*, 1995), Gowda *et al.* (1995) have shown the municipal wastewater could be used for energy production and raising forestry in rural areas. These studies clearly indicate the irrigational and manurial potential of sewage waste for the production of arboreal biomass. In this paper, to study the irrigational and manurial potential of textile effluent on above ground height of *A. ferruginea* which is timber yielder and D. *regia* is fast growing avenue tree species. The statistical analysis showed good performance of above ground height in D. *regia*.

Materials and Methods

The raw and treated effluents were collected from a textile industry situated at Erode, Tamil Nadu. An analysis was made (APHA, 1981) on the physico-chemical parameters, such as pH, electrical conductivity, calcium, magnesium, carbonates, and heavy metals like copper, zinc, iron and lead. Seedlings (16 numbers in each tree species) were transplanted from nursery to the plot in the area of 36 metres. The spacing was made on 2 × 2 metres size. The drip irrigation was done at the rate of litre per day with well water (T_1), raw effluent (T_2) and treated effluent (T_3). The above ground height was measured with the help of centimetre scale from stem at basal ground to the first leaf of tender shoot at an interval of 20 days upto the period of one year (Manivanan and Das, 2002).

Results and Discussion

The collected textile wastewater is analysed for physico-chemical parameter (Manivanan, and Dass, 2002) The higher amount of chemicals in treated effluent is due to the addition of urea and copper sulphate in the raw effluent as a chemical treatment (Table 8.1). The maximum height values are obtained for *A. ferruginea* in treated effluent up to a period of 80 days, from 100 to 140 days, in well water, from 160 to 180 days, in treated effluent, from 200 to 360 days in well water. During the early stages up to a period of 40 days average minimum values were recorded in well water. From 60 to 260 days the lower values were observed in raw effluent and in the later stages (260 to 320 days) lower values were found in treated effluent. From 340 to 360 days period, height was less in raw effluent. Though, there is a fluctuation in the height at various treatments during the study period. The overall maximum above ground height was obtained in well water and minimum in raw effluent.

The maximum above ground height for *D. regia* are observed in treated effluent and minimum in well water for 20 days period. From 40 to 80 days, it was minimum in raw effluent. The lower values were seen in well water irrigation for a period of 100 to 220 days, from 240 to 280 days, it is in raw effluent. Finally, the less height values are recorded in well water during 300 to 360 days period. Though, the maximum above ground height was consistently obtained in treated effluent throughout the study period, lower values of above ground height is fluctuated in well water and raw effluent. The statistical analysis shows that good effluent irrigation with *D. regia* (Table 8.2). Minimum indicated as a initial stage of above ground height and maximum showed final stage of the experiment. With reference to well water irrigation, mode values are 34.5 and 36 in *A. ferruginea* and *D. regia* respectively. Median values are 204.1 and 183.3. Raw effluent and treated effluent irrigations are more than the well water irrigation. The correlation results are less than one for all type of treatments. The data observed for covarience are higher than all values. Kurtosis is negative in all the cases.

Table 8.1 : Physico-chemical Analysis of Textile Effluents Before (1) and After (2) Treatment

Sr. No.	Parameters	Before (1)	After (2)
1.	pH	9.5	8.0
2.	E. Conductivity (Micro mhos)	6.8	5.7
3.	Calcium (mg/l)	0.7	0.5
4.	Magnesium (mg/l)	4.6	4.0
5.	Carbonates (mg/l)	532.0	531.0
6.	Bi Carbonates (mg/l)	1233.0	1340.0
7.	Copper (ppm)	0.34	0.35
8.	Zinc (ppm)	0.09	0.4
9.	Iron (ppm)	0.30	0.52
10.	Lead (ppm)	0.3	.08

Table 8.2 : Statistical Analysis of Plant Species with Irrigated by Well Water, Raw Effluent and Treated Effluent

Parameters	Well Water (&)	Well Water ($)	Raw Effluent (&)	Raw Effluent ($)	Treated Effluent (&)	Treated Effluent ($)
Minimum	34.5	36	38	40.1	40	48.25
Maximum	395.4	352.1	330.9	362.5	355.4	390
Average	209.02	183.33	177.35	185.73	183.71	207.46
Mode	34.5	36	38	40.1	40	48.25
Median	204.1	183.3	177.3556	185.7361	183.7111	207.46
Correlation	0.9940		0.9964		0.9954	
Covarience	14391.94		12126.81		12890.59	
Kurtosis	−1.4358		−1.44225		−1.392	
Skewness	0.1854		0.1295		0.1547	
Standard Deviation	125.53		114.22		119.50	
Standard Errer	13.69818		9.401568		10.42231	

(&) – *Acacia ferruginea*   ($) – *Delonix regia*.

The soil and vegetation act as a wastewater living filter which is called as biopump (Karve, 1996). In the present study, the well water irrigation accelerates the height values in *A. ferruginea* (Fig. 8.1). Raw effluent irrigation shows good potential of nutrient for *Delonix regia* (Fig. 8.2). Figure 8.3 indicated that the treated effluent irrigation also enhanced the height in *Delonix regia*. It is also observed trom the present study that raw effluent did not show any retardation or adverse effect on height in both the tree species. Wastes are now looked upon as a resource since wastewater often provides fair to good quality irrigation water and has an appreciable concentration of nutrients.

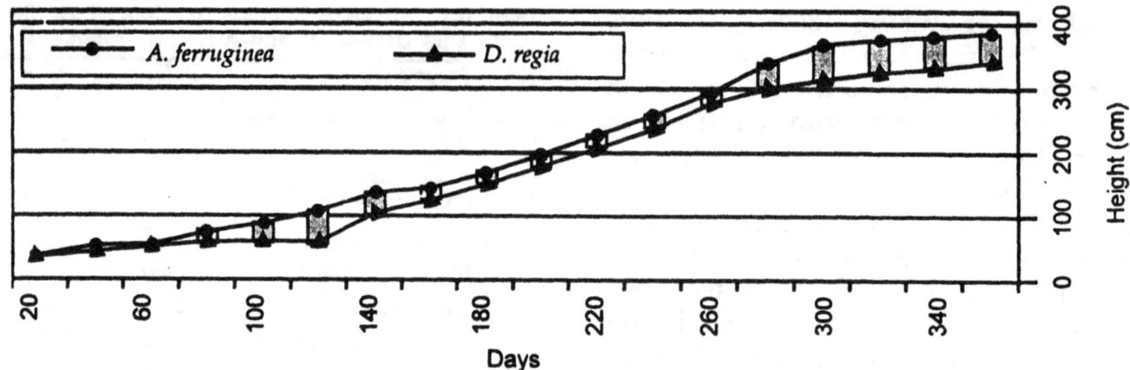

Fig. 8.1 : Comparison of Above Ground Height (cm) in Selected Tree Species with Respect to Well Water.

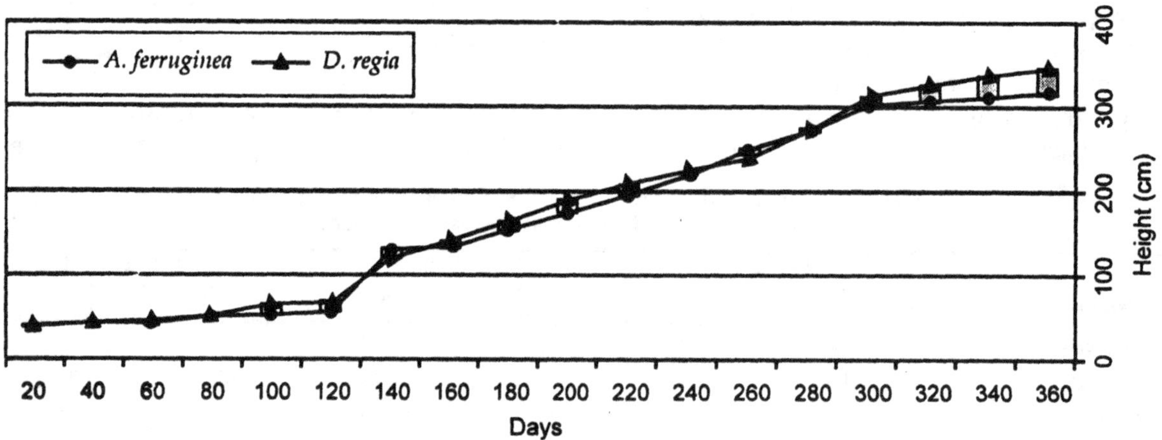

Fig. 8.2 : Comparison of Above Ground Height (cm) in Selected Tree Species with Respect to Raw Effluent.

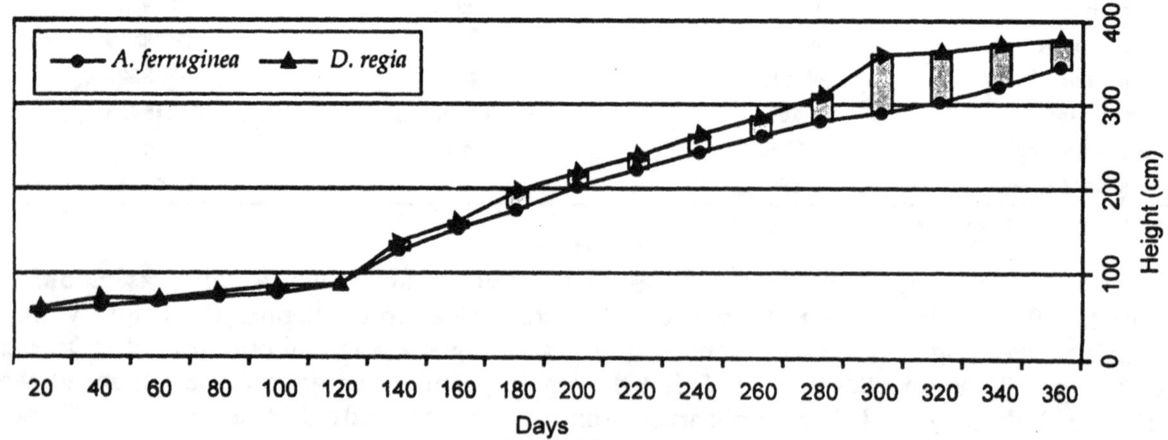

**Fig. 8.3 : Comparison of Above Ground Height (cm) in Selected Tree Species
with Respect to Treated Effluent.**

Similar observations were made with reference to tree species like *Eucalyptus tereticornis*, *Leucaena leucocephala* and *Populus* species. These trees have shown excellent growth in sewage waste as the waste is rich in N.P.K. (Gokate, *et al.*, 1995). Edgar and Stewart (1978) studied wastewater disposal and reclamation using Eucalyptus species. They found to be the best means of wastewater disposal by forest trees. Anon (1990) have shown that forestry crops such as *Eucalyptus terticornis*, *Leucaena leucocephala* and *Populus* can be grown successfully in raw sewage.

Compared to control, distillery effluent at concentrations of 5, 10, 20 and 50% had an inhibitory effect on fresh and dry weight in *oryza sativa* (Behera and Misra, 1981). Rice seeds treated with distillery waste shows retardation in germination speed and seedling growth at higher concentrations (25% and above). Root growth was more adversely affected than the shoot growth at 5% concentration. The adverse effect was attributed to various salts like calcium, chloride, bicarbonates, total nitrogen and organic pollutants. Therefore diluting the effluent to 5% level was suggested to be beneficial for irrigation purposes as a substitute of chemical fertilizers (Sahai *et al.*, 1983). Thomus (1981) have studied the land application of effluents in old growth forests in south Michegan. He found that land application is very useful for forest.

The better growth performance may be related to increased water holding capacity of the soil when irrigated with effluent due to the presence of dissolved as well as suspended solids. Similar observations on increase of water holding capacity by addition of paper mill effluent (Rajannan and Oblisami, 1979), distillery effluent (Swaminathan *et al.*, 1990), steel mill effluent (Archana and Naik, 1992). The enhanced growth in *D. regia* observed during the present studies suggests that tree is able to utilize the minerals and nutrients of the treated textile effluent.

The growth response of *Arachis hypogea* is studied using dyeing effluent, viscose factory effluent (Swaminathan and Vaitheeswaran, 1991 Swaminathan *et al.*, 1992) and glucose factory effluent (Swaminathan *et al.*, 1990). The physico-chemical analysis of above effluents showed high amount of total suspended and dissolved solids. The dyeing factory effluent upto 50% dilution favoured germination and growth while viscose factory effluent retarded plant growth associated with reduction in the number of leaves. This study concludes that viscose factory effluent is more toxic because of high amount of total solids, excess of salt concenrtration and absence of dissolved oxygen. Guerrero and Britto (1995) have shown the orange waste could be used as fertilizers for forest trees.

Disposal of treated wastewater on land is a desirable alternative method rather than surface water (Bower and Chaney, 1974, D'Itri, 1977). Several species and hybrids of *Populus* seems to be ideal for effluent irrigation. Increased height, dry weight and biomass were observed under effluent irrigation. Other trees that have responded favourably for waste-water irrigation are *Fraxinus pennsylvanica*, *Fraxinus americana*, *Liriodendron tulipifera*, *Picea glauca*, *Pinus strobes*, *Pinus resinosa*, *Eichornia crassipes* and *radiat apiru* (Sopper and Kardos, 1973, Cooley A. and Schiess, 1978, Cooley, 1979), De Casabianca (1995) and Cromer (1980).

The notable observations of the present study are, the physico-chemical analysis of the effluent revealed an alkaline pH, with high carbonate and bicarbonate content. In *A. ferruginea*, maximum stem height is reported in well water irrigation and in *D. regia*, treated effluent irrigation showed good results. *D. regia* is found to be more tolerant than *A. ferruginea*. It is observed that wastewater can be used as recycling of wastewater and potential manurial for growth of tree species.

Acknowledgement

The author is greatful to Mrs. V.M. Bendre, Director, CWPRS, Pune for granting permission to publish the paper. The data and work carried out and presented in this paper is Ph.D. work of the author and not necessarily the views of the Central Water and Power Research Station, Pune. The author is grateful to Dr. B.S. Attri, Director (R&D), Department of Environment Forest and Wildlife for financial support in the form of research grant for the project. I am thankful to the Principal and Secretary, Vellalar College for women, Erode for the laboratory help.

References

Agarwal, R.K., Praveen Kumar, Harsh, L.N. and Sharma, B.M. 1994. "Effect of effluents of textile industry on the growth of tree species and soil properties in an arid environment." Indian Forester, 2: 1-60.

APHA. 1981. "Standard methods for the examination of water and wastewater". American Pub. Hlth. Assoc. 1015, 18th Street, N.W. Washington, D.C. 20036, New York.

Anon, 1990. Sewage water, utilisation through Forestry brochure, Pub.: Central Soil Salinity Res. Institute, ICAR, Karnal, pp. 1-15.

Arachana, S. and Naik, M.L. 1992. "Effects of steel mill effluent irrigation on physicochemical characteristics of Kanhar Soil". Jr. Ind. Poll. Cont. 7: 11-16.

Behra, B.K. and Misra, B.N. 1981. "Analysis of the effect of Industrial effluent on growth and development of rice seedlings". Environmental Research, 28(10): 10-19.

Bower, H. and Chaney, R.L. 1974. "Land of Treatment-II water quality and economic aspects of the flushing meadours project". Jr. Wat. Pollut. Control Fedn. 46(5): 844-859.

Coole, D.W. and Schiness, P. 1978. Renovation of wastewater response of forest ecosystems. The pack forest study. In: Proc. State of knowledge in land treatment of wastewater, H.L. Mekim, U.S. Army cold regions research and engineering laboratory, Hanover 63, 755, 2 : 323-331.

Cooley, J.H. 1979. "Growing trees on effluent irrigation sites with sand soil in the upper midwest". Technical Report Data, EPA, Chicago, pp. 127-135.

Cromer, F.J. 1980. Irrigation of radiata piru with wastewater. A review of the potential for tree growth and water renovation. Aust. Forestry, 43(2): 87-100.

Dakshindas, S.D., Faroqui, U.M. and Joshi, V.S. 1995. 'Use of sewage water fertilizer in teak plantations – A case studies,' Proce. IVth regional meeting of silviculturists and research workers of central region, pp. 25-28.

De Casabianca, 1995. *Eichharnia crassipes* Production on petroliferrous wastewater, Effects of salinities, Bioresources Technology, 54: 39-43.

De Casabianca, 1995. Large scale production of *Eichharnia crassipes* on paper industry effluent. Bioresources Technology, 54: 35-38.

D'Itri, 1977. Vegetation management of land treatment of municipal wastewater, Report, Institute of Water Research, East Lansing, Chicago.

Edgar, J.G. and Stewart, H.T.L. 1978. Wastewater disposal and recalamation using Eucalyptus and other trees. Inst. Assoc. on water pollution. Research paper 12 : 11.

Gokate, M.G., Farooqui, V.M. and Joshi, V.S. 1995. Sewage water as potential for the tree growth – A study on teak (*Tectona grandis*) plantation, Indian Forester : 472-481.

Gowda, M.C., Ragahavan, G.S.V., Ranganna, B. and Barrington, S. 1995. Rural waste management waste in south India village a case study, Bioresource Technology, 53: 57-164.

Guerrero, C.C. and Brito, J.C. 1995. Reuse of industrial orange waste as organic fertilizers, Bioresource Technology, 53 : 43-51.

Karve, A.D. 1996. "A report on municipal waste water irrigation in tree species." COSTFORD, IEE-A Report, Pune.

Manivanan, R. and Das S.K. 2002. Phytoremidiation of textile effluent by forest trees (*Acacia ferruginea* and *Delonix regia*). Indian J. Environ. & Ecoplan. 6(2).

Perttu, K.L. 1993. 'Biomass production and nutrient removal from municipal wastes using Willow vegetation filters', Jr. of Sus. For. 1(3): 57-70.

Rajanan, G. and Oblisami, G. 1979. "Effect of distillery effulent on soil crop plants". Indian Jr. Environ. Hlth. 21 : 173-177.

Sahai, R., Jabeen, S. and Saxena, P.K. 1983. "Effect of distillery waste on seed germination, seedling growth and pigment content of rice, Indian Jr. Ecol. 10(1): 7-10.

Sopper, W.E. and Kardos, L.T. 1973. Vegetation responses to irrigation with treated municipal waste-water. *In* : Recycling treated municipal wastewater and sludge through forest and cropland W.E. Sooper and L.T. Kardos, eds. The Pennsylvania State University Press, University Park, Pennsylvania, 16802, pp. 271-294.

Swaminathan, K. and Vaitheeswaran, P. 1991. "Effect of dyeing factory effluent on seed germination and seedling development of ground nut (*Arachis hypogea*)", Jr. Environ. Biol., 12(4) : 353-358.

Swaminathan, K., Manonmani, K. and Sarojini. 1992. "Studies on the toxicity of south India Viscose factory effluent on ground nut (*Arachis hypogea*)", Jr. Environ. Biol. 13 (30) : 252-260.

Swaminathan, K., Arujnan, J. and Gurusamy, R. 1990. Effect of glucose factory effluents on seed germination and seedling development of ground nut. Proce. of 10th session of AEB and symposium on Environ. impact on bio systems, pp. 81-87.

Thomas, M. Burton, 1981. "Studies of land application in old growth forests in Southern Michigan" EPA, Technical data report, Chicago, pp. 147-155.

Vimal, O.P. and Talashilkar. 1985. "Recycling of sewage waste in agriculture prospects and problems". J. Sci. Ind. Res. 14: 126-140.

ENVIRONMENTAL BIOTECHNOLOGY

Edited By :   Professor (Dr.) Arvind Kumar

Published By :   DAYA PUBLISHING HOUSE

9

EFFECT OF PHOSPHATE FERTILIZATION AND VAM ON TWO FLORICULTURAL PLANTS

• *Anjana, S. Kerur and H.C. Lakshman*

Post Graduate Department of Botany,
(Microbiology Lab), Karnataka University, Dharwad, India

Abstract

Different levels of phosphate and vesiecerlar-arbscular mycorrhizal fungal (*Glomus mosseae*) inoculation was carried on two floricultural plants *Tagetes patula*. Plants grew much taller than the non-mycorrhizal plants in 3.0 Mg Rp/kg super phosphate with *Glomus mosseae* treatment. The increased plant height, root length, and reduced per cent root colonization and spore number was observed. But 3.5mg/kg Rp treatment with mycorrhizal inoculation on *Tagetes erecta* influenced in the enhanced plant growth, biomass production and P-uptake in mycorrhizal plants over the non-mycorrhizal plants. The suitable level of rock phosphate and super phosphate treatment with *Glomus mosseae* inoculation has been recommended on these two floricultural plants.

Key Words : Floricultural plants, super phosphate, rock phosphate, per cent root colonization, spore number, Glomus mosseae, vesicular-arbuscular mycorrhizal (VAM) fungi.

Introduction

Proper fertilizer application is an essential method to increase plant production that can be used on infertile soils. Most of the Indian tropical soils lack required phosphorous, which is required for growing many crop plants. It has been well documented that VA-mycorrhizal (VAM) fungi are associated with most of the terrestrial plants. They play a vital role in uptake and translocation of limited nutrients mainly by diffusion of P thereby promoting plant growth (Harley and Smith, 1983). It has been observed that vesicular-arbuscular mycorrhiza fungal external hyphae can be extended as much as 5-8 cm away from the roots, and absorb nutrients from a larger soil volume than the absorption zone surrounding the non-mycorrhizal root system (Raju *et al.*, 1990). Thus absorption of phosphate ions could be possible through some beneficial micro symbionts that are associated with rhizospheric zones of numerous plants. It is also known this fungi that have been associated symbiotically with almost all types of plants including floricultural plants (Allen, 1991).

Rhodes and Gerdmann (1978) studied the influence of phosphorous, sulphur and their nutrition uptake by VA-mycorrhizae infected onion. Smith (1982) reported the inflow of phosphate into mycorrhizal and non-mycorrhizal plants of *Trifojium subterraneum*. Jalali and

Thereja (1985) worked on plant growth response to VA-mycorrhizal inoculation in soils inoculated with rock phosphate. Jones (2000) demonstrated the effect of long-term organic or inorganic fertilization of *Subterraneum* clover with mycorrhiza mediated phosphorous uptake. Since most of the mycotropic plants depend on VAM colonization, when they grow under low external P conditions, their yield can be enhanced either by inoculation of more efficient fungal strains in agronomic practices. Only limited reports are available on the phosphate fertilizer application on floricultural plants to know their effect with mycorrhizal inoculation.

The present study investigates the effect of four different levels of single super phosphate and rock phosphate on two floricultural plants with and without inoculation of VAM (*Glomus mosseae*) have been undertaken.

Materials and Methods

The seeds of experimental plants *Tagetes erecta* and *Tegetes patula* were sown in 15 cm × 15 cm diameter earthen pots containing 04 kg of sterilized soil. Each earthen pot contained sandy loam soil plus sand (3 parts garden soil and 1 part of pure sand). The triplicate sets of pots were maintained with proper controls. Four different levels of super phosphate and super phosphate at the rate of 0.5mg/kg, 1 mg/kg, 1.5 mg/kg and 4.5 mg/kg rock phosphate and super phosphate were provided with 10 gm of VAM (*Glomus mosseae*) mixed inoculum. The inoculum consisted of freshly infected chopped roots pieces, spores, sporocaps, mycelia and rhizospheric soil collected from the host *Soghum sudanese*. The inoculum was placed to each seedling in the vicinity of the root system. The experimental pots were watered on every alternate day. All the pots were maintained in the green house. The following treatments were given for each experimental plant:

1. Non-mycorrhizal plant (Control)
2. Four levels of rock phosphate + VAM (*Glomus mosseae*)
3. Four levels of Super phosphate + VAM (*Glomus mosseae*)

Periodical data were recorded for three harvests in between thirty days intervals. The following parameters such as plant height, dry weight of root and shoot, root/shoot ratio, per cent root colonization, spore number and P content in shoots were determined by vanadomolybdate yellow phosphoric colour (Jackson, 1973). Percentage of mycorrhizal colonization in roots were determined after cleaning the roots in 10% KOH and stained in 0.05% tryphan blue in lactophenol (Phillips and Hayman, 1970). Mycorrhizal spores were recovered by wet sieving and decanting technique (Gerdemann and Nicolson, 1963).

Results and Discussion

Different levels of rock phosphate and super phosphate were given to experimental plants of *Tagtes erecta* and *Tagetes Patula* as shown in Tables 9.1 and 9.2. The VAM (*Glomus mosseae*) inoculated plants grew much taller than non-mycorrhizal plants with P fertilizer application in both of the experimental plants (Tables 9.1 and 9.2). The highest plant height (58 cm) was observed in *Tagetes erecta* when treated with 3.5 mg SP/kg of soil. *Tagetes Patula* showed a higher growth (47 cm) after being treated with 3 gm SP/kg of soil with VAM inoculation. The application SP level (3.0 mg SP/kg soil) influenced the plant height and biomass production with *Glomus mosseae* inoculated plants of *Tagetes Patula*. These plants

Table 9.1 : Effect of Different P levels of VAM (*Glomus mosseae*) Inoculated *Tagetes erecta* for 90 days

Duration	Treatment	Plant Height (cm)	Dry Wt. of Shoot (gm)	Dry Wt. of Root (gm)	Root/Shoot ratio %	% VAM Colonization	Spore No./50g soil	Shoot P (%)	Root P (%)
NM		11.0±1.0	0.91±0.0	0.71±0.0	0.78±0.0	–	–	0.09	0.05
	M+RP0	16.3±4.1	0.81±1.0	0.74±2.0	0.49±0.0	43.1±0.0	51.1±2.0	0.16	0.11
	M+RP1	12.8±2.0	1.65±1.0	0.81±0.0	0.49±0.0	49.6±2.1	50.7±2.1	0.15	0.07
	M+RP2	21.2±2.0	1.10±1.0	0.90±0.0	0.57±0.0	54.5±1.0	60.2±3.1	0.22	0.12
30 days	M+RP3	29.5±2.0	1.73±1.0	1.90±1.0	0.52±0.0	52.5±3.1	50.4±1.1	0.18	0.11
	M+SP0	24.3±1.1	0.96±1.1	0.87±0.0	0.65±0.0	34.5±1.1	38.3±2.0	0.16	0.04
	M+SP1	30.1±2.0	1.80±1.0	0.91±1.0	0.33±0.0	38.5±2.1	49.2±2.1	0.22	0.11
	M+SP2	29.1±2.0	1.20±0.0	0.80±0.0	0.73±0.0	36.3±2.2	42.5±3.1	0.21	0.04
	M+SP3	26.5±1.0	1.62±0.0	0.83±0.0	0.32±0.0	35.2±13.0	47.2±1.1	0.23	0.07
NM		13.2±1.0	1.02±0.0	0.75±1.0	0.73±0.0	–	–	0.11	0.06
	M+RP0	14.3±1.1	0.98±0.0	0.98±0.0	0.47±0.0	45.4±1.0	46.3±2.2	0.15	0.08
	M+RP1	18.3±2.1	1.70±0.0	1.85±0.0	0.50±0.0	51.4±2.1	53.4±2.1	0.18	0.08
	M+RP2	31.2±2.2	1.90±1.0	1.28±0.0	0.39±0.0	55.4±2.0	62.5±2.1	0.23	0.15
60 days	M+RP3	36.1±3.1	1.91±1.0	0.94±0.0	0.49±0.0	54.5±3.1	57.3±1.1	0.21	0.09
	M+SP0	28.5±4.1	1.10±0.0	1.11±0.0	0.31±0.0	34.2±0.0	41.5±0.0	0.21	0.07
	M+SP1	32.5±1.1	1.45±1.0	1.29±0.0	0.37±0.0	36.3±1.1	47.4±1.1	0.25	0.07
	M+SP2	29.5±1.0	1.20±1.0	0.73±0.0	0.61±0.0	35.5±2.1	41.5±2.1	0.22	0.05
	M+SP3	28.2±2.0	1.71±0.0	0.90±0.0	0.30±0.0	35.4±1.0	45.3±3.1	0.24	0.09
NM		15.1±1.0	1.10±0.0	0.80±1.0	0.72±0.0	–	–	0.11	0.07
	M+RP0	10.7±2.1	1.11±0.0	0.73±0.0	0.46±0.0	46.3±1.0	51.1±3.0	0.23	0.11
	M+RP1	21.2±1.0	1.81±1.0	0.92±1.0	0.50±1.0	51.2±1.0	54.5±3.1	0.31	0.12
	M+RP2	33.3±3.2	2.60±1.0	1.32±0.0	0.37±0.0	59.3±2.1	67.4±3.1	0.38	0.18
90 days	M+RP3	47.2±2.1	3.10±0.0	1.00±0.0	0.48±0.0	55.4±0.0	58.3±2.1	0.34	0.13
	M+SP0	26.3±4.1	1.12±1.1	1.10±0.0	0.36±0.0	31.4±1.1	42.2±2.0	0.31	0.10
	M+SP1	32.5±2.1	4.89±1.0	1.40±0.0	0.29±1.0	35.5±1.0	45.3±2.1	0.32	0.11
	M+SP2	31.0±1.0	1.80±0.0	0.90±0.0	0.50±0.0	33.4±0.0	41.2±1.1	0.26	0.09
	M+SP3	30.5±1.1	3.44±0.0	1.25±0.0	0.36±1.0	31.7±1.0	43.1±2.1	0.31	0.07

NM = Non Mycorrhizal, M+RP0 = VAM+0.5 mg rock phosphate/kg of soil, M+RP1 = VAM+1.5 mg rock phosphate/kg of soil, M+RP2 = VAM+3.0 mg RP/kg soil, M+RP3 = VAM+3.5 mg RP/kg soil, M+SP0 = VAM+0.5 mg Super phosphate/kg soil, M+SP1 = VAM+1.5 mg super phosphate/kg soil, M+SP2 = VAM+3.0 mg SP/kg, M+SP3 = VAM+3.5 mg SP/kg of soil.

showed decreased per cent colonization and spore number. Root shoot ratio was drastically decreased in all the inoculated plants over the non-mycorrhizal plants. The P fertilization experiments *T. erecta* indicated that mycorrhizal inoculation alone aids in the effective utlization of rock phosphate than the super phosphate. This is because of changing into available form by mycorrhizal (*Glomus mosseae*) which was later taken up by the plants for their better growth.

Vescicular-arbuscular mycorrhizal fungi clearly produce an increase in absorbing root surface area due to gross changes in morphology of the feeder roots (Hayman and Mosse, 1972, Manjunath and Bhagyaraj, 1984). The date obtained in the present study positively

Table 9.2 : Effect of Different P levels of VAM (*Glomus mosseae*) Inoculated *Tagetes patula* for 90 days

Duration	Treatment	Plant Height (cm)	Dry Wt. of Shoot (gm)	Dry Wt. of Root (gm)	Root/Shoot ratio %	% VAM Colonization	Spore No./50g soil	Shoot P (%)	Root P (%)
NM		13.1±0.0	0.95±0.0	0.72±0.0	0.75±0.0	–	–	0.10	0.05
	M+RP0	14.3±1.1	1.2±0.0	0.63±0.0	0.40±0.0	31.1±2.2	44.1±1.1	0.11	0.05
	M+RP1	16.2±2.0	1.7±1.0	0.85±0.0	0.50±0.0	36.2±5.2	48.2±1.0	0.12	0.07
	M+RP2	22.5±1.0	1.08±1.0	0.9±1.0	0.50±1.0	38.3±2.0	42.5±3.0	0.29	0.11
30 days	M+RP3	18.2±2.0	1.72±1.0	0.88±0.0	0.51±0.0	42.4±3.1	51.4±1.0	0.14	0.12
	M+SP0	19.2±0.0	0.82±0.0	0.78±1.1	0.42±0.0	27.1±1.1	44.2±0.0	0.17	0.12
	M+SP1	23.5±2.0	1.10±1.0	0.91±1.0	0.50±0.0	31.5±1.0	51.4±1.0	0.28	0.15
	M+SP2	27.5±1.0	1.71±0.0	0.86±0.0	0.50±0.0	36.2±1.2	43.3±3.0	0.21	0.12
	M+SP3	36.5±2.0	1.73±0.0	0.88±1.0	0.50±0.0	31.5±2.0	40.2±1.0	0.25	0.14
NM		14.5±0.0	0.97±0.0	0.72±1.0	0.74±0.0	–	–	0.23	0.06
	M+RP0	19.6±1.0	0.79±0.0	0.73±0.0	0.44±0.0	29.2±0.0	47.3±0.0	0.21	0.11
	M+RP1	21.4±0.0	0.82±0.0	0.92±1.0	0.50±0.0	36.4±1.0	51.7±1.0	0.24	0.11
	M+RP2	26.6±1.0	1.81±1.0	0.25±1.0	0.56±0.0	41.5±0.0	52.3±3.0	0.25	0.14
60 days	M+RP3	24.1±1.0	1.95±1.0	0.98±1.0	0.50±0.0	42.3±2.0	54.4±1.0	0.19	0.12
	M+SP0	21.5±1.1	0.92±0.0	0.98±1.1	0.49±1.0	32.2±1.0	37.1±2.0	0.24	0.14
	M+SP1	26.2±2.0	1.80±1.0	1.21±0.0	0.57±0.0	34.4±1.0	40.3±3.0	0.31	0.17
	M+SP2	30.2±2.0	1.98±1.0	1.10±1.0	0.55±0.0	32.5±1.1	41.2±1.0	0.22	0.13
	M+SP3	41.0±1.0	1.81±0.0	1.16±0.0	0.58±0.0	29.3±2.0	41.5±3.0	0.22	0.15
NM		17.5±1.0	1.10±0.0	0.8±0.0	0.72±0.0	–	–	0.23	0.10
	M+RP0	22.6±0.0	1.0±0.0	1.0±0.0	0.51±0.0	36.4±1.1	46.1±3.1	0.16	0.11
	M+RP1	24.2±2.0	2.10±0.0	1.2±0.0	0.57±0.0	39.6±2.0	51.6±3.0	0.18	0.12
	M+RP2	31.1±1.0	3.87±0.0	1.4±0.0	0.50±0.0	47.3±1.1	53.3±1.0	0.28	0.15
90 days	M+RP3	28.2±1.0	1.91±0.0	1.11±0.0	0.55±0.0	45.4±3.1	54.7±1.0	0.21	0.14
	M+SP0	24.1±2.2	1.10±0.0	1.20±0.0	0.31±0.0	28.3±1.1	37.2±2.0	0.31	0.14
	M+SP1	29.1±2.0	1.91±0.0	1.45±0.0	0.35±0.0	31.2±1.0	40.4±3.0	0.35	0.18
	M+SP2	35.1±1.0	1.15±0.0	1.22±0.0	0.56±0.0	32.5±1.0	38.4±1.0	0.24	0.14
	M+SP3	38.2±2.0	2.91±0.0	1.35±0.0	0.35±0.0	24.1±2.0	37.4±1.0	0.32	0.16

NM = Non Mycorrhizal, M+RP0 = VAM+0.5 mg rock phosphate/kg of soil, M+RP1 = VAM+1.5 mg rock phosphate/kg of soil, M+RP2 = VAM+3.0 mg RP/kg soil, M+RP3 = VAM+3.5 mg RP/kg soil, M+SP0 = VAM+0.5 mg super phosphate/kg soil, M+SP1 = VAM+1.5 mg super phosphate/kg soil, M+SP2 = VAM+3.0 mg SP/kg, M+SP3 = VAM+3.5 mg SP/kg of soil.

indicates that the mycorrhizal colonization of the host i.e., floricultural plants were able to utilize soluble phosphates quickly. The super phosphate treatments were more efficient in *Tagetes patula*. The trend of lower per cent mycorrhizal colonization with increasing fertilizer (SP) agrees with the results of previous (Sylvia and Schenck, 1983, Sieverding and Howler, 1985). On the other hand VA-mycorrhizal fungi are known to occur in soils with very high P contents and thus among the species of mycorrhizal fungi, there may be different response to low or high P conditions or to P fertilization treatments (Jones, 2000).

Studies on *Tagetes erecta* treated with 3.5 mg RP/kg soil dosage had the beneficial response with mycorrhizal (*Glomus mosseae*) inoculation. The plants grew much greater than non-mycorrhizal plants. This could be easily explainable that low solubility off source (Rock phosphate) is more effective than the plants treated with super phosphate. Plant roots may increase the rate if dissolution of rock phosphate by lowering the concentration of P in the soil solution by lowering the pH of the soil. The treatment of *Glomus mosseae* with (3.5 mg/kg soil) rock phosphate was most effective on *Tagetes erecta* in promoting plant growth. These findings are, consistent with the earlier workers contributions (Jalali and Thereja 1985, Lakshman, 1992 and 1996).

Plants inoculated with VAM (*Glomus mosseae*) and treated with different levels of super phosphate and rock phosphate grown in green house conditions showed enhanced nutrients in shoot, when compared to non-inoculated plants. These findings are supported by others who reported similar results (Ross, 1971, Powell and Daniel, 1978, Harley and Smith, 1983). Many rhizospheric soils contain organic acids with low calcium, it could be desirable to practice VAM inoculation technique to young nursery plants with different levels of rock phosphate and indigenous mycorrhizal fungi. It may be concluded that mycorrhizal plants absorb 'P' only from the soluble 'P' pools in the soil and that they are unable to utilize source of 'P' to non-mycorrhizal roots or non-mycorrhizal plants.

References

Allen, M. F. 1991. The ecology of mycorrhizae, Cambridge University, New York, pp. 184.

Gerdemann J. W. and Nicolson, T. H. 1963. Spores of mycorrhizal Endogone species extracted from soil by wet-sieving and decanting. Transactions of Brt. Mycol. Soc. 46: 235-244.

Harly, J. L. and Smith, S. E. 1983. Mycorrhizal Symbiosis, Academic Press, London, pp. 483. .

Hayman, D. S. and Mosse, B. 1972 (b). The role of Vescicular-arbuscular mycorrhiza in the removal of Phosphorous from soil by plant root. Review of Ecological and Physiological Science. 9: 463-470.

Jackson, M. L. 1973. Soil Chemical Analysis. Prentice Hall of India Pvt. Ltd., New Delhi.

Jalali, B. L. and Thareja, M. 1985. Plant growth responses to Vescicular-Arbuscular mycorrhizal inoculation in soils incorporated with rock phosphate. Indian Phytopathology. 38(2): 306-310.

Jones, E. J. 2000. The effect of long-term fertilization with organic or inorganic fertilization or mycorrhiza mediated phosphorous uptake in *Sbuterranena Clover*. Biology and Fertility of Soils 32(5): 435-440.

Lakshman, H. C. 1992. Development and Response of *Terminalia bellirica* to inoculation of VA-mycorrhizae. Tropical Forestry. 79: 109-114.

Lakshman H. C. 1996. VA-mycorrhizal studies in some economically important tree species. Ph. D. thesis, Karnatak University, Dharwad, India.

Manjunath, A. and Bhagyaraj, D. J. 1984. Response of pigeon pea and cow pea to phosphate and dual inoculation of vescicular-arbuscular mycorrhizal fungi and *Rhizobium*. Trop. Agric. 61: 48-52.

Phillips, J. M and Hayman, D. S. 1970. Improved procedures for cleaning and staining parasite and vescicular-arbuscular mycorrhizal fungi for rapid assessment of infection. Trans. Brt. Mycol. Soc. 55: 158-162.

Powell, C. L. and Daniel, J. 1978. Growth of white clover in undisturbed soils after inoculation with efficient mycorrhizal fungi. New Zealand journal of Agri Res. 21: 675.

Raju, P. S., Clark, R. B. and Ellis, J. W. 1990. Effects of species of VAM fungi on growth and mineral uptake of *Sorghum* at different temperatures. Plant and soil. 121: 157-164.

Rhodes, I. H. and Gerdmann, J. W. 1978. Influence of Phosphorous nutrition of Sulphur uptake by vescicular-arbuscular mycorrhizal fungi of onion. Soil Biology Biochem 10: 361-364.

Ross, J. P. 1971. Effect of Phosphate Fertilization on yield of mycorrhizal and non-mycorrhizal soybean. Photopathology, 61: 1400-1403.

Sieverding, E., Howler, R. H. 1985. Influence of species of VA-mycorrhizal fungi on cassava field response to phosphorous fertilization. P land and Soil. 88: 213-221.

Smith, S. E. 1982. Inflow of Phosphate into mycorrhizal and non-mycorrhizal plants of *Trifolium subterraneum* at different levels of soil phosphate. New Phytol. 90: 293.

Sylvia, D. M. and Schenck, N. C. 1983. Application of super phosphate to mycorrhizal plants stimulated sporulation of phosphorous tolerant vescicular-arbuscular mycorrhizal fungi. New Phytol. 95: 655-661.

ENVIRONMENTAL BIOTECHNOLOGY

Edited By : Professor (Dr.) Arvind Kumar

Published By : DAYA PUBLISHING HOUSE

10

APPLICATION OF FUZZY LINEAR PROGRAMMING TECHNIQUE IN IRRIGATION MANAGEMENT FOR OPTIMAL CROPPING PATTERN SATISFYING DOWNSTREAM RIPARIAN RIGHT

• *Gokulananda Patel* and Rajendra Gartia**

**Department of Business Administration, Sambalpur University,
Jyoti Vihar, Burla, Sambalpur, Orissa
Department of Statistics, Sambalpur University,
Jyoti Vihar, Burla, Sambalpur, Orissa*

Abstract

The paper addresses the problem of determining a cropping pattern that maximizes the benefits due to irrigation over a water year satisfying the down-stream riparian rights when the inflow to the reservoir is uncertain. Objective function, cropping pattern and land utilization is set by fuzzy linear programming whereas the release to the down-stream refers to the deterministic model.

Key Words : Fuzzy linear programming, Irrigation management, Riparian right.

Introduction

The demand for water resources is ever increasing with the explosive population growth, advancement of civilization, increased agricultural and industrial productivity and other inter-racial systems. Water resources are limited compared to its ever growing needs. India is predominantly an agrarian country with more than 70% of the population depending directly on agriculture. Agriculture needs more water than domestic and industrial consumption. Agricultural utilization in our country accounts for more than 80% of the water resources. Further, in our country most of the monsoon fall is concentrated in four monsoon months, with uncertain space and time distribution, resulting in flood and drought. All these factors place tremendous responsibility on water resource planners and decision makers.

Irrigation management is the process by which naturally available water for controlled application in agriculture is planned, organized and implemented. The purpose of an irrigation system is to service and sustain an agricultural system in a timely and adequate manner so that water, together with other inputs and infrastructure creates favourable conditions for achieving high level crop production. Due to design parameters and operating policies of a project, expected return from an irrigation system cannot match with those actually achieved. Design parameters being relatively difficult to change, the project performance can be improved by improving the operating policies.

To determine the operating policy of a reservoir becomes complex because the inflow to the reservoir is uncertain; data available for a limited period and downstream riparian right. Maass *et al.* (1962) and Allan Manne (1962), have developed LP model and solved it for a small hypothetical problem, while Hufschmidt and Fiering, (1966), developed and analysed a simulation model for a real life problem. Chatterjee and Kalro (1984), used mathematical programming and regression for reservoir optimal control. An LP technique has been applied by Thandaveswar (1992), to maximize the net benefits from crops in the commands of Cauvery basin. In the report, Kolavalli, and Kalro (1993) have described the operating policy of Meshwo reservoir system by a econometric model.

This paper aims at finding an optimization technique satisfying a strongdown-stream riparian right in an uncertain inflow through Linear programming (LP) and fuzzy linear programming (FLP).

Deterministic Model

To develop the mathematical model, we use the following notations:

I	:	The set of crops (indexed by $i \in I$)
J	:	The set of months (indexed by $j \in J$)
NB_i	:	Net benefit from i^{th} crop
$x_{ij} = x_i$	:	Area of ith crop in ha in the direct command in j^{th} month
$x_{ij} = 0$	:	If i^{th} crop has not been cultivated in j^{th} month
TL_j	:	Total land available in j^{th} month in ha
S_j	:	Storage in reservoir at the beginning of month j in million cubic meter (mcm)
Inf_j	:	j^{th} month inflow into reservoir in mcm
RC_j	:	Release from reservoir to canal in j^{th} month in mcm
RR_j	:	Release from reservoir to river in j^{th} month in mcm
DR_j	:	Minimum down-stream release required in j^{th} month in mcm
S_{max}	:	Maximum capacity of reservoir in mcm
S_{min}	:	Dead storage of reservoir in mcm
W_{ij}	:	Water required by i^{th} crop in j^{th} month in mcm. per ha
E_j	:	Evaporation loss in j^{th} month in mcm.

The objectie function and various constraints are described as follows :

Objective Function : The objective function is to maximize the total net benefits from all crops commanded by the reservoir in the direct command.

$$\text{Max} \quad \sum_{i \in I} NB_i x_i$$

Land Availability Constraints : The total irrigated area in each month is less or equal to the possible maximum area available for irrigation.

$$\sum_{i \in I} x_{ij} \leq TL_j \, \forall \, j \in J$$

Storage Continuity Constraints : Based on the balance within reservoir during each month, storage continuity constraints can be written as

$$S_{j+1} = S_j + Inf_j - RC_j - RR_j - E_j \ \forall j \in J$$

Live Storage Constraint : The live storage during any month 'j' in the reservoir should not exceed the maximum live storage capacity of the reservoir.

$$S_j \leq S_{max} \ \forall j \in J$$

Dead Storage Constraints : The storage during any irrigated month 'j' in the reservoir should not be less than the dead storage capacity.

$$S_{min} \leq S_j \ \forall j \in J$$

Down-stream Release Constraints : The down-stream release into the river in any month 'j' should not be less than the down-stream release required.

$$RR_j \geq DR_j \ \forall j \in J$$

Crop Water Requirement Constraints : The total amount of water required by all crops during any month 'j' should be less than the water released into the canal during the same month.

$$\sum_{i \in I} w_{ij} x_{ij} \leq RC_j \ \forall j \in J$$

Non-negativity Constraints : All decision variables greater than or equal to zero.

The LP model that emerges from the objective function, land availability for irrigation, storage continuity constraints, live storage constraints, dead storage constraints, down-stream release constraints and crop water requirements constraints is as follows :

$$\text{Max} \quad \sum_{i \in I} NB_i x_i \tag{1}$$

Subject to

$$\sum_{i \in I} x_{ij} \leq TL_j \ \forall j \in J \tag{2}$$

$$S_{j+1} = S_j + Inf_j - RC_j - RR_j - E_j \ \forall j \in J \tag{3}$$

$$S_{min} \leq S_j \leq S_{max} \ \forall j \in J \tag{4}$$

$$RR_j \geq DR_j \ \forall j \in J \tag{5}$$

$$\sum_{i \in I} w_{ij} x_{ij} \leq RC_j \ \forall j \in J \tag{6}$$

All decision variables ≥ 0 (7)

Based upon the data available in the report of Kolavalli *et al.* (1993), on inflow to Meshwo reservoir in Gujarat, actual cropping pattern, month-wise crop plantation, water requirements, net benefits from crops and storage continuity constraints (Appendices A, B and C), the model (1) - (7) is run for different water years and the detail solution is presented in Table 10.1.

Regression Models

To avoid cumbersome calculations for arriving at decision, three models are derived from the above results on regression.

We use the following notations for the models.

Table 10.1 : Cropping Pattern in Deterministric Model

Year	Sept_B (mcm)	Sept_I (mcm)	Oct_I (mcm)	Nov_I (mcm)	Kharif Groundnut	Rabi Groundnut	Rabi Wheat	Profit in Cr(Rs)
1972-73	25.00	00.00	00.00	00.00	2063.0	6880.0	–	2.437
1973-74	53.00	24.00	00.00	00.00	6880.0	2560.0	4320.0	6.895
1974-75	42.00	00.00	00.00	00.00	6880.0	6448.0	0432.0	4.996
1975-76	36.00	26.00	00.00	00.00	6880.0	4212.0	2668.0	6.088
1976-77	77.25	06.00	00.00	01.00	6880.0	1774.0	5106.0	7.278
1977-78	77.25	10.00	00.00	00.00	6880.0	1444.0	5436.0	7.493
1978-79	66.00	10.00	00.00	00.00	6880.0	2686.00	4194.0	6.833
1979-80	50.00	00.00	00.00	00.00	6880.0	5565.0	1315.0	5.427
1980-81	40.00	00.00	00.00	00.00	6880.0	6669.0	0211.0	4.888
1981-82	53.00	00.00	00.00	00.00	6880.0	5234.0	1646.0	5.589
1982-83	50.00	00.00	00.00	03.00	6880.0	5218.0	1662.0	5.597
1983-84	55.00	07.00	09.00	00.00	6880.0	3212.0	3668.0	6.576
1984-85	77.25	00.00	00.00	00.00	6880.0	2559.0	4321.0	6.895
1985-86	23.14	00.00	01.90	00.00	2098.0	6880.0	–	2.454
1986-87	20.42	00.00	00.00	00.00	0376.0	6880.0	–	1.615
1987-88	09.55	00.00	00.00	00.00	0242.0	6880.0	–	1.549
1988-89	53.24	01.30	00.00	00.00	6880.0	5063.0	1817.0	5.672
1989-90	23.67	01.39	00.00	00.00	2089.0	6880.0	–	2.450
1990-91	56.53	17.49	00.00	00.00	6880.0	2896.0	3984.0	6.731
1991-92	42.75	00.50	01.74	00.00	6880.0	6112.0	0768.0	5.160

Notations used in Table 10.1 and the following regression models.

Sept_B : Reservoir balance on 1st September in mcm.

Oct_B : Reservoir balance on 1st October in mcm.

Nov_B : Reservoir balance in 1st November in mcm.

GN = : Area for groundnut to be cropped in ha.

SF = : Area for safflower to be cropped in ha.

WH = : Area for wheat to be cropped in ha.

RR = : Down-stream release during the month of Sept. in mcm.

Model-I

This model considers the September balance of the reservoir for Kharif and November balance for Rabi.

Kharif

If $Sep\_B \geq 27.58 + 1.01 * RR$

$\qquad GN = 6880$

else

$\qquad GN = \min (-7141.22 + 368.47 * Sep\_B, 6880)$

endif

Rabi

If $Nov\_B \geq 8.576$

$SF = 7805.03 - 113.73 * Nov\_B$

$WH = 6880 - SF$

else

$SF = 6880$

$WH = 0$

endif

The mode is 99% explainable.

Model-II

This model considers the September balance for Kharif and October balance for Rabi

Kharif

If $Sep\_B \geq 27.58 + 1.01 *RR$

$GN = 6880$

else

$GN = min(-7141.22 + 368.47 * Sep\_B, 6880)$

endif

Rabi

If $Oct\_B \geq 8.73$

$SF = 7699 - 110.77 *Oct\_B$

$WH = 6880 - SF$

else

$SF = 6880$

$WH = 0$

endif

The mode is 97% explainable.

Model-III

The decision of Rabi crops at the beginning of September can explain only up to 67 per cent.

Down-stream Release Rule

Let $\quad DR = \sum_{i \in I} DR_i$ and $DR \geq 10.4 \, mcm$

Assuming that the people on the down-stream has legal right of at least of 10.4 mcm water, the rule can be written as

if $Sep\_B \leq 27.58 + 1.01*RR$

$RR = DR$

else

$RR = DR + ((Sep\_B - (27.58 + 1.01*DR))/2)$

endif

As long as river release $RR \leq Sep\_B - 38.08$ *mcm*, the net benefits change at the rate of Rs. 0.03 crore per mcm. (Table 10.1) of RR and the change in benefits is at the rate of Rs. 0.20 crore per mcm. if $RR > Sep\_B - 38.08$ *mcm*.

Treatment of Uncertainty

In practice, real situations are very often not deterministic and complete description of real system often requires more data to be processed and understand the problem. The available data has range, mean and standard deviation as per Table 10.2.

Due to highly variable environment we chose to find the targets by solving a fuzzy linear programming. Let us depart from the linear programming (1) - (7).

Table 10.2 : Mean Range and Standard Deviation in mcm

	Range		Mean	Standard Deviation
	Min.	Max.		
Balance on 1st Sept.	9.55	77.25	46.6	19.11
Inflow of Sept.	0	26	5.18	8.09
Inflow of Oct.	0	9	0.63	1.99
Inflow of Nov.	0	3	0.2	0.68

(a) Let us not maximize the objective function (1) and reach some aspiration level.

(b) Secondly, instead of satisfying the inflow, down-stream release in strictly mathematical sense, accept the smaller violations in any constraint if it can not be adequately be approximated by a crisp constraint.

(c) Finally, attach different degree of importance to violations of different constraints as per aspiration level set to achieve.

Now the classical linear programming (CLP) problem

$$\max f(x) = c^T x$$

subject to

$$Ax \leq b$$
$$x \geq 0 \tag{8}$$

can be stated as

Find x such that

$$c^T x \gtrsim z$$
$$Ax \lesssim b \tag{9}$$
$$x \geq 0$$

Where z is the aspiration level for the achievement. $\lesssim$, $\gtrsim$ are fuzzified version of $\leq$, $\geq$ respectively. By substituting

$\begin{pmatrix} -c^T \\ A \end{pmatrix}$ *with B and* $\begin{pmatrix} -z \\ b \end{pmatrix}$ *with d, FLP* can be written as

Find x such that

$$Bx \lesssim d \tag{10}$$
$$x \geq 0$$

For our problem we use the following membership function (Zimmermann, 1994) over the tolerance interval p_i,

$$\mu_i(x) = \begin{cases} 1 & \text{if } B_i x \le d_i \\ 1 - \dfrac{B_i x - d_i}{p_i} & \text{if } d_i < B_i x \le d_i + p_i \\ 0 & \text{if } B_i x > d_i + p_i \end{cases} \tag{11}$$

The membership function of the fuzzy set decision is

$$\mu_D(x) = \min_i \mu_i(x) \tag{12}$$

$\mu_i(x)$ can be interpreted as the degree to which x fulfils the fuzzy inequality $B_i x \le d_i$. Introducing λ for min $\{\lambda, (x)\}$ we arrive at

max λ such that

$$\lambda p_i + B_i x \le d_i + p_i \tag{13}$$
$$x \ge 0$$

If (λ, x_0) is the optimal solution of (14), then x_0 is the maximizing solution of (8) assuming membership function specified in (11). To have the problem in the form of (13), we estimated the following parameters.

Table 10.3 : d_i (eq. 10) and p_i in Low Rainfall and Normal rainfall

Constraints	Low rainfall (< 46.5 c.m)		Normal rainfall (> 46.5 c.m)	
	d_i	p_i	d_i	p_i
Benefits	3.5 crore	0.5 crore	6.5 crore	1 crore
Sept_B	20 mcm (mean)	10 mcm (SD)	61 mcm (mean)	11 mcm (SD)
Sept_Inf	0.25 mcm (mean)	0.5 mcm (SD)	15 mcm (mean)	7.5 mcm (SD)
Down-stream Release	10 mcm	5 mcm	20 mcm	5 mcm

On solving (13) with above d_i and p_i s and (1)-(7) with same p_i s, we get the following results.

Solutions are mentioned under fuzzy and non-fuzzy columns respectively separately for low rainfall and normal and above rainfall. When the meteorology department forecasts rainfall as normal or below normal, targets of cropping pattern and land utilization for the direct command area can be set on solving fuzzy linear programming problem with appropriate parameters. Table-10.4 contains both the solutions for LP (Non-fuzzy) and fuzzy linear programming considering the parameters given in Table 10.3 and based on of Kolavalli *et al.*, through appendix-A, B and C. If we would not have considered FLP, our targets for utilization of land would have been the solution of non-fuzzy problem having less benefits (Table 10.4)

Table 10.4 : Fuzzy and Non-fuzzy Solutions

Crops	Low		Normal and above normal	
	Fuzzy	Non-fuzzy	Fuzzy	Non-fuzzy
Groundnut	5441 ha.	3996 ha.	6880 ha.	6880 ha.
Safflower	6880 ha.	6880 ha.	2267 ha.	5421 ha.
Wheat	–	–	4613 ha.	1458 ha.
Z	4.08 crore	3.37 crore	7.04 crore	5.49 crore

Conclusions

The approach followed in this paper is general enough to replicate on other projects and can be extended to multi purpose project to consider hydro power, domestic use of water, allocation to industries for their use, flood control operation, irrigation distribution system. Thus, there remains independent fields or integrated fields for detailed future investigation.

References

Chatterjee, A.K. and A.H. Kalro, 1984. "Derivation of an Operation Policy for Reservoir Operation in a Multi Purpose Project", Opsearch, 21(2), 102-112.

Hufschmidt, M. and M. Fiering, 1966. "Simulation Technique for Design of Water Resources System", Cambridge Harvard Univ. Press.

Kolavalli, S., A.H. Kalro, 1993, "Management of Irrigation Systems : The case of Meshwo Reservoir in Gujarat", CMA, IIMA.

Maass, A. 1962. "Design of Water Resource System", Cambridge, Harvard Univ. Press.

Manne, A.S. 1962. "Product Mix Alternative : Flood Control, Electric Power and Irrigation", Intl. Econ. Review, 3, 30-59.

Thandaveswar, B.S. 1992. "Modelling an Overdeveloped Irrigation System in South India", Water Resources Development 8(1), 17-29.

Zimmermann, H.J. 1994. "Fuzzy Set Theory", Kluwer Academic Press.

Appendix A
Cropping Pattern in deterministic model

Year	Sept_B (mcm)	Sept_I (mcm)	Oct_I (mcm)	Nov_I (mcm)	Kharif (ha)	Rabi (ha)	River Release (mcm)
1972-73	25.00	00.00	00.00	00.00	0825.59	0000.00	22.86
1973-74	53.00	24.00	00.00	00.00	1020.02	2313.14	08.48
1974-75	42.00	00.00	00.00	00.00	1742.85	0000.00	22.01
1975-76	36.00	26.00	00.00	00.00	1152.85	2692.05	20.78
1976-77	77.25	06.00	00.00	00.00	0472.85	2509.90	08.31
1977-78	77.25	10.00	00.00	00.00	0790.35	2339.30	00.42
1978-79	66.00	10.00	00.00	00.00	0895.70	2382.82	14.78
1979-80	50.00	00.00	00.00	00.00	1535.85	0000.00	22.32
1980-81	40.00	00.00	00.00	00.00	1464.17	0000.00	37.18
1981-82	53.00	00.00	00.00	00.00	2087.81	2508.68	21.39
1982-83	50.00	00.00	00.00	03.00	2515.67	0000.00	31.49
1983-84	55.00	07.00	09.00	00.00	1920.80	3259.02	08.42
1984-85	77.25	00.00	00.00	00.00	1417.46	3495.90	13.14
1985-86	23.14	00.00	01.90	00.00	1779.63	0000.00	01.05
1986-87	20.42	00.00	00.00	00.00	1616.78	0605.98	03.45
1987-88	09.55	00.00	00.00	00.00	0000.00	0000.00	03.37
1988-89	53.24	01.30	00.00	00.00	1207.69	3252.76	11.06
1989-90	23.67	01.39	00.00	00.00	1723.49	0000.00	14.14
1990-91	56.53	17.49	00.00	00.00	0000.00	3420.61	04.79
1991-92	42.75	00.50	01.74	00.00	2202.27	0000.00	06.85

Appendix B

Monthwise Crop Plantation, Water Requirement (cubic metre) per hectarare and not Benefits from crop/ha

Crop	July	August	September	October	November	December	January	February	March	Net benefit (Rs)
Jowar(x1)			1896							1805
Bajra (x2)			1896							2881
Maize (x3)			1896							2592
Groundnut (x4)			2689							4875
Tur (x5)			1613	4160	3380	2510	2710			9048
Castor (x6)			1613	4160	3380	2510	2710			3862
Cotton (x7)			1613	4160	3380	2510	2710			4178
Mustard (8)					2000	1070	2520	2840		5872
S.flower (x9)							440			2080
Wheat (x10)					2030	1220	2600	2440	440	6963

Appendix C

Storage Continuity Constraints

In the absence of the record of humidity, temperature, reservoir storage and related evaporation loss, we have taken evaporation loss as the rate of change of storage level for different months. Simplifying storage continuity constraints from September to March, we arrive at

$$1.0086S_2 - 0.99S_1 - I_1 + RC_1 + DR_1 = 0$$
$$-0.99S_2 + 1.0093S_3 - I_2 + RC_2 + DR_2 = 0$$
$$-0.9899S_3 + 1.013S_4 - I_3 + RC_3 = 0$$

11

BIOMASS : A SUSTAINABLE ENERGY SOURCE

• *Prof. (Dr.) M.H. Fulekar and Prof. U.S. Bagde*

Life Sciences, University of Mumbai,
Kalina, Vidyanagari, Santacruz (E), Mumbai

Introduction

India, in the early 1970's recognized the importance of increasing use of renewable energy sources for achieving a sustainable energy. Indian scientists have developed solar cooker and water heaters in the fifties. Thereafter Indian scientists have been making efforts to design, develop and induct on a large scale a variety of renewable source of energy devices. The commission for additional sources of energy set up in 1981 and the Department of Non-conventional Energy Sources (DNES) set up in 1982. Thus, India earned the distinction of being the only country to have an exclusive Ministry for Non-conventional Energy Sources (MNES). The MNES now has sectoral groups of (a) rural energy, (b) urban/industrial energy, and (c) power generation. Through the restructuring, emphasis shifted towards policies, planning and institutional linkages to promote Renewable Energy Technology's (RET's) within each sector. Each such sector now consists of integrated programmes to serve different energy needs; for instance, cooking energy is now comprehensively dealt with, under the rural energy group rather than individual technologies being implemented separately. The change in the structure has resulted in significant changes in the focus of the programmes.

In recent years, the rationale has been further buttressed by the environmental imperative. Local and regional environmental problems associated with the generation of conventional energy have provided a strong argument for enhancing the role of renewable energy within the broad energy development plans of the country. More recently, the Kyoto Protocol, agreed at the Conference of Parties to the Framework Convention to Climate Change, in December 1997, adds a global perspective to the environmental imperative. It has been directed, last decade-and-a half, for promotion of wind, biomass, and solar energy technologies (and of other RETs) in the Indian energy economy. This has provided a great deal of empirical knowledge about strategies for successful commercialization.

Biomass-Sustainable Energy : Future

Unlike coal, oil and natural gas, whose reserves are limited, sources like the sun, wind, and vegetative 'waste' can be used to generate energy in a sustainable way. The sun's energy can be used to generate electricity, and heat as well as cool buildings cheaply over a long period of time without creating pollution. The wind energy is also used to generate electricity. Biogas plants can utilise human and animal waste to produce fuel for cooking and other uses, reducing the dependence on fossil fuels. It is estimated that the country has potential of 100,000 MW renewable energy (Padmanabhan, 1999)[1]. However, the share of renewable energy sources is 1378 MW, a mere 1.5% (exclusive of hydro power) of the total grid power generating capacity in the country (90,000 MW). It has often been pointed out that an important reason for the slow rate of diffusion of Renewable Energy Technology's is the high front-end cost. This will no longer be there as the fossil fuel are expected to reach their maximum potential and their prices will become higher than the renewable energy options. It is expected that the setting of clean technology in the coming years will facilitate channelling of funds from the developed countries to support renewable energy sources development in developing countries due to issues, such as, climate change becoming urgent. Thus finance is no longer a constraint.

Renewable source of energy other than hydropower e.g. solar, wind and geothermal sources, currently provide only a small fraction of global energy use. The most prevalent source of energy is biomass. Biomass fuels include wood, logging wastes and sawdust, animal dung, and vegetable matter consisting of grass, leaves, crop residues and agricultural waste. Globally, biomass fuels account for around 12% of total energy requirements. In developing countries, however, biomass accounts for 36% of all energy used (Smith, 1987, DeKoninge *et al.*, 1985). In India, the biomass programmes are mainly targeted to meet the needs of rural and remote areas and have helped in reaching electricity to the hitherto unreached section of the population. Besides, Renewable Energy provides a non-polluting, health hazards-free energy required for cooking, heating and drying. But statistically looking, there is long way to go to tap the potential to the extent desired.

One of the reasons for slowdown of installation/commissioning of biomass-renewable energy generation is said to be inadequacy of the input material. To overcome this, attempts are being made to use alternatives to cattle dung like poultry droppings, sericulture wastes, press mud, wastes from sago industry, bagasse from sugar mills and likewise. Since biomass based energy system can help to reduce carbon-dioxide emissions, a project on carbon emission reduction through biomass energy for rural India, prepared by the Centre for Application of Science and Technology in rural areas, in the Indian Institute of Science, is proposed to be posed to UNDP/GEF for multilateral funding.

Estimates indicate that if all forms of biomass were taken into account, their carbon dioxide emission reduction potential would be equivalent to about 150 million tones by the year 2010 (Sharma, 1999). But this will only be possible once biomass is used as a source of energy. Mere a forestation may not balance out excess carbon for an extended period of time. Because uncontrolled burning and decay of the mature plantations will bring back a sustainable quantity of carbon, back into circulation. The increased production of carbon dioxide in developing countries should be offset by greater energy conservation. Efforts to make renewable sources of energy less costly and more widely available should continue; practicable methods should be developed for waste incineration as a source of energy.

Fig. 11.1 : Biomass Programmes

Target/Programmes	Potential / Capacity	Installed
Bio-gas plants	12 million	< 3 million
Biomass Programme- Biomass Gassifer System (BGS)	Under- 31 (BGS)	3,085 KW
Around 28 projects- on Biomass and Bagasse	19,500 MW	141 MW
Another 27 projects on Biomass and Bagasse		180 MW (under- Execution)
Urban and Municipal Waste	1,000 MW	7 MW (completed)
Industrial waste	700 MW	4 MW (under-Execution)

Technology for Resourceful Waste (Case Study)

Given the success of an experimental kitchen waste biogas plant in Delhi, kitchen refuse may no longer remain an obnoxious waste. The 2 cubic metre improvised Deenbandhu biogas plant fed on a kitchen waste is functioning successfully for over a year now. Thanks to the initiative by All India Women's Conference (AIWC) and the innovation by Action for Food Production, the Plant is able to put the kitchen waste from AIWC working women hostel to some profitable use.

AIWC hostel accommodates 350 working women and the total organic waste generated from the hostel kitchen amounts to over 70 kilograms every day. Over the years, the hostel administration has been wrestling with the problem of disposing this waste. Towards the beginning of 1997, the problem was brought to Mrs. Lalita Balakrishnan's attention. Mrs. Lalita Balakrishnan heads the Rural Energy Units at the AIWC, which has been implementing a bio gas programme for past four years or so. It occurs to her that if a biogas plant could digest cattle dung, it may well be effective in digesting organic waste. Armed with this simple logic she approached AFPRO, who were quick to depute their biogas specialist Mr Ishrat Hussain for the task (AFPRO, 1999).

In place of a usual inlet tank, this plant has the inlet pipe directly linked to the digester. Kitchen waste is pushed down this pipe into the digester with the help of bamboo pole. A stirrer has been installed inside the digester, which can be easily rotated by a handle placed on the top of the dome. These two simple improvisations on the existing Deenbandhu model have worked well in effective functioning of the plant.

The family of a staff member residing in the campus is using the gas thus generated from the plant. The manure has been found useful for the potted plants in the hostel and the administrative block. What used to be problem some months ago is now a resourceful waste for the AIWC.

Wardha-based Centre of Science for Villages (CSV) is one of the agencies identified by the MNES for providing technical backup support on biogas technology within a given area. Over past 8-10 years, CSV has trained at least 200 technicians on biogas technology. The Government of Maharashtra has cared to engage only a handful of these trained persons in furthering biogas technology through the Zilla Parishad authorities.

Biomass-pollution Control : Strategies

The combustion of unprocessed biomass dominates rural energy combustion in the developing world and may also be important in urban communities. As many as two billion people, particularly women and children, may be exposed to indoor pollution resulting from the use of an open fire for cooking and heating, with inadequate ventilation. Concentration of particulates and oxides of sulphur and nitrogen substantially exceed proposed health norms. The most important effects are respiratory, ranging from predisposition to acute infections in children to chronic obstructive pulmonary diseases in adults. As many as 700 million women in developing countries may be at risk of developing such a serious disease. In addition to these direct effects on health, the environmental degradation resulting from the unsustainable use of Biomass may compromise the food-producing capabilities of rural communities (WHO-Commission, 1992).

Mitigation of indoor air-pollution can be achieved by the use of processed biomass (charcoal, biogas, or methanol) and the adoption of simple ventilation measures and improved stoves. Hopefully there will be a discernible trend towards this in developing countries together with the extension of local processing technologies. The introduction of appropriate species of vegetation, to provide a renewable source of biomass should be planned as a part of environmentally sound development (Kazuhisa, 1997).

Fig. 11.2 : Biomass Resources

	Starch Sugar Crops	– Grains (Rice, Wheat) – Sugarcane – Corn – Potatoes
	Aquatic Plants	– Sea weeds – Water-hyacinth, chlorella
Crops	Rubber Plants	– Para rubber tree – Guayule rubber tree
	Oil Plants	– Palms, etc.
	Petroleum Plants	– Heliopora – Yukar
	Woods	
BIOMASS		
	Agricultural wastes	– Rice straw, chaff, wheat straw – Sugarcan Bagasse – Corn, Stover
Unused Resources	Forestal Wastes	– Saw Dust – Pulp waste – Thinned woods
	Municipal Wastes	
	Industrial Wastes	
	Livestock Wastes	– Excreta of Livestock

Wider uses of improved technologies for the local conversion of raw biomass into better, more efficient types of fuel, such as biogas, are needed throughout the developing world. This results in little pollution if equipment is properly maintained, and contributes to a cleaner indoor environment. It will also be necessary to promote renewal of biomass vegetation, in order to prevent environmental degradation with loss of the agricultural land essential for the survival of the rural communities. Renewal of biomass also promotes a balance between carbon dioxide productions during fuel combustion and its uptake by the biomass vegetation during photosynthesis.

References

AFPRO, "Resourceful Waste-Technology" Rural Energy Journal, Vol. 8, p.13, 1999.

De Koning, H. W., *et al.*, "Biomass fuel combustion and health, Bulletin of the World Health Organization" 63 (1): pp. 11-26, 1985.

Kazuhisa Miyamoto, "Renewable biological system for alternative sustainable energy production" FAO Agricultural Services Bulletin, 128, Rome, pp. 1-108, 1997.

Padmanabhan, B. S., "RENEWABLE ENERGY-Into the reckoning" The Hindu - Survey of Indian Industry, pp. 144-146, 1999.

Sharma, S., "Biomass - The Burning Issue" Rural Energy Journal Vol. 8, p. 3, 1999.

Smith, K. R., "Biofuels, air pollution and health - a global review". New York, Plenum Press, 1987.

WHO-Commission, "Health and Environment - Report of the Panel on ENERGY" WHO/EHE/92.3, Geneva, pp. 1-155, 1992.

ENVIRONMENTAL BIOTECHNOLOGY

Edited By : **Professor (Dr.) Arvind Kumar**

Published By : **DAYA PUBLISHING HOUSE**

12

INVESTIGATION ON *IN VITRO* AND *IN VIVO* EFFECT OF *ALLIUM SATIVAM* AND *ALLIUM CEPA* ON MULTIDRUG RESISTANT *ENTEROBACTERIACEAE* MEMBERS AND COAGULASE POSITIVE *STAPHYLOCOCCUS AUREUS*

• *S. Muthukumaravel*, *K.T.K. Anandapandian and C. Rathika*

PG Unit of Microbilogy, V.H.N.S.N. College, Virdhunagar, Distt. Virdhunagar

*PG Unit of Microbiology & Zoology, Thiagarajpur College, Theppakulam (PO), Madurai

Abstract

Antimicrobial activity of the garlic was examined by using *Enterobacteriaceae* members and *Staphylococcus aureus*. Garlic extracts were prepared by mascerating the cloves in a blender and then filter sterilizing it. Anti-microbial spectrum was also studied by *In vivo* test using induced infection of *Staphylococcus aureus* in Balb/c strain mice. Based on the result, Garlic water was more active than the alcohol extract of Garlic.

Key Words: Enterobacteriaceae, Staphylococcus aureus, Allium Sativam, Allium cepa.

Introduction

Treating of affliction with chemical ingredients has been practised for centuries beginning from Louies Pasteur (Brewster, 1994). The first being the detection that sulphonamide compounds could be used for the cure of certain bacterial diseases and the second was the detection of a innovative and potent class of antibacterially strong chemotherapeutic agents, named antibiotics. Chemical agents were formulated in the chemical laboratory and most of them were derived from micro-organism and some plants. The sulphonamide drugs have been the development of chemotherapeutic agent with expanded anti-microbial activities in host animal (Frazier, 1989). For centuries meat and pisces were cooked with many food additives like Garlic and Onion. Countless ingredients in the garlic act as restraining substances. The heat stable components of these garlic and onion has been appraised for their anti-microbial animation since early 1900s. (Dorothy *et al.*, 1995). Thus these provocative anti-microbial activities of garlic instigated us to strive on this discussion.

Materials and Methods

Preparations of Garlic water extract and Alcohol extract

Garlic cloves were soaked separately in sterile distilled water and alcohol (g/ml) and kept for 24 hours for imbibitions. Imbibed cloves were ground to paste by blender (Usha

Anand Rao *et al.,* 1988). The resulting paste was let to be soaked separately in the sterile distilled water and alcohol for 48 hours. The paste is squeezed and filtered through whatmann's No.1 filter paper. Filtered extract is sterilized by seitz filter (less than 0.02μ) and stored at 4°C.

In vitro Test

Minimum inhibitory concentration value of garlic, and onion against certain multi-drug resistance of Enterobacteriaceae members were done by agar incorporation methods (Herald, 1987). Anti-bacterial activities were investigated by agar dilution method in Muller-Hinton Medium. The concentrations of garlic and onion extracts to that of the Muller-Hinton agar were prepared from 1: 1 0, 1 :20,– 1: 100 dilution. In every dilution the concentration of garlic and onion extracts were constant. Multi-drug resistant *Salmonella chloraesuis, Enterobacteriaceae agglomerans, Escherichia coli,* and *Enterobacter clocae* belong to *Enterobacteriaceae* family were used for testing the anti-microbial effect of garlic and onion (Victor, 1990). Resistance pattern of test organisms were determined by Kirby-Bauer Method.

In vivo Test

BALB/c strain of mice were used and caged into four batches. Each batch containing four mice. The mice were depilated by using the hair remover (Victor, 1990). Warm water was used for cleaning the surface. Mice weighed 20 to 21g, and then it was injected subcutaneously with *staphylococcus aureus* and the effects of extracts were observed by topical application of them on the infected site of the mice. During all the batch sterile broth, alcohol and water extract of garlic were used. Simultaneously control was treated in all the batch.

Recovery of *Staphylococcus aureus* From Infected Sites

After 24 hours incubation all the animals were sacrificed under sterile precautions the skin was cut open in the abdomen area and swabs dipped in sterile nutrient broth were used to take swabbing under the skin. Immediately the swabs were swabbed on mannitol salt agar.

Result and Discussion

The minimal inhibitory value of alcohol and water extract of garlic and onion were determined and it was shown in Table 12.1. The plate incorporation methods shown 1 : 29 dilution is the minimal inhibitory concentration of garlic water extract. Garlic water extract is more active than alcohol-Garlic extract.

Table 12.1 : Minimal Inhibitory Concentration of Strains Against Different Extracts

Sl.No.	Extracts	% value of effective dilution
1.	Garlic Water Extract	1 : 29
2.	Garlic Alcohol Extract	1 : 21
3.	Garlic Water Extract	1 : 3
4.	Garlic Alcohol Extract	1 : 1

The *Enterobacteriaceae* strains were also checked with various drugs to find out their sensitivity pattern by Kirby-Bauer method and the sensitivity percentage of the test organisms were determined, it was shown in Table 12.2.

The survival rate of *Staphylococcus aureus* was determined by *in vivo* method and it was shown in Table 12.3. Here also the garlic water extracts was more active than any other extracts. The survival rate of organism in garlic water extract is only 6.36%.

Table 12.2 : Resistance Percentage of Test Organisms

Sl. No.	Organism	Resistance Pattern	Percentage of Resistance
1.	*Salmonella chlorasusis*	15	83.3%
2.	*Enterobacter agglomerans*	9	50%
3.	*Escherichia coli*	8	44.4%
4.	*Enterobacter cloacae*	8	44.4%

Table 12.3 : Effect of Garlic Extract on *Staphylococcus aureus* Infection (*in vivo* method)

Sl. No.	Batch of Animals	Average colony forming unit	% of reduction rate	Survival rate of *S. aureus* %
1.	If-Garlic Water extract	56	93.64	6.36
2.	If-Garlic Alcohol extract	194	77.96	22.04
3.	If-Topical Treatment	380	56.4	43.06

Note : Animals treated with *Staphylococcus aureus* alone forming 880 CFU. This should be treated as control when compared to other extracts.

If-Infection, The pyogenic bacteria such as *Staphylococcus aureus* were inhibited by aqueous extract of garlic and it is also positively documented that onion extracts displayed an inferior growth inhibiting activity. The 4% aqueous extract of garlic were totally inhibited the growth of *Salmonella typhosa*, *Shigella dysenteriae*, *Staphylococcus aureus* and *Escherichia coli*. The crude extract was exceptionally effective against *Staphylococcus aureus*. The onion juice has only a feeble bacteriostatic impact. Onions were not effective against gastrointestinal bacteria. The component from garlic acts actively and acrolein components inhibit the bacterial growth (Alfred Larry Branen *et al.*, 1983).

References

Alfred Larry Branen, Michael Davidson, P., 1983. Antimicrobials in Foods pp. 374-376.

Brewster, J.L and Wellesbouene, UK 1994. Onion and Other Vegetable Alliums: Chapter 1 pp. 1-18.

Dorothy, W. Sanders, Paul Wetherwax and L.S. McClung 1945. Antibacterial substances from plants collected in Indiana. Journal of Bateriology V. 49 pp. 611-615.

Fraizer, W.E., Westhoff D.C. 1989. Food Microbiology VIII edition pp. 163.

Herald J. Bensen 1987. Microbiological App. PB2-133 v ed.

Usha Anand Rao. K. V. Murthy 1988. Effect of Garlic extract on Vibrio Parahaemolyticus. Journal of biomedicine. Vol.15/8/pp. 1-12.

Victor Lorian, M.D. 1990: Antibiotics in Laboratory Medicine III ed topical treatment.

ENVIRONMENTAL BIOTECHNOLOGY

Edited By :   **Professor (Dr.) Arvind Kumar**

Published By :   **DAYA PUBLISHING HOUSE**

13

EVALUATION OF PROVENANCES OF *PUNICA GRANATUM* L. (WILD POMEGRANATE) USING BIOCHEMICAL AND FRUIT TRAITS

• *K.S. Pant, Arun Handa*, Pankar Panwar & C.V. Saraswat*

Department of Silviculture and Agroforestry,
University of Horticulture and Forestry, Nauni, Solan (H.P.)

Introduction

Large scale plantations demand large amount of planting material and it is necessary to start with superior planting material in order to achieve high yields of quality produce. Further, study of variation is the first step for any breeding programme. Effective tree breeding depends on an understanding of tree variation in nature and preserving such variation for future use. In natural population of the species the presence of genotypic differences cause variation, which may be genetically associated. To study available variation in entire range of species distribution and delineation of the provenances capable of providing best adapted tree, parameters like seedling performance in nursery are suggested which often take long time to accomplish. The quick and easy alternatives are the methods which use biochemical and fruit traits to evaluate variations among provenances. Thus in the present study variations in the provenances of wild pomegranate were ascertained using biochemical and fruit traits.

Materials and Methods

A pilot survey of population of *Punica granatum* in Himachal Pradesh was undertaken to identify the sites, where this species occurs in wild stands. The sampling procedure includes delineation of the whole area, under the species into a number of sites depending upon the altitudes, aspect and variation in morphological characters. In this way fifteen provenances were selected in five districts of Himachal Pradesh. The name and geographical location of the selected sites is given in Table 13.1. The selected provenances were confined to 32°52′4″ to 32° 32′ 18″ N latitude and 76°06′ 11″ to 77°52′18″ E longitude. These localities fall under sub-tropical to sub-temperate climate. Five naturally occurring trees of wild pomegranate, which were mature, having approximately same age, crown size and mean diameter of 15-25 cm were selected within each site as per method adopted by Savinin, 1967; Pozdnjakov, 1969 and Dumitriu-Tataranu, 1970. These trees were marked for the

Table 13.1 : Geographical Locations of Different Provenances of Wild Pomegranate (*Punica granatum* L.) in Himachal Pradesh

Code Number	Provenances	District	Altitude (m)
S1	Oachghat	Solan	1200
S2	Kandaghat		1150
S3	Darlaghat		1250
S4	Daro Deoria	Sirmour	1000
S5	Narag		1250
S6	Bagthan		1300
S7	Basantpur	Shimla	950
S8	Ghanati		1050
S9	Shoghi		950
S10	Sundernagar	Mandi	900
S11	Rewalsar		800
S12	Sarkaghat		900
S13	Kullu	Kullu	1300
S14	Banjar		1250
S15	Nagwain		1300

Source : Toposheets of Sermour, Solan, Mandi, Kullu and Shemla districts of Himachal Pradesh.

collection of fruits. Eight fruits from each of the selected trees were collected, depending upon their direction i.e two each from south, east, north and west, and were mixed together. Average weight of fruit per tree was determined. The fruit rind was removed and dried and its average weight was estimated. The thickness of the rind was recorded with digital vernier calliper. Biochemical parameters viz. total soluble solids (TSS), acidity, total sugar, reducing and non-reducing sugar were estimated. Total soluble solid of the seeds was determined with the help of hand refractometer and was expressed in Brix units. The titrable acidity was estimated by titrating a known aliquot of sample against N/10 NaOH solution using Phenolphthalein as an indicator. The titrable acidity was expressed as per cent citric or acetic acid as described by A.O.A.C., 1980. Sugars (reducing and total) were determined by Lane and Eynons volumetric method (A.O.A.C., 1980) by titrating the samples against Fehlings solution (before and after hydrolysis) and expressed in per cent. Non-reducing sugar was calculated by subtracting the contents of reducing sugar from that of total sugar and multiplying the difference by 0.95 factor.

Results and Discussions

Variations in biochemical traits as affected by source of fruit is presented in Table 13.2. The data reveals that total sugar content ranged between 5.74 to 8.07 per cent. Highest sugar contents (8.07 per cent) was recorded for the fruits collected from S3 (Darlaghat) provenance, which was followed by S1 (Oachghat) and S4 (Daro Deoria). But differed significantly from the remaining provenances. S13 (Kullu) provenance recorded the least value of 5.74 per cent for total sugar, followed by S6 (Bagthan) provenance but differed significantly from that of the remaining provenances. Coefficient of variation gave a value of 5.42 per cent. Percentage of reducing sugar varied from 3.10 to 6.14. Heighest value of 6.14 per cent was registered for S7 (Basantpur) provenance which was statistically at par with S3 (Darlaghat) and S4 (Daro Deoria) provenances, but showed significant variations with the remaining twelve provenances. Lowest value of 3.10 per cent was registered for S14 (Banjar) provenance, which was statistically at par with S13 (Kullu) and S9 (Ghanati) provenance but differed significantly from S15, S11, S10, S6 and S12. Coefficient of variation for reducing sugar was 8.28 per cent non-reducing sugar showed highly significant variation and varied between 1.39 to 3.07 per cent and recorded 20.09 per cent coefficient of variation. High value of non-reducing sugar was found in S15 (Nagwain) provenance. Lowest value of non-reducing sugar (1.39 per cent) was recorded from S7 (Basantpur) which was statistically at par with S2 (Kandaghat), S5 (Narag) and S12 (Sarkaghat) provenances. Total soluble solids ranged from 10.60 to 14.96 Brix among 15 provenances with 7.13 per cent coefficient of variation. Total soluble solid were heighest (14.96 Brix) in S3 (Darlaghat) provenance which was statistically at par with that of S1 (Oachghat) and S4 (Daro Deoria) provenances. Lowest value of 10.60 Brix was noticed from that of S13 (Kullu) provenance which was statistically at par with S6 (Bagthan) and S14 (Banjar) provenance but showed significant variation from that of remaining provenances. Per cent acidity ranged between 5.48 to 7.05. High acid content, 7.05 per cent, was recorded for S9 (Shoghi) provenance which did not differ significantly from S6 (Bagthan), S13 (Kullu), S14 (Banjar) and S11 (Rewalsar) provenances. Lowest 5.48 per cent acidity was recorded from the Oachghat provenance. Coefficient of variation gave a value of 6.83 per cent among different provenances.

Table 13.2 : Variation in Biochemical Traits Among Different Provenances of *Punica granatum* L.

Provenances	Total Sugar (%)	Reducing Sugar (%)	Non-reducing Sugar (%)	TSS (Brix)	Acidity (%)
S1	7.99	5.49	2.36	14.72	5.48
S2	6.94	5.39	1.46	12.88	6.64
S3	8.07	5.78	2.17	14.96	6.58
S4	7.71	5.70	1.90	14.32	6.16
S5	6.79	5.07	1.63	12.48	6.72
S6	6.04	3.96	1.97	11.28	7.04
S7	7.61	6.14	1.39	13.86	6.80
S8	6.44	3.68	2.61	11.96	6.76
S9	6.27	3.52	2.61	11.72	7.05
S10	6.43	3.82	2.47	11.88	6.65
S11	6.72	3.80	2.76	12.44	6.89
S12	6.40	4.40	1.89	11.84	6.57
S13	5.74	3.33	2.30	10.60	7.03
S14	6.22	3.10	2.95	11.11	6.99
S15	7.01	3.77	3.07	13.04	6.37
$CD_{0.05}$	0.46	0.47	0.57	0.86	0.57
Coefficient of Variation (%)	5.42	8.28	20.09	7.13	6.83

Sugar content more or less follows the pattern of fruit growth and is directly related with the development phase and ripening of fruit (Grebensky, 1936 and Dzamic *et al.*, 1967) for pulm varieties. Increase in fruit total solid might be associated with increased translocation of organic assimilates from leaves to hormonal stimulation (Hansen, 1967). Total soluble solid as the indicator of positive correlation with total sugar showed a significant variation among the different provenances. These findings are in line with the findings of Sharma, 1990. Various factors like hormonal stimulation of assimilates may contribute to depression or enhancement of acid content in fruit. Present observations on acidity are in line with the findings of Sharma, 1992.

Data on variation fruit traits viz. fruit size, weight, rind thickness and rind dry weight of different provenances are presented in Table 13.3. It was observed that fruit collected from Kandaghat provenance (S2) was having heighest fruit length (7.44 cm) showing non-significant differences with Oachghat (6.95 cm), Basantpur (6.77 cm) and Nagwain (6.93 cm)

Table 13.3 : Variations in Fruit Traits in *Punica granatum* Among Different Provenances

Provenances	Fruit length (cm)	Fruit diameter (cm)	Fruit weight (g)	Rind thickness (cm)	Rind dry weight (g)
S1	6.95	6.27	77.48	0.57	11.67
S2	7.44	6.37	87.80	0.44	11.50
S3	5.14	4.42	91.25	0.45	15.12
S4	5.59	4.67	91.57	0.37	9.22
S5	5.96	4.82	85.87	0.69	9.17
S6	6.05	5.29	101.00	0.69	12.75
S7	6.77	5.48	80.60	0.48	13.12
S8	6.44	5.84	99.75	0.71	11.00
S9	6.59	5.72	70.50	0.70	9.37
S10	6.10	5.13	60.65	0.51	8.27
S11	5.68	5.22	59.77	0.36	7.75
S12	5.94	5.39	62.35	0.54	13.40
S13	6.43	5.31	63.75	0.41	7.47
S14	6.48	5.91	70.30	0.53	8.45
S15	6.93	6.12	65.17	0.34	7.30
$CD_{0.05}$	0.77	0.80	14.63	0.14	2.72
Coefficient of Variation (%)	9.68	11.52	14.82	21.15	20.73

provenances. Fruit of Darlaghat provenance (S3) gave the least fruit length (5.14 cm). Following the same trend, fruit diameter of Kandaghat provenance was largest having a value of 6.37 cm which was at par with Ghanati (5.84 cm), Shoghi (5.72 cm), Banjar (5.91 cm) and Nagwain (6.12 cm). The least fruit size was of Darlaghat (5.14 cm) provenance. Fruit obtained from Bagthan provenance were heaviest in weight (101 g). Fruits of Kandaghat (87.80 g), Darlaghat (91.25 g), Daro Deoria (91.57 g) and Ghanati (99.75 g) showed statistically no difference among them and with Bagthan provenance. Ghanati provenance registered maximum thickness of rind (0.71 cm) which was at par with Oachghat (0.57 cm), Narag (0.69 cm), Bagthan (0.69 cm) and Shoghi (0.70 cm). Least rind thickness was recorded in Nagwain provenance (0.34 cm). Fruits harvested from Darlaghat provenance registered highest (15.12 g) rind dry weight which was statistically at par with Sarkaghat provenance (13.40 g) and showed significant variations from all other provenances. The least rind dry weight (7.75 g) was recorded in fruits collected from Rewalsar provenance. The coefficient of variation among fruit characters was highest for rind thickness (21.15 per cent) followed by rind dry weight (20.73 per cent), fruit weight (14.82 per cent) and least for fruit length (9.68 per cent).

The results indicate that the fruit size was highest in Kandaghat provenance with mean fruit length and width of 7.48 cm and 6.42 cm, respectively. However, highest fruit weight was recorded for Bagthan provenance. The increase in fruit might be due to the enhanced synthesis of carbohydrates and water uptake and their movements into the fruits (Miniraj and Shanmugavelu, 1987). The increase in fruit size might also be due to an enlargement of the cells in the fleshy parts of the fruits as has been observed by Turkey and Young, 1939. The maximum mean rind thickness of 0.71 cm has been recorded from Ghanati provenance and lowest 0.31 cm for Daro Deoria provenance. These findings are in line with the findings of Sharma, 1990 who reported that rind thickness lies between 0.33 - 0.72 cm in wild pomegranate. Since the fruits were collected from trees of similar age, size and crown exposure, therefore, variations observed in fruit traits may be attributed to genetically variable sources as a result of adaptation of diverse environment throughout their distribution (Salazar and Quesada, 1987).

References

A.O.A.C., 1980. Official methods of analysis, Association of Official Analytical Chemists, Washington, D.C, USA.

Dumitriu- Tataranu, I. 1970. Studies with a view to selection of some provenances and forms of larch occurring naturally in Rumania. For. Abst. , 37, 792 1976.

Dzamic, M., Popovik, Z. and Pantelic, M. 1967. Changes in the glucose, fructose and sucrose contents of the fruits in the pulm variety pozegaca. zborn.Rad. , Poliouriv Fak. Beograd., 15 (451).

Grebensky, S.O. 1936. Biochemical investigations of pulms. Bull. Appl. Bot. Liningr., Ser. 11 (15): 31-34.

Hansen, P. 1967. C-Studies in apple trees. 1 The effect of fruit on the translocation and distribution of photosynthates. Physiol. Plant, 20 : 382 - 391.

Miniraj, N. and Shanmugavelu, K.G. 1987. Studies on the effect cf triacanthol on growth, flowering yield, quality and nutrient uptake in chillies (*Capcicum annum* L.), South Indian Hort., 35 : 362-366.

Pozdnjakov, A.A. 1969. Morphological characters for the cisual determination of age generations of *Abies siberica*. For. Abst., 32,1243.

Salazar, R. and Quesda, M. 1987. Provenance variation in *Guazuma ulmifolia* L. in Costa Rica. *Commonwealth For. Rev.* 66(4) : 317 - 324.

Savnin, A.G. 1967. Age estimation by external features in spruce and fir. For. Abst., 27, 2607.

Sharma, J.C. 1990. Taxanomic status of some wild fruits. Ph.D Thesis, Deptt. of Fruit Breeding, UHF, Nauni, Solan (H.P.) India.

Sharma, S. D. and Sharma, V. K. 1990. Variation for chemical characters in some promising strains of wild pomegranate (*Punica granatum* L.), *Euphytica*, 49 (2): 131 -133.

Turkey, H.B and Young, J.O. 1939. Histological study of the developing fruits of the sour cherry, *Bot. Gaz.*, 100: 723-749.

ENVIRONMENTAL BIOTECHNOLOGY

Edited By :   Professor (Dr.) Arvind Kumar

Published By :   DAYA PUBLISHING HOUSE

14

CERAMIC TILE INDUSTRIAL WASTE TREATMENT PLANT – RESIDUE USAGE IN THE PRODUCTION PROCESS

• *B. Kotaiah*, I.V. Ramana Reddy** & S. Sreedhar Reddy****
* Professor & Head, Deptt. of Civil Engg., S.V.U. College of Engg., Tirupathi
** Assoc. Professor, Deptt. of Civil Engg., S.V.U. College of Engg., Tirupathi
*** Lecturer, Deptt. of Civil Engg., Vellore Institute of Technology, Vellore

Abstract

The paper deals with the feasibility of usage of treatment plant residue of ceramic tile manufacturing industrial waste water in the production process. The experimental results indicate that the treatment plant residue can be mixed with the basic raw materials at a proportion of 1 : 10 to 1 : 20 without any deterioration to the quality of tiles standards. A proportion of 1 : 10 increases the bending stress of tile and decrease considerably the compressive strength. Water absorption was decreased for all admixtured residue tiles.

Key Words : Reuse, waste treatment plant residue, Ceramic tile industry.

Introduction

Recovery and reuse of waste constituents from industrial waste is a cost effective solution in the control of environmental pollution. Ceramic tile industries produce liquid wastes with high inorganic colloidal matter through floor, equipment and process operation. The treatment of these wastewaters by alum coagulation (dose 500-600 mg/lit) and clarification produce about 3600 to 3800 kg of chemical sludge (on dry weight basis) per 40 tonnes of tiles produced (Kotaiah, 1990). The residue contains silt and clay minerals, metal oxides (glazing substances) colour stains etc., some of which are toxic and post-pollution problems, if discharged into the environment. One way of solving this problem is to reuse it in the production process.

The present paper examines the concept of reuse of the treatment plant residue in the production process and consequent impact of these on the quality of tile standards.

Materials and Methods

Recovered treatment plant residue from wastewater was first oven dried at 100°C for 24 hours. Then it was mixed in various proportions (1 : 10 to 1 : 20) with the body materials. The mixture was fed into the production process (slip-preparation, spray drying, glazing and gas-fired (kilning) as usual processes). The obtained tiles along with others (unmixed) were subjected by various standard tests of tiles (IS 13630:1992). In each test 10 randomly selected tiles were used.

Results and Discussion

The characteristics of tiles of both normal and waste residue mixed with the body material are presented in Table 14.1. Both normal and different fractions of waste mixed with the body material are compared (IS 13630 : 1992).

Table 14.1 : Characteristics of Tiles

S.No.	Parameter	Normal Tiles (Unmixed)	Residue Mixed Tiles with Different Proportions			Italian Standards
			1:10	1:15	1:20	
1.	Flatness of the Surface Deviation (%)	± 0.4 (max)	± 0.4 (max)	± 0.4 (max)	± 0.4 (max)	± 0.5 (max)
2.	Compressive Strength kgf/cm²	1000 - 1250 (1100)	900 - 1020 (1010)	950 - 1100 (1075)	1010 - 1200 (1070)	700
3.	Transverse Strength kgf/cm²	280 - 340 (310)	330 - 360 (338)	320 - 345 (328)	310 - 330 (314)	200
4.	Water Absorption (%)	4 - 4.8 (4.6)	3.6 - 3.8 (3.68)	4.0 - 4.4 (4.16)	4.3 - 4.6 (4.38)	3 - 6
5.	Hardness (MOH's Scale)	5	5	5	5	5 (min)

Note : 1. 10 Nos. of tiles are collected randomly for each test.

2. Total number of batch-feeding for each proportion of residuce as admixture-3 (at different time intervals)

3. Average Value in parenthesis.

The results indicate that water absorption of the tiles decreased with waste residue admixtured as compared to unmixed tiles. This is possibly due to the presence of lower sized colloidal matter in the residue (metal oxides) which might have graded the body material pore volume. This consequently decreased the total pore volume of the tile with a lesser water absorption. A slight increase of flexural strength is observed in all the tiles produced by admixtured residue. This may be due to the increase of plasticity in the body material of the tile in the presence of metal oxides and hydroxides, and other $Al(OH)_{3(S)}$ species as additives. A slight decrease in compressive strength of admixtured tiles as compared with normal tiles is not quite apparent in this investigation. Further investigations are needed to understand the reasons for such decrease in compressive strength.

Conclusions

Based on the present investigation, the following conclusions can be drawn :

1. Recovery and reuse of chemical sludges in the production process is an economic and feasible solution for all the ceramic tile industries.
2. An admixture of 1 : 10 to 1 : 20 proportion of treatment residue with body material do not deteriorate the quality of tile to the standards.

References

IS 13630: 1992. Specification for Ceramic Tiles. Bureau of Indian Standards, New Delhi.

Kotaiah, B. 1990. "Management of Liquid Waste of Stiles India Limited, Unpublished Project Report.

ENVIRONMENTAL BIOTECHNOLOGY

Edited By :   Professor (Dr.) Arvind Kumar

Published By :   DAYA PUBLISHING HOUSE

15

SEASONAL INCIDENCE AND IMPACT OF ABIOTIC FACTORS ON ORIENTAL FRUIT FLY, *BACTROCERA DORSALIS* (HENDEL) ON MANGO (*MANGIFERA INDICA* L.) AT JAMMU

• *R.K. Arora*

Division of Entomology, Faculty of Agriculture,
Sher-e-Kashmir University of Agricultural Sciences & Technology-J,
Udheywalla, Jammu-180002

Abstract

Studies on the seasonal incidence of *Bactrocera dorsalis* Hendel with methyl eugenol traps in mango orchard at Jammu during 1999-2000 revealed that the population build-up of the pest commenced from 10th standard week (4 -10 March, 2000). Maximum number of flies (482.33 and 472.66) were trapped during 30th (22-28 July) and 31st (29 July-4 August, 2000) standard weeks when the mean temperature 35.5°C (max.) and 25.6°C (min.) coupled with 86.9 per cent relative humidity and total rainfall 319.62 mm were found contributing towards highest fly population coinciding with the ripening of Dashehari cultivar of mango. Simple correlations between weekly trap catches and abiotic factors recorded a highly significant and positive correlation with maximum temperature ($r = 0.5494$), relative humidity ($r = 0.98$) and rainfall ($r = 0.7122$).

Key Words : Mango, Fruit fly Bactrocera dorsalis, seasonal incidence, abiotic factors.

Introduction

Mango (*Mangifera indica* L.) is the most important fruit grown in tropical and sub-tropical regions of the world. In Jammu region of Jammu and Kashmir State in India, mango occupies an area of approximately 7190 hectares with an annual production of 7838 tones (Department of Horticulture, 1997). One of the major constraints in the production of mango is insect-pests causing damage to all parts of the plant at all stages of its growth and development (Ramasubbarao, 1994). Among them, the oriental fruit fly, *Bactrocera dorsalis* Hendel is one of the most destructive, widely distributed and economically important pests of mango (Veeresh, 1989). The pest has been reported to be active throughout the year in South India, whereas, in northern cooler areas, it over winters in pupal stage and adult flies appear when the ambient temperature persists (Kapoor, 1993, Srivastava, 1997). The adults survive under unsuitable weather conditions by feeding on exudations and the honeydews of various other insects (Panwar, 1995). A number of studies have been conducted on the incidence and management of this pest elsewhere in the country. However, no information was available so far on the actual time of appearance of this pest, its peak population and effect of various weather factors on its population build-up at Jammu. Therefore, the present studies were carried out to monitor the onset of its activity in mango orchard using methyl eugenol traps.

Material and Methods

Studies were conducted in a twenty years old mango orchard at Jammu which lies in humid sub-tropical tract (274.3 m above mean sea level and 32°N latitude and 74°E longitude). One hundred and ten mango trees of Dashehari variety were selected for observations planted at a variable distance of 18 to 20 feet. The design of the trap was slightly modified after Steiner (1957) and the plastic shoot fly trap designed by Mekongsee *et al.* (1981). Each trap was filled with 100 ml solution containing 1.0 ml methyl eugenol, 0.5 ml dichlorvos and 98.5 ml of water. The quantity of methyl eugenol and insecticide were in accordance with the concentrations standardized by Singh (1993). Trap catch counts were taken at weekly interval by emptying the traps in the laboratory and replenishing them again with fresh solution. The numbers of flies so trapped were counted separately for each trap and the catch record was maintained regularly. For linear association among various weather factors, weekly mean population counts of *B. dorsalis* were recorded from 40th standard week of 1999 to 39th standard week of 2000 (1st Oct., 1999 to 29th Sept., 2000). The data were subjected to statistical analysis to determine the effect of various weather factors on the build-up of fly population. To study the degree of their association, simple correlation coefficients were worked out.

Results and Discussion

From 52 weekly receipts (Table 15.1), it was observed that the fly population significantly fluctuated over the period of study. Lowest mean trap catches (63.0-2.33 flies/trap) were obtained during the cooler months from 40th standard week (1-7 October, 1999) to 4th standard week (21-27 January, 2000) with weekly mean temperature and relative humidity ranging between 33.8°C (max.) to 5.9°C (min.) and relative humidity 88.57 to 36.0 per cent, respectively. A constant mean population ranging between 7-8 flies/week was recorded from 5th standard week (28th January - 3rd February, 2000) to 9th standard week (25th February - 3rd March, 2000) at an average temperature of 20.28°C (max.) and 8.6°C (min.) and relative humidity 79.42 to 42.28 per cent. Thereafter, an orderly increase in the mean number of flies entrapped was registered from 10th standard week (4-10 March, 2000) to 22nd standard week (20-26 May, 2000) ranging from 15.33 to 112.33 flies/trap. Weather parameters also depicted similar increasing pattern when weekly mean temperature [25.9°C (min.), 43.8°C (max.)] and relative humidity ranging between 37.3 to 50.3 per cent were found conducive for the fly population build-up. Trap counts recorded from 23rd standard week (3-9 June, 2000) to 27th standard week (23-30th June, 2000) registered an unabated sharp increase and attained peak in 30th standard week (22-28th July, 2000) maintained almost a similar trend in 31st standard week (29th July -4th Aug.) and started declining up to the end of experimentation. The present findings conform to the observations of Bagle and Prasad (1983) who recorded the lowest weekly trap catches during December (42 flies/trap) and January (71 flies/trap) and maximum during March to June with monthly average catch/trap being 270,416,487 and 1268, respectively. In another study, it was estimated that most favourable period with maximum fly catches in India was 2nd week of June to middle of August with more than 300-500 flies/trap (Rahman, 1990). Reports of some other works

Table 15.1 : Monitoring of Adult Flies of *Bactrocera dorsalia* (Hendel) in Relation to Abiotic Factors

Standard week	Mean number of adult flies trapped	Temperature (°C)		Relative Humidity (%)		Rainfall (mm)	Wind velocity (km/h)
		Maximum	Minimum	Morning	Evening		
40	63.00	33.88	21.55	66.57	48.71	Nil	12.42
41	50.66	33.81	17.54	61.71	42.14	Nil	16.14
42	38.66	32.30	14.90	58.71	39.85	Nil	11.57
43	36.00	31.00	14.15	60.71	36.28	Nil	11.14
44	34.66	30.00	14.00	71.14	42.57	Nil	11.42
45	6.33	29.24	15.01	58.85	37.00	5.00	20.42
46	10.66	27.80	12.57	72.71	47.71	Nil	11.71
47	9.33	26.88	9.08	66.28	36.00	Nil	15.14
48	7.66	25.17	10.61	64.28	42.85	Nil	13.28
49	8.00	26.44	8.65	74.28	41.42	Nil	11.57
50	9.33	23.64	7.82	77.57	46.14	Nil	10.85
51	8.33	21.52	6.11	79.28	50.71	Nil	10.28
52	3.66	22.07	5.95	66.85	37.71	Nil	10.85
1	2.66	22.90	7.91	61.42	41.71	Nil	12.85
2	3.00	20.48	8.18	81.14	57.85	51.10	14.85
3	2.33	17.82	8.05	88.57	67.00	6.50	17.42
4	2.33	20.21	7.38	78.28	52.85	1.20	13.57
5	6.33	20.98	7.70	79.42	49.28	Nil	24.28
6	6.66	21.61	9.05	73.00	52.14	0.20	21.85
7	8.00	21.08	8.60	67.00	42.28	0.30	26.14
8	8.66	20.28	8.87	73.85	48.42	37.70	25.42
9	12.66	24.27	9.11	69.42	45.57	Nil	19.42
10	15.33	25.92	11.05	68.42	43.85	Nil	12.71
11	22.66	28.55	14.17	68.71	46.28	Nil	17.85
12	28.33	29.97	14.98	64.85	49.14	2.80	27.42
13	39.00	31.00	17.17	63.14	44.28	Nil	20.00
14	46.33	28.58	14.85	61.14	45.28	22.40	28.42
15	42.66	30.70	15.88	38.85	38.42	Nil	24.42
16	72.00	30.92	17.27	45.71	37.42	1.00	19.81
17	80.00	33.30	17.81	51.71	36.71	11.70	23.00
18	76.33	36.84	22.65	43.42	34.28	4.27	23.85
19	86.66	36.40	21.02	41.85	31.00	9.60	33.00
20	99.00	37.77	22.48	46.28	32.85	6.00	29.57
21	104.66	41.95	25.07	30.71	25.85	0.20	33.28
22	112.33	42.85	27.70	29.71	20.71	Nil	14.00
23	165.66	43.44	27.48	23.85	16.42	0.50	15.57
24	280.00	38.44	25.25	35.28	26.42	0.20	29.58
25	320.00	43.58	29.40	37.28	28.28	Nil	27.71
26	327.00	36.57	27.112	71.14	58.57	30.20	25.71
27	274.66	33.64	25.62	82.85	69.85	147.84	26.85
28	407.00	33.42	26.27	84.71	74.28	97.79	26.42
29	363.66	33.24	25.45	86.14	80.57	190.75	21.42
30	482.33	33.51	25.67	85.57	75.85	64.26	17.42
31	472.66	33.34	25.34	88.28	82.14	255.36	23.85

Contd...

Table 15.1 – Contd...

Standard week	Mean number of adult flies trapped	Temperature (°C)		Relative Humidity (%)		Rainfall (mm)	Wind velocity (km/h)
		Maximum	Minimum	Morning	Evening		
32	347.00	32.14	24.31	92.42	85.57	236.74	16.57
33	232.66	32.78	24.77	86.42	75.42	165.06	14.42
34	114.33	33.40	25.44	86.14	77.57	64.05	14.28
35	122.66	33.54	24.78	85.85	72.71	31.78	14.71
36	128.33	32.68	24.81	83.14	72.57	27.37	15.71
37	152.00	33.00	23.30	79.57	67.71	35.00	24.57
38	109.00	33.54	21.62	75.14	57.57	Nil	12.85
39	94.66	32.30	21.35	77.42	49.20	9.00	18.71

conducted over the years on *B. dorsalis* (*Dacus dorsalis* Hendel) have shown that both biological and environmental factors influence the distribution and demography of the fruitfly population. However, temperature, moisture and availability of host appear to be the most important factors and are primarily inter-related to the number of adult flies (Balasubramaniam *et al.*, 1972, Lakshamanan *et al.*, 1973). In the present study also, the peak period of activity of the pest coincided with the 3rd and 4th stages of maturity of mango fruits in the experimental orchard which was visually assessed as described by Singh (1990) i.e. shoulders outgrown from the stem end and colour of the fruit between olive green and light olive green. It was, therefore, concluded that methyl eugenol traps admixed with an appropriate insecticide could be employed in mango orchard to control the pest effectively.

Correlation between fruitfly population and weather parameters

The results of simple correlation coefficients between weekly fruitfly catches per trap and weather parameters are presented in Table 15.2. The population of *B. dorsalis* was found to have been influenced significantly as there existed a positive correlation between weekly

Table 15.2 : Simple Correlation Coefficients of Fruitfly Population Trapped Per Standard Week with Abiotic Factors

	Temperature (°C)		Relative Humidity (%)		Rainfall (mm)	Wind velocity (km/h)
	Maximum X_2	Minimum X_3	Morning X_4	Evening X_5	X_6	X_7
X_1	0.5494**	–0.0493	–0.0720	–0.0231	0.7122**	0.3325
X_2		–0.0216	–0.1682	–0.1119	0.1418	0.3352
X_3			0.9818**	0.9874**	–0.0395	–0.0637
X_4				0.9970**	0.0176	–0.1417
X_5					0.0488	–0.1047
X_6						0.1140

* Mean number of flies trapped per week;
** Significant at 5%.

population counts and maximum temperature (r = 0.5494). These findings are in conformity with Clark *et al.*, (1967), Arai (1976) and Bagle and Prasad (1983). Shukla and Prasad (1985) had also established a significant positive correlation between fly population and temperature, responsible for timing of population process and their synchronization with changes in environment. However, Verghese and Sudha Devi (1998) suggested considering the minimum temperature as a constant variable for the development of a forecasting model using methyl eugenol bait traps to help in IPM decision making. In the present studies, relative humidity (morning-evening) was also found to play an important role in the population build-up of fruit fly as it acquired a highly significant positive correlation (r = 0.98) with mean fly catches. During the peak period of activity, weekly average rainfall (819.62 mm) also favoured the fly population significantly as the correlation matrix obtained under the present investigation revealed a highly significant positive correlation (r = 0.7122) resulting in increased relative humidity and directly affecting the rise in population density. These findings are in agreement with Nishida (1963) who reported expansion in population when rainfall was adequate. Shukla and Prasad (1985) also established a positive correlation with maximum relative humidity and significant negative correlation with minimum relative humidity and weekly rainfall and number of *B. dorsalis*. The differences in the correlations of relative humidity and rainfall obtained in the present findings could be attributed to the differences in the weekly average relative humidity and rainfall data of the two different places of study. Tan and Serit (1994) established that peak numbers of *Bactocera dorsalis* males and host fruit occurred entirely within wet season of the year. There was a negative correlation between average weekly wind velocity and trap counts of fruitfly. Relatively higher number of fruit flies was captured mostly during the weeks of lower wind velocity. Trapping fewer flies during the weeks of lower wind velocity may be due to hindrance in their movement and flight. These results are in conformity with Bagle and Prasad (1983) who obtained negative correlation between wind velocity and fly population.

The fly population was observed to decline after 4th August, 2000 with the trap catches lower in number. This could be attributed to heavy rainfall thereafter and water stagnation in the orchard, which mediated the extrication of adults from pupae. It could, thus, be copiously inferred that the range of environment is extremely broad to which fruit fly *B. dorsalis* is exposed and there was no single environmental component that could stand out to determine its abundance. The principal components on which the rates of increase or decrease depended were temperature, relative humidity and rainfall that governed the fruitfly population in mango orchard.

References

Arai, T. 1976. Effect of temperature and light dark circles on the diel rhythm of emergence in the Oriental fruitfly, *Dacus dorsalis* Hendel (Diptera: Trypetidae) Jpn. J. Ento. Zoo., 20(2): 69-76.

Bagle, B.G. and Prasad, V.G. 1983. Effect of weather parameters on population dynamics of oriental fruitfly, *Dacus dorsalis* Hendel. J.ent. Res., 7 (2): 95-98.

Balasubramaniam, G., Abraham, E.V.S., Vijayaraghvan, S., Subramaniam, T.R., Santharaman, T. and Gunasekaran, C.R. 1972. Use of male annihilation technique in the control of Oriental fruitfly, *Dacus dorsalis* Hendel. Indian J. Agric.Sc, 42 (11): 975-977.

Clark, L.R., Geier, P.W., Hughes, R.D. and Morris, R.F. 1967. The ecology of insect populations in theory and practice, London, Methuen. pp.232.

Department of Horticulture. 1997. Annual Report. Directorate of Horticulture, Jammu & Kashmir Government.

Kapoor, V.C.1993. Indian Fruitflies (Insecta: Diptera: Tephritidae). Oxford & IBH Publishing Company Pvt. Ltd., New Delhi. pp.228.

Lakshamanan, P.L., Balasubramaniam, G. and Subramaniam, T.R. 1973. Effect of methyl eugenol in the control of oriental fruitfly *Dacus dorsalis* Hendel on mango. Madras Agric. J., 60(7): 628-629.

Mekongsee, B., Chawanapong, M., Sangkasuwan, U. and Poonyathaworu, P. 1981. The biology and control of sorghum shoot fly, *Atherigona soccata* Rondani, in Thailand. Insect Sci. Appl, 2 (1-2): 111-116.

Nishida, T. 1963. Zoogeographical and ecological studies of *Dacus cucurbitae* in India. Hawaii Agri. Expt. St. Tech. Bull, 54:28.

Panwar, V.P.S. 1995. *Agricultural insect pests of crops and their control.* pp. 265. Kalyani Publishers, New Delhi.

Ramasubbarao, V. 1994. New record of blossom thrips *Megaluothrips distalis* on mango (*Mangifera indica*). Indian J.Agric.Sci., 64(4): 417-418.

Rahman, S.J. 1990. Ecology and management of *Dacus dorsalis* Hendel and *Dacus cucurbitae* Coquillett. Ph.D. Thesis. G.B.Pant University of Agriculture and Technology, Pantnagar, India. pp.224.

Shukla, R.P. and Prasad, V.G. 1985. Population fluctuations of the oriental fruitfly, *Dacus dorsalis* Hendel in relation to hosts and abiotic factors. Trop. Pest Mgmnt. 25 (3): 389.

Singh, R.N. 1990. *Mango.* I.C.A.R., New Delhi. pp. 95-96.

Singh, T. 1993. Management studies on oriental fruitfly, *Dacus dorsalis* Hendel on Guava, *Psidium guajava* L. M.Sc. (Ag.) Thesis, SKUAST, Jammu & Kashmir.

Srivastava, R.P. 1997. Mango Insect Pest Management. International Book Distributing Co. Lucknow, India. pp.I-34

Steiner, L.F. 1957. Low cost fruitfly trap. J. Econ.Entomol., 50:508-509.

Tan, K.H. and Serit, M. 1994. Adult population dynamics of *Bactrocera dorsalis* (Diptera: Tephritidae) in relation to host phenology and weather in two villages of Penang Island, Malaysia: Environ. EntomoL, 23 (2): 267-275.

Veeresh, G.K. 1989. Pest problems in mango: World situation. Acta Horti.2 (231): 551-565.

Verghese, A. and Sudha Devi, K. 1998. Relation between trap catch of *Bactrocera dorsalis*, Hendel and abiotic factors. Proc. Natnl. Symp. Pest Mgmnt. in Horti. Crops, , Bangalore. In Advances in IPM for Horti. Crops (Eds. P.P. Reddy, N.K. Krishna Kumar and A. Verghese). Association for Advancement of Pest Management in Horticultural Ecosystems, Bangalore. pp. 15-18.

ENVIRONMENTAL BIOTECHNOLOGY

Edited By :   Professor (Dr.) Arvind Kumar

Published By : DAYA PUBLISHING HOUSE

16

OPTIMIZATION STUDIES ON THE EXTRACELLULAR LIPASE PRODUCING *PSEUDOMONAS* SP

• *A. Sahul Hameedu*, K.T.K. Anandapandian**,*
*Venkatesh Babu, J.*** & R. Rajesh Kumar**

* P.G. Unit of Microbiology, V.H.N.S.N. College, Virudhunagar, Tamil Nadu
** P.G. Unit of Microbiology, Thiagarajar College, Madurai, Tamil Nadu
*** P.G. Unit of Microbiology, Srimad Andavar College, Trichy, Tamil Nadu

Abstract

Lipolytic *Pseudomonas* species was isolated from oil effluent and it was identified and confirmed as per Bergy's manual. The organism was subjected to various carbon sources, pH, temperature, and time of incubation and looked for its optimum enzyme production. The best carbon source was given by olive oil and the optimum range of pH was 6, the temperature was 50°C and the incubation time for the lipase production was 72 hrs. Also the enzyme was partially purified by salt precipitation method and dialysis technique.

Key Words : Pseudomonas sp., lipase.

Introduction

Lipases find a number of potential applications in detergent industry, oleo chemical industry, and paper manufacturing industry, organic chemical processing, biosurfactant synthesis, nutrition, cosmetics, pharmaceuticals and agrochemical industry. Lipases (Triacyl glycerol acyl hydrolase's) are one of the versatile classes of enzymes, which catalyze the hydrolysis of triacylglycerols into glycerol and free fatty acids (Sidhu *et al.*, 1998). Many bacteria and fungi, which can grow on fatty acids or on triacylglycerols, will perforce produce a lipase and, indeed, induction of lipase synthesis is achieved by growing micro-organisms on oils or fats or even of hydrocarbons. The advantages of Lipase catalysed hydrolysis over the conventional high temparature, high pressure steam splitting are that a product is obtained which is of higher quality in terms of odour and colour and that of energy requirement is less (Suzuki *et al.*, 1988). Lipases can thus occur both intra-and extra-cellularly. Many bacterial lipases are extra-cellular enzymes, catalyzing the hydrolysis of fatty acid esters to free fatty acids. Many *Pseudomonas* species produce, extracellular lipases and some of these have been purified and characterized, including those from *Pseudomonas flourescens* (Sztajer *et al.*, 1991); *Pseudomonas tragi* (Nishio *et al.*, 1987, Pabai *et al.*, 1991) and *Pseudomonas aeruginosa* (Gilbert *et al.*, 1991). In this study an attempt has been made to optimize the condition, which is suitable for the lipase enzyme production by *Pseudomonas* species.

Materials and Methods

The oil refinery effluent samples were collected from an oil mill at Virudhunagar.

Enrichment and isolation

Enrichment was carried out by adding 10 ml of sample to olive oil (1.5% v/v) in minimal medium (Glucose 2.0 g; KH_2PO_4 1.0 g; K_2HPO_4 3.0 g; NH_4Cl 5.0 g; $MgSO_4.7H_2$ 0.1 g; . pH 7.0). All the flasks were incubated at 50°C for 72 hours. Transfers (2% v/v) were carried out two times at daily intervals (Bradoo *et al.*, 1999). Finally one ml from each flask approximately diluted, and purified and maintained in nutrient agar. These plates were incubated at 50°C for 24 hrs, the colonies were purified and maintained in nutrient agar slants at 4°C.

Screening for lipolytic activity

The isolates were streaked on tributyrin agar and the zone of hydrolysis was noted after 48 hrs at 50°C. The best zone production *Pseudomonas* species was selected for further process. Newly isolated *Pseudomonas* species was checked by tributyrin agar assay (Lawrence *et al.*, 1967). Lipolytic activity of *Pseudomonas* sp was assayed by titrimetric method by using 1.5% olive oil as substrate at 50°C and pH 7.0. One unit of lipase was defined as the amount of enzyme that liberates 1 µ mole of equal fatty acids/min/ml under assay condition (Naka *et al.*, 1992)

Partial purification of lipase

Lipase was precipitated from the culture supernatant by adding ammonium sulphate to 85% saturation. After this concentrate was left overnight at 4°C. The resultant precipitate was collected by centrifugation, dissolved in 5 ml of phosphate buffer (0.01µ, pH 7.0) and dialysed against the same buffer overnight at 4°C. The protein concentrations in various enzymatic fractioned were determined according to a modification of the Lowry Method (Bradoo *et al.*, 1999) using BSA as the standard for calibration. Characterization of the lipases with reference to various carbon sources (olive oil, tributyrin, tween 20 and tween 80), various time of incubation (24-120 hrs), pH tolerance (4.0-9.0), and temparature (15-70°C) tolerance was examined using the titrimetric standard assay.

Table 16.1 : Effect of Different Carbon Substrates on Enzyme Activity After 72 hrs at 37°C

S.No.	Carbon source	Enzyme activity (µ/ml)
1.	Olive oil	6.2
2.	Tributyrin	5.0
3.	Tween 20	4.2
4.	Tween 80	4.0

* 30 min incubation at 50°C.

Result and Discussion

Fifteen per cent of *Pseudomonas* strains were able to hydrolyze tributyrin. The strain that showed the highest activity was chosen for subsequent studies. Lipase from *Pseudomonas* species was found to be inducible in nature and its yield was significantly affected by the carbon source used. The amount of lipase production was dependent on the composition of carbon sources and the presence of lipids and the conditions for optimal lipase production was differed for each bacterial strain (Brune *et al.*, 1992). In the present study olive oil (triglycerides) is observed as a best inducer for producing extra-cellular lipase (Table 16.1).

Table 16.2 : Effect of pH on Enzyme Activity

S.No.	pH	Enzyme activity (µ/ml)	Biomass production (660nm)
1.	4	8.4	0.4
2.	5	14.0	1.3
3.	6	14.9	1.5
4.	7	13.0	1.1
5.	8	11.2	0.9
6.	9	10.0	0.6

* 30 min incubation at 50°C.

Table 16.3 : Effect of Temperature on Enzyme Activity

S.No.	Temperature	Enzyme activity (µ/ml)	Biomass production (660nm)
1.	15	2.2	0.6
2.	27	6.7	0.9
3.	37	9.8	1.5
4.	50	14.5	2.5
5.	60	11.0	1.8
6.	70	7.5	1.0

* 30 min incubation at 50°C.

Table 16.4 : Effect of Time Duration for Enzyme Activity

S.No.	Time	Enzyme activity (µ/ml)	Biomass production (660nm)
1.	24	10.4	1.9
2.	48	12.2	2.2
3.	72	15.6	2.8
4.	96	12.06	2.5
5.	120	11.5	3.2

* 30 min incubation at 50°C.

pH tolerance

Effect of pH and lipase enzyme activity of *Pseudomonas* species was identified and the result was shown in Table 16.2. Similarly the lipase activity of the partial purified enzyme fraction from *Pseudomonas putida*, *Aspergillus niger* and *Rhizopus oryzae* were measured from pH 5.0-10.0, the activity of the *Pseudomonas putida* increased with increasing pH being optimal at pH 7.0 (Pabai *et al.*, 1995).

Temperature tolerance

The temperature tolerance of *Bacillus sterothemophilus*, enzyme lipase which retained 40% of its activity at 100°C, even after 30 min at pH 6.0 (Bradoo *et al.*, 1999). The different temperature range of the *Pseudomonas* species for enzyne production being optimal at 50°C and it was shown in Table 16.3.

Time of incubation

Time of incubation play the main role in any aspect of enzyme production. Lipase activity decreased sharply however after 24 hrs which was probably due to the presence of proteases in the culture filterate (Baillargeon *et al.*, 1989). In our study various time duration (24, 48, 72, 96 and 120 hrs) were used. The maximum enzyme production was observed at 72 hrs and it was shown in Table 16.4.

References

Baillargeon, M.W., Bistline, R.G., and Sonnet, P.E. 1989. Evaluation of strains of *Geotrichum candidum* for lipase production and fatty acid specificity. Applied Microbiology and Biotechnology. 30: 92-96.

Bradoo, S., Saxena, R.K. and Gupta., R.K. 1999. World Journal of Microbiology and Biotechnology. 15: 97-102.

Brune, K.A. and Gotz, F. 1992. Degradation of lipids by bacterial lipases. In microbial Degradation of Natural Products, Ed Winkelmann, G. pp. 243-266.

Gilbert, E.J., Drodzd, J.W. and Jones, C.W. 1991. Physiological regulation and optimization of lipase activity *Pseudomonas aeruginosa* EF2. Journal of General Microbiology. 137: 2215-2221

Lawrence, R.C., Frer, T.F. and Reiter, B. 1967. Rapid method for the quantitative estimation of Microbial lipases. Nature. 213: 1264 - 1265.

Naka, Y. and Nakamura, T. 1992. The effects of serum albumin and related amino acids on pancreatic lipase and bile salts inhibited microbial lipases. Bioscience, Biotechnology and Biochemistry 56: 1066 - 1070.

Nishio, T., Chikano, T. and Kamimura, M. 1987. Purification and some properties of Lipase produced by *Pseudomonas* 22.39 B Agriculture and Biological chemistry 51: 181-186.

Pabai, F., Kermasha, S. and Morin, A. 1999. World Journal of Microbiology & Biotechnology 15: 97-102.

Siddhu, Prabhdeep Sharma, Rohit, Soni, S.K., and Gupta, J.K. March 1998. Indian Journal of Microbiology. 38: 9-14.

Sztajer, H. and Maliszewska. J. 1998. The effect of cultural conditions on lipolytic productivity of Micro-organisms. Biotechnology letters 10: 199-204.

Suzuki Takahiro, Mushiga Yashinao., Tsuneo Yamane and Shoichi Shimizu. 1987. Appl. Microbiology and Biotechnology. 27: 417- 422.

Vulfson, E.N. 1994. Industrial application of lipases. In lipases structure, Biochemistry and application. 271-28.

ENVIRONMENTAL BIOTECHNOLOGY

Edited By :   Professor (Dr.) Arvind Kumar

Published By :   DAYA PUBLISHING HOUSE

17

GREEN MANURE CROP FOR NUTRIENT MANAGEMENT

• *N.K. Bohra*

Arid Forest Research Institute, Jodhpur (Rajasthan)

Abstract

Green manures both non-grain legumes and grain legumes improve the productivity especially in rice field. Their value in substituting inorganic N and also improving soil fertility are now well established. It is the need of the present era to overcome some of the constraints in its use. There is a scope for land improvement and nutrient management by using green manure.

Key Words : Legumes, green manure, and fertilizer

Introduction

Fertilizers and manures are used in agriculture to supplement the nutrients that the plants can obtain from the soil alone. The result is usually an increase in yield. The role of green manure in improving soil fertility and supplying a part of the nutrient requirement, of crops is well known.

Green manure refers to fresh plant matter that is added to the soil largely for supplying the nutrients contained in its biomass. Such biomass can both be grown *in situ* and incorporated or grown elsewhere and brought in for incorporation in the field to be manured. Their use in crop production is recorded to have been practised in China as early as 1134 B.C. These are one of the main components of integrated nutrient supply system along with inorganic fertilizers and biofertilizers. Green manures can meet a part of the nutrient needs (Particularly N) of crops for optimum production and to that extent can result in fertilizers saving. In India, an estimated 7 million ha. were green manured during 1990-91 (nearly 4.5% of the net sown area). Andhra Pradesh and Uttar Pradesh accounted for 60% of green-manured area. The six states of A.P., Karnataka, Madhya Pradesh, Orissa, Punjab and U.P. comprised 88% of total area (FAI, 1990).

Plants for Green Manure

An ideal green manure should possess early establishment and high seedling vigour, tolerance to drought, shade, flood and adverse temperature, Possess early onset of N fixation and its efficient sustenance. It must have an ability to accumulate large biomass and N in 4-6 weeks, easy to incorporate, quickly decomposable, and tolerant to pests and diseases. (FAG 1977, IRRI 1988. Cosico 1990).

Green manures may be either non-grain legumes or leguminous plants. Non-grain legumes include Crotolaria, *Sesbania Centrosema Stylosanthes* and *Desmodium* while leguminous green manure plants include pigeon pea, green gram, cowpea, Soyabean and Perrenial woody legumes such as *Leucaena leucocephala* (Subabul), *Gliricidia sepium* and *Cassia siamea* etc.

Table 17.1 : Green manure crops

Field crop	Recommended Green Manure
Rice	Sun hemp (*Crotolaria juncea*) Sesbania & Wild indigo (*Indigofera tinctoria*)
Sugar cane	Sun hemp (*Crotolaria juncea*)
Finger millet	Sun hemp (*Crotolaria juncea*)
Wheat	Sun hemp (*Crotolaria juncea*)
Sorghum	Sun hemp (*Crotolaria juncea*) Subabul (*Laucaena leucocephala*)
Banana	Leaves of *Gliricidia sepium*
Potato	Lupin (*Lupinus albus*), Sun hemp (*Crotolaria juncea*) Cowpea (*Vigna sinensis*) Guar (*Cyamopsis tetragonoloba*) Buckwheat (*Fagopyrum* sp.)

Green manure crops suitable for some field crops were listed by Krishnamurthy (1978) and Sharma *et al.* (1991).

Leguminous green manures

These differ widely in nitrogen conc. and yield. Among 86 species of leguminous green manure crops used in India their nitrogen content was found from 2 to 4.9% (Vachhani and Murthy, 1964, Ghai *et al.*, 1985). Their utility in rice field was studied by several workers. Some common green manure crops for rice fields and their potential nitrogen contribution is given in Table 17.2.

Table 17.2 : Potential of various green manure crops in rice

Local name	Botanical Name	Growing season	Output in 45-60 days green matter (t/ha)	Nitrogen Contribution (Kg/ha)
Sun hemp	*Crotolaria juncea*	Wet	21.2	91
Dhaincha	*Sesbania aculeate*	Wet	20.2	86
Pillipesara	*Phaseolus trilobus*	Wet	18.3	201
Green Gram	*Vigna radiata*	Wet	8.0	42
Cowpea	*Vigna sinensis*	Wet	15.0	74
Guar	*Cyamopsis tetragonoloba*	Wet	20.0	68
Senji	*Melilotus alba*	Dry	28.6	163
Barseem	*Trifolium alexandium*	Dry	15.5	67
Khesari	*Lanthyrus sativus*	Dry	12.3	66

(Abrol & Palaniappan, 1988)

Non-grain legume

Evaluation of some nof the promising green manure crops at Coimbatore indicated the potential of Dhaincha and the newly introduced stem nodulating S rostrate (Table 17.3).

The exotic stem nodulating *S. rostrate* of Senegalese origin has much promise for low land rice especially with adequate irrigation. Another promising stem nodulating introduction from Medagascar is Aeschynomene afraspera, which is capable of withstanding water stress to some extent. Its potential under Indian conditions is under study *Tephrosia purpurea* and *Phaseolus trilobus* grow rather slowly and accumulate much less nitrogen than sesbania and are more adopt to drought (Palaniappan *et al.*, 1990). Among them Tephrosia purpurea has the additional advantage of self-sowing and is not browsed by cattle. Wild Indigo is another drought tolerant green manure crop suitable for rainfed rice, and it can contribute 62-182 Kg N/ha. to the succeeding rice crop depending on soil, climatic and cropping conditions. (Garrity and Flinn, 1988).

Table 17.3 : Evaluation of green manures for rice based cropping system at Coimbatore

Green Manure	Output in 60 day			Special feature
	Biomass	N accumulation		
		Kg/ha	Kg/ha/day	
Crotolaria juncea (Sun hemp)	16.8	159	2.65	Quick growing, easy seed production
Sesbania aculeata (Dhaincha)	26.3	185	3.08	High biomass accumulation, wide adaptability
Sesbania rostrata	24.9	219	3.65	Stem nodulating tolerant to water logging
Tephrosia purpurea (Kolinji)	16.8	115	1.92	Drought tolerant self-seedling
Phaseolus trilobus	17.6	126	2.10	Green manure-cum fodder-cum-cover crop

(Palanippan *et al.*, 1990)

Grain legume

Some annual grain legume crops are also used as green manure after all or part of the grain is harvested. Green gram Stover incorporated in the soil after pod harvest contributed about 60 Kg. N/ha to the succeeding rice crop (Rekhi and Meelu, 1983). The green gram yielding about 1 ton grain and 2.5 ton dry matter/ha equivalent to 50 Kg. N/ha (Meelu *et al.*, 1986). It is concluded that green gram, black gram (Phaseolus mungol) and cowpea could provide about 50-60 kg N/ha for the succeeding rice crop. (Kulkarni and Pandey, 1988). Green manuring is a very old established practice, known to Romans in the early centuries A.D. using leguminous crops such as vetches and lupins. The essential characteristics of green manure crops are rapid growth, vigorous root development and abundant tops. Cheap, reliable and rapidly germinating seed is essential and the crop must not present difficult husbandry problems. The main functions of green manures include addition to the soil of readily mineralizable plant nutrients particularly nitrogen, supply of organic matter to the soil, provision of soil cover in areas where erosion is a problem, prevention of leaching of nutrients by recycling them, control of weeds by smothering them. Green manures used properly should either increase humus content or increase the immediate supply of available nitrogen and other nutrients, but cannot effectively do both at the same time. If the main aim of growing a green manure crop is to increase the supply of nutrients, with emphasis on nitrogen, a legume should be grown and ploughed in while immature. These, partly because of the N_2 fixing symbiotic bacteria in the root nodules, give material with a very low carbon-nitrogen ratio. After ploughing in, rapid decomposition takes place with release of available nitrogen for the following commercial crop. Much of the organic matter is lost to the atmosphere as carbon dioxide, so that the humus content of the soil is not greatly increased. For this suitable leguminous crops are red clover (*trifolium pratense*), sweet lovers (*Melilotus* spp.), common vetch (*Vicia sativa*), trefoil (*Medicago lupulina*) and yellow lupin (*Lupinus luteus*).

If the main aim of growing the green manure crop is to maintain or increase soil organic matter, it is preferable to grow a non-leguminous crop and allow it to grow further towards maturity, though still green, before ploughing in. This material will have a higher carbon-nitrogen ratio and will contribute more persistent humus to the soil. Suitable species are mustard (Sinapis alba), rye (secale cereale), rape (Brassica napush) turnip rapes (Bras-sica rapa and crosses) and ryegrass (lolium spp.). Sowing leguminous green manure crops by, for example, direct drilling into stubble after early-harvested cereals followed by autumn ploughing - in, which can supply 40-60 kg/ha of available nitrogen for the following crop. Benefits of green manuring are generally interpreted in terms of their capacity to provide N or substitute fertilizer N. The role of green manures in enhancing the availability of other macro and micro-nutrients has been under study. Several workers reported increase in N and P contents in agricultural crops by using green manures. Enhanced availability of iron upon green manuring has been reported (Takkar and Nayyar, 1986).

Economics of green manuring

Green manuring increased rice yield by 78% in low fertile compared to 22% in high fertile soils (Gu and Wen, 1981). In a rice-rice-mung bean system, inclusion of black gram or cowpea in the pre-rice summer season increased the productivity level of the system over the conventional rice-rice-mung bean system (Siddeswaram, 1992). Besides these the residual effect of green manuring in cropping system improve nutrient status in the soil. However, it depends on soil types, environment condition, quality and quantity of green manure application, etc. (Meelu and Morris, 1984). It is now well established fact that incorporatation of grain legume haulms as organic source gave net returns comparable to that with green manuring. Combined application of green manure and recommended fertilizer N to rice produced high net returns (Srinivasaulu Reddy, 1988, Sivabal, 1989, Padmavathy, 1992, Siddeswaram, 1992).

Wider adoption of green manure technology at the farm level and its practical and economic feasibility are critical issues. Tillage and seeding operations are major components of the labour and cash outlay for farmers who grow green manure crops (Garity and Flinn, 1988). The financial analysis seeks to compare costs and benefits, between the combined use of green manure and inorganic N and inorganic N alone. Overall an added cost to produce a green manure crop is estimated to be about Rs. 700/ha to 1000/ha that is equivalent to nearly 100 kg. urea and without any harmful effects as a plus point.

Constraints

Besides many benefits green manure crops have several constraints hampering its wide spread adoption. They are -

1. In intensively cropped areas, farmers do not wish to set apart 6-8 weeks exclusively for growing a green manure crop with no direct cash benefit.
2. When rice is grown after wheat, farmers find it difficult to do the farm operations in the intense heat of May and June.
3. The most costly item in green manuring is the seed. Inadequate availability of quality seeds of desired species at reduced cost is one of the major problems in the adoption of green manuring practice.
4. Benefits of green manuring are not perceptible to the farms because these are not directly visible as for fertilser N.
5. The inter-cropped green manures may compete for resources with the main crop. The present day investigation should be focused to develop viable solution to overcome these constraints.

Further Research Needs

The cost of fertilizers and their harmful effects on environment, growing awareness of ecological impact of agro-chemicals, integrated nutrient management with organic, inorganic and biofertilizer sources have become imperative for sustainable crop production. To make green manure technology more viable and feasible following efforts are required in future.

1. Collection, conservation, cataloguing and evaluation of green manure species that show promise for different ecosystems and their screening for N accumulation patterns under controlled field situations.
2. Identification of the optimum combination of fast and slow N-release green manure species. A combination of two types of legumes may provide the best agro-ecosystem management.
3. Research to improve the establishment and stand of green manures particularly on the aspects of seed treatments for early and uniform germination.
4. Contribution of nutrients other than N in different soils and environments with long-term economic advantages including sustainability facts.
5. Multi-purpose trees and short season legumes with pests and disease tolerance that can be grown on bunds or wastelands. Legumes that provide both forage and green manure or fuel would be more appropriate. Research on legume species with efficient rhizobium strains for symbiotic N fixation even in adverse soil environments.
6. Design of cropping systems including green manure crops without displacement of any economic crop in the system.

Summary

Although green manuring is a 3000 years old practice it is still relevant in today's context and must be made use of wherever feasible and economical. For the proper rotting of the green manure, it is necessary that the green material should be succulent and there should be adequate moisture to the soil. Plants at the flowering stage contain the greatest bulk of succulent organic matter with a low carbon/nitrogen ratio. The incorporation of the green manure crop into the soil could enforce fertility of the soil by 30 to 50%.

References

Abrol, I.P., Palaniappan, S.P. 1988. Green manure crops in irrigated and rainfed lowland rice based cropping system in South Asia Proc. symp. on sustainable agriculture. The role of green manure crop in rice farming systems. IRRI, Phillipines, 71-82.

Cosico W.C. 1990. Studies on green manuring in the Philippines. Extension Bulletin No. 314, FFTC, Taiwan. pp.14.

FAI. 1990. Fertilizer. statistics.

FAO. 1997. Recycling of organic waste in agriculture, Report on FAO/UNDP tour to Republic of China. Soils Bulletin. FAO, Rome.

Garity, D.P., Flinn, J.C. 1988. Farm level management system for green manure crops in Asian rice environments. Proc. symp. on sustainable agriculture. The role of green manure crops in rice farming systems, IRRI, Phillipines, 111-129

Ghai, S.K. et.al. 1985. Comparative study of the potential of *Sesbanias* for green manuring. Tropical Agric. (Trinidad) 62:52-56.

Gu, R. and Wen, Q. 1981. Cultivation and application of green manure in paddy fields of china. pp. 207-219. Proc. Of symp. on paddy soil. Beijing, China, Institute of Soil science, Academia, Sinica and Science Press.

IRRI (International Rice Research Institute). 1988. Symposium recommendations, pp. 1-9, Green manure in rice farming; Proceeding of a symposium on sustainable agriculture, 25-29 May, 1987, Los Banos, Philippines. IRRI.

Krishnamurthy, K.K. 1989. Plantation crops - Research and development in the 21st century. JO. plantation crops 16 (supplement). 1-6.

Krishnamurthy R. 1980. A manual on compost and other organic manures. Today and Tomorrow Publishers, New Delhi, 85-119.

Kulkarni, K.R., Pandey, R.K. 1988. Annual legumes for food and as green manure in a rice based cropping system. Symposium on Sustainable agriculture. The role of green manure crops in rice farming systems, IRRI, Philippines, 289-299.

Meelu, O.P. 1986. Green manuring in low land rice. Paper at planning meeting workshop of international network on soil fertility and evaluation research. Hangzhon, China.

Meelu, O.P. and Morris, R.A. 1984. Integrated management of plant nutrients rice and rice based cropping system. Fertilizer news 29(2); 65-75.

Padmavathy, V.K. 1992. Effect of green manure and green gram on the economy of nitrogen management in rice based cropping system. M.Sc. (Ag.) Thesis. Tamil Nadu Agricultural University, Coimbatore, India. pp.133

Palaniappan, S.P. 1990. Green manure evaluation for rice farming system. Proc. Int. Symposium on notational resources management for sustainable agriculture.

Rekhi, R.S. and Meelu, O.P. 1983. Effect of complementary use of mung straw and inorganic fertilizer N on the nitrogen availability and yield of rice. Oryza 20: 125-129.

Sharma, R.C. et al. 1991. Nitrogen management in potato. Tech. Bull. 32, CPRI, Shimla pp. 74.

Siddeswaran, K. 1992. Integrated nitrogen management with green manure and grain legumes in rice based cropping systems. Ph.D. Thesis. Tamil Nadu Agricultural University, Coimbatore, India. pp. 204.

Sivabal, M.K. 1989. Integrated nitrogen management practices with green manures for low land rice. M.Sc. (Ag.) Ph.D. Thesis. Tamil Nadu Agricultural University Coimbatore, India.

Srinivas Reddy, P. 1988. Integrated nitrogen management in rice -based cropping system. Ph.D. Thesis. Tamil Nadu Agricultural University, Coimbatore, India. pp. 327.

Takkar, P.W., Nayyar, V.K. 1986. Integrated approach to combat micro-nutrient deficiency. Proc. FAI seminar on growth and modernization of fertilizer industry, New Delhi. PS III/22/1-16.

Vachhani, M.V. and Murthy, K.S. 1964. Green manuring for rice. ICAR Research report serial No. 17, New Delhi, India : Indian Council of Agricultural Research. pp. 50.

ENVIRONMENTAL BIOTECHNOLOGY

Edited By :   **Professor (Dr.) Arvind Kumar**

Published By :   **DAYA PUBLISHING HOUSE**

18

THE MYCOFLORA OF DUSTS FROM DIFFERENT NICHE IN RICE MILLS IN IMPHAL – A SOURCE OF FUNGAL BIOPOLLUTANTS AND BIODETERIOGENTS POPULATION IN INDOOR ENVIRONMENT

• *S.R. Singh, N. Sangbanbi Devi Seram*, N. Bimola Devi*

Post Graduate Studies Centre, HRDRI, Canchipur

*Department of Botany, Imphal College, Imphal

Abstract

The mycoflora of dust from different niche of rice mills in present day post-harvest technologies; depends on the different types of the dust. The quantitative fungal species are high in dust from floor, followed by bottom of containers dust from the mill wall, dust from the field bag, jute bags etc. Fungi potentiality pathogenic to rice were eventuated dust from different sources. These fungal types are known biodeteriogents, and allergenic to human beings and act as sources of hazardous bio-pollution in the indoor environment.

Key Words : *Niche, Dust, Mycoflora, Indoor environment, hazardous bio-pollution.*

Introduction

It is well known fact that more than two million people in India, working in the grain industry, come in contact with a variety of potentially hazardous substrates including airborne pollutants, dust, fungi, zoonatic microbes and other particulate matters (PM) (Karve *et al.*, 1989). New system of rice milling with high tech machines cause dust of different sizes including microsize particulate matters inside the milling room (Chakraverty and De, 1981, Mukherjee and Bose, 1981). The dust when inhaled causes respiratory disorders to workers (Lacey, 1980)

Moreover, the respiratory disease caused by rust is high atopic since the confirmation of allergenic action of pathogenic rust and smut fungi is well known (Cadham, 1924, Harris, 1939, Gimenez *et al.*, 1995). The information about mycoflora of dust is also very scanty (Konn *et al.*, 1963, Grub *et al.*, 1965). Only a few studies have been reported on dust from mouldy hay, mill and grain harvesting (Gregory and Lacey, 1963, Harris, 1939, Warren *et al.*, 1974).

Further, post-harvest technology being inter-disciplinary it spreads to different subjects of drying, processing, milling, methods of utilization of by products etc and all these section have specific influence to the mycoflora inside a rice mill by inducing the substrates and contaminations.

The present study attempts to determine the mycoflora of dust from different niche within the rice mill and its relevance to bio-pollution, bio-deterioration, and ecological adaptation of the biopollutants.

For detection of the source of fungal spores in the rice mill, dust samples were swept from the scrapped of bamboo baskets, crates, bottom of containers, mill walls, mill floor and cultured on sterilized petridish containing Czapek's Dox Agar (CDA) and Potato Dextrose Agar (PDA) media. Then the culture plates were incubated at room temperature for about 7 to 20 days in an upside down position. The colonies developed were examined, counted and identified. The sampling was done for 18 months at 1 month's interval following the method of Barkai Golan, 1966 (Lacey, 1980 and Singh and Singh, 2000).

The investigation in the dust of rice mill showed that the dusts are contaminated with a number of fungal spores (Table 18.1). Some of the dominant types are *Aspergillus flavus, Aspergillus niger, Aspergillus nidulans, Alternaria alternata, A. solani, Chaetomium funicola, Cladosporium herbarum, C. cladosporioides, Curvularia geniculata. Curvularia lunata, Fusarium avenaceum, Fusarium monoliforme, Fusarium oxysporum, Fusarium pallidorosearum, Fusarium gramenearum. Helminthosporium oryzae* (≡*Bipolaris oryzae*), *Humicola mucormucedo, Nigrospora oryzae, Penicillium digitatum, Phoma. oryzae, Rhizopus arrhizus, Rhizopus nigricans, Rhizoctonia solani, Trichoconis padwickii, Trichoderma viridae,* white sterile, yellow sterile etc.

Aletrnaria alternata, A. solani, A. spp., Aspergillus clavatus, Aspergillus flavus, Aspergillus nidulans, A. spp., Cladosporium herbarum. C. cladosporioides, Chaetomium funiculosa, Chaetomum spp., Curvularia lunata, Curvularia geniculata, Fusarium monoliforme, Fusarium graminearum, Fusarium oxysporum, Fusarium pallidorosearum, Fusarium avenaceum, F. spp., Geotrichum candidus, Helminthosporium oryzae, Nigrospora oryzae, Humicola oryzae, Penicillium digitatum, Phoma oryzae, Phoma spp., Rhizopus arrhizus, Rhizopus nigricans, Rhizopus spp., Rhizoctonia solani, Sclerotium rolfsii, Trichoderma spp., Trichoconis padwickii, Verticillium spp., black sterile etc. were freely isolated from the dust samples taken from the mill walls. The finding was in agreement with that of various workers (Singh, 1988, Singh, 1990, Singh and Singh, 2000). In this regards the expected fungal types may come from fungi growing on walls, paintwork, floor etc. Spores may be accumulated in house dust and grow if the humidity is high enough (Davies, 1960).

Table 18.1 : Fungal Types Found in Dust Samples From Different Locations for the Rice Mill During the Investigation Period (March 2001 to Sept. 2002)

Fungal types	Location			
	Field bags, Jute bags Field flat mat etc.	Bottom of sorting containers	Mill walls	Mill Floor
a	b	c	d	e
1. *Aletrnaria alternata*	++	++	++	++
2. *A. solani*	+	+	+	++
3. *A. spp.*	−	+	+	+
4. *Aspergillus flavus*	++	+	+	+
5. *Aspergillus clavatus*	−	++	++	+
6. *Aspergillus niger*	+	+	−	+
7. *Aspergillus nidulans*	+	+	+	++
8. *Aspergillus* spp.	+	−	+	−

Contd...

Table 18.1 – *Contd...*

a	b	c	d	e
9. *Absidia* sp.	+	−	−	+
10. *Actinomucor* sp.	−	−	−	+
11. *Cephalosporium* sp.	−	−	−	+
12. *Chaetomium funicola*	−	−	+	++
13. *Chaetomium* spp.	+	+	−	−
14. *Circinella* sp.	−	−	−	+
15. *Cladosporium herbarum*	+	+	++	++
16. *Cladosporium cladosporiodes*	−	+	++	++
17. *Cladosporium* spp.	−	+	+	−
18. *Cuninghmella elegans*	+	+	−	+
19. *Curvularia lunata*	+	++	++	++
20. *Curvularia geniculata*	++	+	++	++
21. *Fusarium monoliformae*	−	+	++	++
22. *Fusarium graminearum*	++	++	++	++
23. *Fusarium oxysporum*	−	+	+	+
24. *Fusarium pallidoroseum*	−	+	+	++
25. *Fusarium avenaceum*	−	+	+	+
26. *Fusarium* spp.	−	−	+	−
27. *Geotrichum candidus*	−	−	+	+
28. *Helminthosporium oryzae*	+	++	++	++
29. *Humicola* sp.	+	+	+	++
30. *Mucor mucedo*	+	+	−	++
31. *Mucor racemosus*	+	+	−	+
32. *Mucos pusillus*	+	+	−	−
33. *Nigrospora oryzae*	+	+	++	++
34. *Penicillium digitatum*	++	+	++	++
35. *Penicillium italicum*	+	−	−	+
36. *Penicillium* spp.	+	+	−	−
37. *Pullularia* spp.	+	−	−	−
38. *Phoma oryzae*	+	++	++	++
39. *Phoma* spp.	−	−	+	−
40. *Rhizopus arrhizus*	++	++	++	++
41. *Rhizopus nigricans*	+	−	++	++
42. *Rhizopus* spp.	−	−	+	−
43. *Rhizoctonia solani*	−	+	++	++
44. *Sclerotinia* spp.	+	+	−	−
45. *Sclerotium rolfsii*	−	+	·+	+
46. *Trichoderma viridis*	−	−	−	++
47. *Trichoderma* spp.	+	+	+	+
48. *Trichoconis padwickii*	+	+	++	++
49. *Verticillium* spp.	+	+	+	+
1. White sterile spp.	−	+	−	++
2. Black sterile spp.	+	−	++	−
3. Yellow sterile spp.	−	−	−	++
	31	36	35	41

Aletrnaria alternata, A. solani, Aspergillus flavus, Aspergillus niger, Aspergillus nidulans, A. spp., Absidia sp., Chaetomium spp., Cladosporium herbarum, Cuninghmella elegans, Curvularia lunata, Curvularia geniculata. Fusarium graminearum, Helminthosporium oryzae. Humicola sp., Mucor mucedo, Mucor racemosus, Mucos pusillus, Nigrospora oryzae, Penicillium digitatum, Penicillium italicum, Penicillium spp., Pullularia spp., Phoma oryzae, Rhizopus arrhizus, Rhizopus nigricans, Sclerotinia sp., Trichoderma spp., Trichoconis padwickii, Verticillium spp., Black sterile etc., were isolated from the sample of field bags, jute bags, field flat mat. The finding indicates the key role of containers in becoming a genuine source of fungal types inside the rice mill. Similar finding was reported by Singh and Singh (1989). The boxes with dust or mouldy fruits in the packinghouse enriched the atmosphere with spores. The conveyer belts, the sorting containers, the floor and walls of the packinghouse etc. are the main sources of spores within the air of packinghouse.

Aletrnaria alternata, A. solani, A. spp., Aspergillus flavus, Aspergillus clavatus, Aspergillus niger, Aspergillus nidulans, Chaetomium spp.. Cladosporium herbarum, Cladosporium cladosporiodes, Cladosporium spp., Cuninghmella elegans, Curvularia lunata, Curvularia geniculata. Fusarium monoliformae, Fusarium graminearum, Fusarium oxysporum. Fusarium pallidoroseum, Fusarium avenaceum, Helminthosporium oryzae, Humicola sp., Mucor mucedo. Mucor racemosus, Mucos pusillus, Nigrospora oryzae, Penicillium digitatum, Penicillium spp., Phoma oryzae, Rhizopus arrhizus, Rhizoctonia arrhizus, Rhizoctonia solani, Sclerotinia spp., Sclerotium rolfsii, Trichoderma spp., Trichoconis padwickii, Verticillium spp., white sterile etc. were predominantly isolated from the dust samples taken from the bottom of sorting containers. The present finding was in agreement with that of the opinion given by Mehrotra (1981) that the principal sources of inoculums of diseases comes after harvest as containers used to transport or store the harvested products, and other cleaning materials. Similar finding was reported from the rice grains store-cum-sale shops in Imphal by Singh and Singh (1989).

It is clear from the present study that dust from different sources of rice mills has always been found surcharged with certain kind of fungal spora. In deep analysis on the number of fungal types, the highest number is isolated in the dust from floor (41) which is followed by dust from the bottom of container (36), dust from the mill wall (35) and dust from the field bags, jute bags (31). The present finding of mode of fungal types contaminated dust was in agreement with that of other workers (Singh, 1990, Lacey, 1980). Various fungal types in the dust apart from causing diseases to the rice, including seed infection, deterioration and their discolouration, might be responsible for causing allergic human diseases to the susceptible workers, likers, millers, helpers, sellers and customers.

A number of fungal types that detected from the present investigations apart from causing diseases on field crops and stored grains but also responsible for causing allergic human diseases to susceptible workers or mill or grain handlers (Eyre, 1972, Pirie, *et al.* 1971, Dennis, 1973, Wiseman *et al.*, 1973, Agarwal and Shivpuri, 1974, Gamble and Lewis, 1996). One of the remarkable events is that Tse *et al.* (1973) reported 75% of the workers exposed to dust are grain elevators showed the similar symptoms. Further, dock workers handling grains suffering from respiratory allergic diseases (Dunner *et.al.*, 1964). In this connection Lacey (1980) reported that harvest dust was an important factor in allergic diseases and storage dust was responsible for both allergy and infection.

Out of the 49 fungal types isolated from the different niche inside the rice mill, fourteen fungi namely, *Alternaria alternata, Aspergillus sp., Curvularia geniculata, Curvularia lunata, Fusarium avenaceum, Fusarium monoliformae, Helminthosporium oryzae, Nigrospora oryzae, Penicillium spp, Phoma oryzae, Rhizoctonia arrhizus, Rhizoctonia solani, Trichoconis padwickii* were reported to be found consistently associated with discolouration of seeds of different varieties of paddy viz. China 1039, China 988, Huikap phou, Leimaphou, Norin 18 etc. (Singh, 2000).

The present investigation clearly shows that many of the fungal types detected from the dust of different niche in rice mill are well known allergens e.g. *Cladosporium, Aletrnaria, Puccinia* that cause type I immediate allergy. *Aspergillus clavatus, A. fumigatus* cause Type III delayed allergy, other such as *Aspergillus* spp., *Penicillium* spp. *Absidia* sp. and *Mucor pusillus* may cause infection (Dutkiewiez, 1973).

The present finding also clearly indicates that dust from different niche in rice mill are the resourceful source of various fungal types that are closely associated with the different parts of paddy, the essential raw material used in rice mill. A number of fungal types isolated from the dust of different niche were also isolated from husk, endosperm, and embryo of discoloured rice grains with high dominance of *Curvularia lunata, Fusarium monoliforme and Helminthosporium oryzae*. The occurrence of fungi in all grain parts of the test varieties, *C. lunata,* ranged to 30, 40, and 30% on husk, endosperm and embryo of China 1039. *A. alternata, C. geniculata, C. lunata, F. avenaceum, F. graminearum, F. moniliforme, H. oryzae* and *R. solani* were abundantly detected from the endosperm (Singh, 2000).

Acknowledgements

The authors are thankful to Director CABI, Mycological Institute for confirming the identity of fungi.

References

Agarwal, M.K. and Shivpuri, D.N. 1974. Fungus spores - their role in respiratory allergy. pp. 78-128. In advances in pollen spore Research 1.

Barkai-Golan, R. 1966. Reinfestation of circus fruits by pathogenic fungi in the packinghouse. Israel J. Agric. Res.

Cadham, F.T. 1924. Asthma due to grain rusts. JAMA 83:27-16 133-138.

Chakraverty, A. and De, D. S. 1981. Post-harvest technology of cereals and pulses. Oxford and IBH Publishing House Pvt. Ltd., New Delhi.

Davies, R.R. 1960. Viable mould in house dust. Trans. Br. Mycol. Soc. 43: 617-630.

Dennis, C.A.R. 1973. Health hazards of grain storage. pp. 367-387. In: Grain storage part of a system (Eds. By R.C. Singh & W.E. Muir). A VI Publishing Company, West Port, Connecticut.

Dunner, L., Herman, R. and Bengnell, D.J.T. 1964. Pneumoconiosis in dockers dealing with grain and seeds. Brit. Journal of Radiology, 19:506-511.

Dutkiewiez, J., Uminski, J., Dutkiewiez, E., Smerdel, C. and Krys Inska-Tragezyk, E. 1973. Allergenic action of grain dust II Immunological examination of workers from grain elevators and mills. Med Wie iska, 8:147-156.

Eyre, P. 1972. Equine Pulmonary emphsema: a bronchopulmonary mold allergy. Veterinary Record 91:134-140.

Gamble, J.F. and Lewis, R.J.F. 1996. Health and respirable particulate (PM10) of air pollution. Environ Health Perspective 104(8) 839.

Gimenez, C., Fouad, K., Choudat, D., Laurillard, J., Bouscaillaw, and Leib, E. 1995. Chronic and acute respiratory effects among grain mill workers. Int. Arch. Occup. Environ. Health, 67(5):311-316.

Gregory, P.H. and Lacey, M.E. (1963). Mycological examination of dust from mouldy hay associated with farmer's lung disease. Journal of General Microbiology 30-75.

Grub, W., Rollo, C.A. and Howes, J.R. 1965. Dust problems in poultry environments. Trans. Ann. Soc. Agric. Engrs. 8 : 338.

Harris, L.H. 1939. Allergy to grain dusts and smuts. J. Allergy 10 : 327-336.

Karve, R.B., Niphadkar, P.V., Papadia, R.M., Jain, R.K. and Ghamande, P.B. 1989. Nasobronchial allergy among wheat grain handlers. XXIII ICAAI Convention. Abstract.

Konn, J., Howes, J.R. Grub, W. and Rollo, C.A. 1963. Poultry dust. origin and composition. Agric. Engg. 44 : 608.

Lacey, J. 1980. The microflora of graindust. In, Occupational pulmonary disease: focus on graindust and health. (Eds. J.A. Doseman and D.A. Cotton) Academic Press, N. York, pp 417-440.

Mehrotra, R.S. 1981. (Ed.) Plant Pathology 771. Tata McGraw, New Delhi.

Pirie, H.M., Dawson, C.O., Breeze, R.G. Wiseman, A. and Hamilton, J. 1971. A bovine disease similar to farmer's lung, extrinsic allergic alveolitis, Veterinary Record. 88 : 346-351.

Singh, N.I. 2000. Present status of rice grain discolouration in Manipur. Silver Jublee Souvenir, ICAR Complex for NEH Region, Manipur Centre, Imphal.

Singh, N.I. and Singh, S.R. 1989. Indoor fungal airspora of rice grains store cum sale shops in Imphal. Canadian and Pan American Symp. On Aerobiology and Health Canada. June 7-9 No. 62.

Singh, S.R. 1988. Studies on the air spora of Imphal. Ph. D. Thesis, Manipur University, Canchipur. p. 555.

Singh, S.R. and Singh, N.I. 2000. The mycoflora of dusts from different niche in fruits and vegetables store-cum-sales shops in Imphal - A source of fungal bio-pollutants population in indoor air. Poll. Res. 19(4):517-521.

Singh, S.R. 1990. A survey for fungal diseases of fruits and vegetables found in the major markets of Manipur. Final Technical Report DSTE, Manipur.

Tse, K.S., Warren, P., Janusz. M., McCarthy, D.S. and Cherniack, R.M. 1973. Respiratory abnormalities in workers exposed to grains dust: Archieves of Environmental Health 27 : 74-77.

Warren, P., Cherniack and Tse, K.S. 1974. Hypersensitivity reactions to grain dust. Journal of Allergy and Clinical Immunology 53: 139.

Wiseman, A., Selman, I.E., Dawson, C.O., Breezer, R.G. and Pirie, H.M. 1973. Bovine farmers lung: A clinical syndrome in a herd of cattle. Veterinary Record 93 : 410-417.

ENVIRONMENTAL BIOTECHNOLOGY

Edited By :   **Professor (Dr.) Arvind Kumar**

Published By : **DAYA PUBLISHING HOUSE**

19

Pseudomonas fluorescens as Biocontrol Agent Against *Fusarium oxysporum* and *Sclerotium rolfsii*

• *P. Prema, P.S. Dheenan and P. Balaji*

Post Graduate Department of Microbiology, V.H.N.S.N. College,
Virudhunagar

Abstract

Four fluorescent *Pseudomonas* isolates were designated as PARB, GNRB, TORB, and ORFRB were isolated from rhizosphere soil samples of agricultural field showed invitro antibiosis against two fungal pathogens such as *Fusarium oxysporum* and *Sclerotium rolfsii*. All the four isolates showed positive (+++*) HCN production. But among the four isolates, ORFRB showed moderate level (++* *) of HCN production. All the four isolates produced siderophores in two media. In vitro antagonism as well as HCN and siderophore production revealed highest suppressive performance against fungal pathogens of diseased crops.

Key Words : *Flourescent Pseudomonas, Biocontrol agent, Fungal pathogens, Siderophores, HCN production.*

Introduction

Biological control has been described as a non-hazardous strategy to reduce crop damage caused by plant pathogen (Weller, 1988). Fluorescent *Pseudomonas* have received much attention in recent years as biological control agent for suppression of root pathogenic fungi (Schippers *et al.*, 1988). In recent years various plant, root colonizing Pseudomonas species have been shown to be potent microbiological control agent in various plant pathogens (Thomashaw and Weller, 1995).

Stem rot caused by *Sclerotium rolfsii* is a potential threat to groundnut production and is of considerable economic significance for groundnut grown under irrigated conditions. The disease causes severe damage to the crop at any stage of crop growth and yield losses of over 25% have been reported (Mayee and Datar, 1988). *Fusarium* wilt disease is responsible for important yield losses in number of crops. *Fusarium oxysporum* is an important pathogen of lentil with potential. Biological control of *Fusarium* wilt by antagonistic *Pseudomonads* has been often attributed to competition for certain nutrient such as iron (Van Peer *et al.*, 1989). The application of micro-organism as agent for biocontrol of plant diseases in agriculture is now considered as an important alternative to the use of chemical fungicide. Certain *Pseudomonas* species have been shown to effectively control fungal pathogen like

Fusarium oxysporum and *Sclerotium rolfsii* (Hebbar *et al.*, 1992). There is no previous report in the use of biocontrol agents as a strategy to reduce crop disease. In this work, we report the isolated *Pseudomonas fluorescens* that suppress the growth of fungal pathogens.

Materials and Methods

Isolation and Characterization of Fluorescent *Pseudomonas*

The fluorescent Pseudomonas designated as PARB, GNRB, TORB and ORFRB isolated from rhizosphere soil samples of different agricultural field at Thirumangalam. Totally 135 isolates were collected and pure cultures were obtained. The isolated organisms were capable of growing on King's B medium and characterized as fluorescent *Pseudomonas*. Identification of the organisms were done according to Bergey's Manual of Systematic Bacteriology (Krieg and Holt, 1984).

Isolation and identification of pathogenic fungi

The fungal pathogens such as *Fusarium oxysporum* and *Sclerotium rolfsii* were collected from the Indian Type Culture Collection, Indian Agriculture Research Institute (IARI), New Delhi. They were identified by Lacto phenol cotton blue staining and for their mycelial growth. The pathogens were maintained on PDA plates.

Invitro antagonistic studies

Dual culture test, as described previously (Dileep Kumar and Dube, 1991) was used to examine the antagonism of the bacterial isolates on KB medium. The fluorescent *Pseudomonas* isolates were screened for invitro antagonism against two fungal pathogens such as *Fusarium oxysporum* and *Sclerotium rolfsii*. Streaking the bacteria as 4cm line on KB medium tested invitro antogonism of the fluorescent *Pseudomonas* isolates. Mycelial plugs, 5mm in diameter of the fungal pathogens were transferred to the most distal point from the bacteria on the plate. Plates were replicated and repeated thrice. Plates were incubated at 28 ± 2°C and zones of inhibition were recorded after 8 days. Inhibition zone (mm) was calculated by taking the difference of the pathogens growth in control and treated plates.

Production of HCN

Production of HCN was determined as described by Wei *et al.* (1991). Bacteria were grown on TSA supplemented with 4.4 g/l of glycine tubes. Then filter paper strips soaked in picric acid solution (2.5 g of picric acid, 12.5 g of $Na_2 CO_3$ 1 litre of water) were placed in tubes. Then the tubes were sealed with paraffin film and incubated for 2 days at 28°C. HCN production was indicated by the presence of coloured zone around the bacteria. Reactions are scored as weak (Yellow to light brown), moderate (Brown), and strong (Reddish brown) for each of the bacterial strains. The experiment was repeated thrice.

Siderophore production

Two different media used were KB medium and Succinate medium. Fluorescent *Pseudomonads* were inoculated with 10^6 cells per ml, obtained from 18 hours old cultures at 28-30°C. 48 h old culture was centrifuged at 10,000 rpm for 10 min. and the cell free supernatant was examined for extracellullar siderophore production by standard assay as given below.

FeCl$_3$ test

This test was determined as described by Jalal and Dick van der Helm (1990). 0.5 ml of cell free culture supernatant was added to 0.5 ml of 2% aqueous FeCl$_3$ solution. Appearance of orange or reddish brown colour indicated the presence of the siderophore.

CAS assay

CAS assay (Chrome Azurol Sulphonate) was carried out according to Schwyn and Neilands (1987). 0.5 ml of CAS solution was added to 0.5 ml of cell free culture supernatant. The change in colour of the blue dye to orange indicated the presence of siderophores.

CAS agar plate test

CAS agar plate test was done according to Schwyn and Neilands (1987). Fluorescent *Pseudomonas* isolates were streaked on CAS agar plates for 48 h at 30°C. Formation of orange halos around the colonies indicated the presence of siderophores.

Spectrophotometric assay

Spectrophotometric assay for siderophore production was determined as described by Mayer and Abdallah (1978). Cell free culture supernatant were examined for their absorption maxima in Shimadzu UV-vis 160 A spectrophotometer. A peak at or near 405 nm indicated the presence of siderophores.

Chemical nature of Siderophores

Hydroxamate, Catecholate and Carboxylate nature of the siderophore was examined by the following tests.

Hydroxamate nature

Neiland's spectrophotometric assay

According to Neiland's (1981) 1 ml of cell free supernatant was added to 5 ml of freshly prepared 2% aqueous FeCl$_3$ solution, and absorbance between 400-600 nm was noted. A peak between 420-450 nm indicated the hydroxamate nature of the siderophores.

Catecholate nature

Neiland's spectrophotometric assay

According to Neiland's (1981), 5 ml of freshly prepared 2% aqueous FeCl$_3$ solution was added to 1 ml of the test sample. Formation of wine coloured complex was read at 495 nm indicated catecholate nature of siderophores.

Carboxylate nature

Spectrophotometric assay

According to Shenker *et al.* (1992), 1 ml of cell free supernatant was added to 1 ml of 250μM CUSO$_4$ and 2 ml of acetate buffer (pH 4.0). Copper complex was observed for absorption maximum between 190-280 nm. Entire wavelength was scanned to observe the peak of absorption for siderophores.

Results and Discussion

Isolation and identification of *Pseudomonas*

Mainly four isolates were identified and designated as PARB, GNRB, TORB and ORFRB. *P. fluorescens* was gram-negative motile rods. They showed fluorescence under UV illumination. They were able to hydrolysis casein but not starch. They were positive for citrate, catalase and oxidase production. But they were negative for Indole, MR, VP, Lactose and Sucrose fermentation (Table 19.1).

Isolation and identification of pathogenic fungi

The pathogenic fungi such as *Fusarium oxysporum* and *Sclerotium rolfsii* were isolated from the infected root of tomato and groundnut plant. They were identified by Lacto phenol cotton blue staining and based on mycelial growth.

Table 19.1 : Biochemical Tests for the Identified Organism (*P. flourescens*)

Biochemical tests	Results
Indole	Negative
MR	Negative
VP	Negative
Citrate	Positive
Casein hydrolysis	Positive
Starch hydrolysis	Negative
Catalase	Positive
Oxidase	Positive
Gelatin liquefaction	Positive
Carbohydrate fermentation	
Glucose	Negative
Lactose	Negative
Sucrose	Negative

In-vitro antagonistic studies

The four antagonistic bacteria showed good inhibition zone against phytopathogen (Table 19.2). Flouroscent *Pseudomonas* (ORFRB and TORB) were more effective against *F. oxysporum* (52 and 48 mm inhibition) and *S. rolfsii* (42 and 39 mm inhibition zone), followed by these GNRB and PARB were effective against both pathogens (36 to 45 mm zone of inhibition). Kloepper *et al.* (1980) demonstrated the siderophore-mediated competition as a mechanism of biological control. Dileep Kumar and Dube (1991) reported that fluorescent *Pseudomonas* owed their inhibition to the siderophore produced in iron deficient King's B medium is strongly suggested by the fact that the antagonism was reduced when iron was induced in the growth medium. Similarly here the siderophore produced by *Pseudomonas* possess antagonism against pathogenic fungi. It revealed that an effective antagonism showed in PDA plate compared to the PDA plate with $FeCl_3$ solution.

Table 19.2 : Inhibition Zone (mm) of Fungal Pathogens by *P. fluorescens*

S.No.	Pathogens	Inhibition zone (mm)			
	Flourescent *Pseudomonas*	PARB	GNRB	TORB	ORFRB
1.	*Fusarium oxysporum*	42	45	48	52
2.	*Sclerotium rolfsii*	40	36	39	42

Table 19.3 : HCN Production of Fluorescent *Pseudomonas* Isolates

S.No.	Fluorescent *Pseudomonas* isolates	HCN Production
1.	PARB	+++*
2.	GNRB	+++*
3.	TORB	+++*
4.	ORFRB	++*

HCN Production

The HCN production by fluorescent Pseudomonas isolates recorded are given in Table 19.3. Among the four isolates, PARB, GNRB and TORB exhibited good response (strong) than the ORFRB antagonist (Moderate). Production of HCN indicated that the isolates showed suppressive nature against pathogens. HCN production is reported to play a role in disease suppression (Wei *et al.*, 1991). Similar trend was observed in the present experiment. Here *P. fluorescence* produced HCN helped the suppression of pathogenic fungi.

Siderophore production

Siderophore production by *Pseudomonas* isolates on different media (succinate medium and KB medium) and their spectrophotometric assay were given in Table 19.4. Yeole *et al.* (2001) suggested that the hydroxamate nature of siderophore was evident by their absorption maximum between 402-412 nm. In this experiment, the chemical nature of siderophore such as catecholate and carboxamate were produced by fluorescent *Pseudomonas*. Absorption maximum of the chemical nature of siderophore produced in media was ranged from 408 to 425 nm (Table 19.5).

Table 19.4 : Siderophore Production by Four Fluorescent *Pseudomonas* Isolates in Two Different Media After 48 hrs. of Growth

Sl. No.	*Pseudomonas* isolates	Media	FeCl$_3$ test	CAS assay	CAS agar plate test	Spectro photometric assay λ max (nm)
1.	PARB	1	+	+	+	408
		2	+	+	+	407
2.	GNRB	1	+	+	+	404
		2	+	+	+	406
3.	TORB	1	+	+	+	404
		2	+	+	+	405
4.	ORFRB	1	+	−	−	−
		2	+	−	−	−

Media : 1. Succinate medium; 2. King's B medium.

Table 19.5 : Detection of Chemical Nature (Hydroxamate, Catecholate and Carboxylate) of Siderophores Produced by Four Fluorescent *Pseudomonas* Isolates

Sl. No.	Flourescent *Pseudomonas* isolates	Hydroxamate (Neiland's Spectrophotometric assay)	Catecholate (Neiland's Spectrophotometric assay)	Carboxylate (Neiland's Spectrophotometric assay)
1.	PARB	408	−	−
2.	GNRB	425	−	−
3.	TORB	412	−	−
4.	ORFRB	410	−	−

Based on the data recorded in the present result, several rhizosphere bacteria have the potential to control various root foliage and post-harvest disease of agricultural crops caused by plant pathogenic fungi, bacteria and viruses. These bacteria are not only the appealing candidates for bio control of plant disease, but many of these play an important role in promoting plant growth and yield. These bio control agents are ecofriendly in nature.

References

Dileep Kumar, B.S. and Dube, H.C. 1991. Plant growth promoting activity of fluorescent *Psedumonas* from tomato rhizoplane. Indian J. Exp. Biol. 29: 366-370.

Hebbar, K.P., Davey, A.G. and Dart, P.J. 1992. Rhizobacteria of maize antagonistic to *Fusarium moniliforme*, a soil-borne fungal pathogen: Isolation and identification. Soil Biol. Biochem. 24: 979-987.

Jalal, M.H.F. and Dick van der Helm. 1990. Isolation and spectroscopic identification of fungal siderophores. In: Hand Book of Microbial iron chelates (Ed. Winkelmann, G.), CRC Press, Boca raton. pp 235.

Kloeppper, J. W., Leong, J., Teintze, M. and Schroth, M.N. 1980. Enhanced plant growth by siderophores produced by plant growth promoting rhizobacteria Curr Microbiol. 4: 317.

Krieg, N.R. and Holt, J.G. 1984. Bergey's Manual of Systematic Bacteriology. 1: 1-964.

Mayer, J.M. and Abdallah, M.A. 1978. The fluorescent pigment of Pseudomonas flourescens: Biosynthesis, purification and Physico-chemical properties. J. Gen. Microbiol. 107: 319.

Mayee, C.D. and Datar, V.V. 1988. Disease of groundnut in the tropics. Rev. Trop. Plant Pathol. 5: 85-118.

Neilands, J.B. 1981. Microbial iron transport compounds (siderophores) as chelating agent. In: Development of iron chelators for clinical use (Eds Martell, A.E., Anderson, W.J. and Badman, D.G.), Elsevier Press, North Holland, Amsterdam. p.13.

Schwyn, B. and Neilands, J.B. 1987. Universal chemical assay for the detection and determination of siderophores. Anal. Biochem. 160: 47.

Schippers, B., Bakker, A.W. and Bakker, P.A.H.M. 1988. Interactions of deleterious and beneficial, micro-organisms and the effect of cropping practices. Annual Rev. Phytopathol. 25: 339-358.

Shenker, M., Oliver, I., Helmann, M., Hadar, Y. and Chen, Y. 1992. Utilization by tomatoes of iron mediated by a siderophore produced by *Rhizopus arrhizus*. J. Plant Nutr. 15: 2173.

Thomashaw, L.S. and Weller, D.M. 1995. Current concepts in the use of introduced bacteria for biological disease control mechanisms and antifungal metabolites. *In:* Plant microbes interaction (Eds. Stacey, G. and Keen, N.) Chapman and Hall, New York. 1:187-235.

Van Peer, R., Rahink, H. and B. Schippers. 1989. Biological control of cornation wilt caused by *Fusarium* oxysporum f. sp. dianth in hydroponic system. In: Soilless culture, Proc. Integr. Soilless culture (Eds. Stainer, A.A. and Vittien, J.J.), Secretariat ISOSC, Wageningen, Netherlands. 361-373.

Wei, G., Kloepper, J.W. and Sadik, T. 1991. Induction of systemic resistance of Cucumber to *Collectotrichum orbiculate* by selected strain of plant growth promoting rhizobacteria. Phytopathol. 81: 1508-1512.

Weller, D.M. 1988. Biological control of soil borne plant pathogen in the rhizosphere with bacteria. Ann. Rev. Phytopathol. 26: 379-407.

Yeole, R.D., Dave, B.P. and Dube, H.C. 2001. Siderophore production by fluorescent *Pseudomonas* colonizing roots of certain crop plants. Indian J. Expt. Biol. 39: 464-468.

ENVIRONMENTAL BIOTECHNOLOGY

Edited By : Professor (Dr.) Arvind Kumar

Published By : DAYA PUBLISHING HOUSE

20

RETRIEVAL OF TREATED TANNERY EFFLUENT CAUSED DECREASE ON GROWTH AND METABOLISM OF *HARDWICKIA BINATA* ROXB. BY SPIC CYTOZYME

• *Mariappan, V.*

Department of Biology, Gandhigram Rural Institute,
Deemed University, Gandhigram-(Tamil Nadu)

Abstract

The paper deals with the retrieval of tannery effluent caused decrease on growth and metabolism of *Hardwickia binata* Roxb. Tannery effluent shows brought about a considerable reduction in growth characters of plants treated with it. Almost all the growth and biochemical parameters analyzed except proline in treated plants were reduced. Increase in proline level as its action as osmo-regulatory agent clearly indicates the suffering of plants subjected to effluent treatment. Retrieval effect of cytozyme is understood with gradual increase in growth and biochemical characters. The decline in proline level after the cytozyme indicates the relief of plants from stress.

Key Words : Treated tannery effluent, Hardwickia Roxb, Growth, and metabolism, cytozyme.

Introduction

Environment is defined as an aggregate of all social, biological, physical and chemical factors of the surroundings. Man has been continuously affecting the environment by his activities. Macro and micro-development activities have led to rapid industrialization for economic progress. Industrialization has brought with it the hazards of environmental pollution. Industrial discharges have large quantities of liquid, solid and gaseous wastes which pollute the air, water and soil. This damages the over all ecosystem. Most of the industrial effluents are highly toxic to the growth of plant system. Only few studies have proved that the effluents can be utilized for agriculture and other processes. (Khambatta and Kethkar, 1960, Bishop, 1983, Swaminathan *et al.*, 1989). Dindigul is an industrial town with 61 tanneries located in around it. Tanneries are attractive because of its foreign exchange earning potential and its pollution to agro-ecosystem. The quantity of effluent released from the tanneries is about 30-35 litres per kilogram of leather produced. This wastewater produces serious consequences of pollution in freshwater streams and lands used for agriculture (Geetha and Vembu, 1998). Usually the wastewater affects ground water and it is not suitable for drinking and agricultural purposes (Wolstenholme, 1974 and Apporao and Karthikeyan, 1990).

Cytozyme the phytohormone used in the present study belongs to the group cytokinins. These are different types such as zeatin and kinetin. Kinetin primarily promotes the cell division, cell enlargement, tissue differentiation, dormancy breaking and promotes fruits and flower development. Many plant growth regulators have been used to study their effect on growth and yield in different crop plants (Foda *et al.*, 1973, Yadava and Sreenath, 1975). Among the growth regulators SPIC cytozyme a biologically derived growth promoter supports the nutritional needs of plants during foliar spray of it since it contains auxins and cytokinins (Ryan *et al.*, 1982).

Hardwickia binata Roxb a tree species is of greater economic value for its timber, has finy leaves and the tree canopy in not dense which is quite favourable for crop production and its interspaces. Alsō *H. binata* is nitrogen firing tree species has numerous advantages (Singh *et al.*, 1990, Singh, 1982). Hence it is programmed to study the retrieval of treated tannery effluent caused decrease on growth and metabolism of *Hardwickia binata* by SPIC cytozyme.

Material and Methods

I. Collection of Effluent Sample

For the present study the treated tannery effluent were collected from TALCO Tannery Environ. Contol Pvt. Ltd. Dindigul, Dindigul district, Tamil Nadu. The physico-chemical parameters of the effluents were made as per the standard procedures (APHA, 1990). The water quality Index were calculated by Tiwari and Manzoor Ali (1987). Effluent was collected on weekly basis and the physico-chemical parameters were studied. The effluent was then used for irrigation. After standardization, 10% concentration was used for further studies. The effluent was analyzed for its chemical constitution to see their effect on growth and metabolism.

II. Cultivation of Plant

Healthy, uniform and dry seeds of *Hardwickia binata* Roxb. were collected and pre-treated as per the procedures of Oddukkam seed centre, Palani Hills Conservation Council, Nallampatty, Dindigul district, Dindigul. The seeds were surface sterilized with 0.1% mercuric chloride for one minute and washed with running tap water followed by distilled water. They were soaked in distilled water for about 3 hrs and sown in earthern pots containing garden soil.

The soaked seeds were sown in earthen pots of 2 kg capacity filled with garden soil and effluent of 10% (already standardized) concentration was added. The soaked seeds were sown in 3 pots and named A, B, C and they were further used for D, E and F also description of which is given hereunder.

A. Control - Treated with water for 305 days.

B. Treatment with effluent for 30 days.

C. Treatment with SPIC Cytozyme for 30 days.

D. Control - Treatment with water for 60 days.

E. Treatment with effluent for 30 days followed by SPIC Cytozyme for another 60 days.

F. Treatment with SPIC Cytozyme for 60 days.

Growth parameters (Shoot length, root length, fresh weight and dry weight) were estimated. The biochemical parameters such as Total chlorophyll, (Arnon, 1949), Total soluble protein (Lowry's *et al.*, 1952), Total sugar (Jeyaraman, 1981), Proline (Bates *et al.*, 1973), Catalase (Addy and Goodman, 1978) were quantitatively estimated.

Results and Discussion

The physico-chemical analysis of treated tannery effluent (Table 20.1) shows that it was much toxic in nature. This is due to the presence of high hardness with the ingredients of sodium and chloride. The pH of the treated tannery effluent slightly alkaline in nature. The electrical conductivity of the treated tannery effluent was (10,000 micromhos) found to have more of ionic substances. The soluble sodium percentage of treated tannery effluent indicates the high degree of pollution due to the presence of organic chemicals arising from the discharge of a number of organic chemicals in tannery water (Rajan and Rukmani, 2000 and Mariappan *et al.*, 2001) certain wastewaters (Sreenivasan and Soundarraj, 1964). WQI (Water Quality Index) was calculated and a measure of over all extent of pollution was found to be 2120.74 for the treated

Table 20.1 : Physical-chemical Analysis of Treated Tannery Effluent

Sl. No.	Parameters	Value
1.	Temperature (°C)	34.2
2.	pH	7.91
3.	Odour	smell of H_2S
4.	Colour	Dark brown
5.	Electrical conductivity	10,000
6.	Total solids	19.615
7.	Total dissolved solids	18.660
8.	Total suspended solids	0.955
9.	Total hardness	1800
10.	Sodium	175
11.	Potassium	38
12.	Calcium	15
13.	Magnesium	32
14.	Sulphate	375
15.	Chloride	8364
16.	Carbonate	–
17.	Bicarbonate	3.8
18.	Phosphate	3.2
19.	Salinity	22
20.	Soluble Obsorption Capacity	130.87
21.	Soluble Sodium Percentage	8.87
22.	Oil and grease	0.016
23.	Nitrogen	2.8
24.	Dissolved Oxygen	5.3
25.	Dissolved Carbon dioxide	39.7
26.	Chemical Oxygen Demand	256
27.	Biological Oxygen Demand	82.5
28.	Water Quality Index	2120.74

All the values are expressed in mg/l. except EC (micromhos).

tannery effluent. This indicate treated tannery effluent are highly polluted to a higher extent and therefore it is not suitable for any human use without further treatment (Trivedy *et al.*, 1990 and Apparao and Karthikeyan, 1990).

Effect of treated tannery effluent on growth in 30 days old seedlings of *H. binata* were shown in Table 20.2. The treated tannery effluent has brought about considerable reduction in growth of the plants in terms of shoot length, root length, fresh weight and dry weight which are good indicators of stress. (Rajan and Rukmani, 2000, Mariappan *et al.*, 2001). This reduction in total biomass is attributed to a condition which may seriously affect and control permeability to water and synthesis of toxic metabolites retarding the normal growth (Strogonov, 1965 and Soma Sekhar *et al.*, 1984). Reduced growth due to increased Na^+ and Cl^- concentration Na^+/K^+ ratio and decreased concentration of K^+ (Srivastava, 1991).

Table 20.2 : Effect of Treated Tannery Effluent on Growth in 30 Days Old Seedlings of *Hardwickia Binata* Roxb

S. No.	Parameters	Control A	Effluent B	SPIC Cytozyme C	% of Suppression B over A	% of enhancement after SC C over A
1.	Fresh weight (mg)	2.402 ± 0.01 (100.0)	0.672 ± 0.004 (28.0)	3.102 ± 0.02 (129.14)	72	29.14
2.	Dry weight (mg)	0.972 ± 0.01 (100.0)	0.212 ± 0.008 (51.0)	0.492 ± 0.005 (120.0)	49	20.0
3.	Shoot length (cm)	18.6 ± 1.44 (100.0)	12.3 ± 0.68 (66.0)	22.6 ± 0.21 (121.0)	34	21.0
4.	Root length (cm)	10.72 ± 0.33 (100.0)	8.6 ± 0.36 (80.2)	16.12 ± 0.16 (150.0)	20	50

Figures in braces indicate percentage;

± indicates standard error.

Effect of treated tannery effluent on metabolism in 30 days old seedlings on *H.binata* were shown in Table 20.3. All the biochemical parameters analyzed in our studies remain affected due to effluent treatment. The chlorophyll, protein, sugar remain affected. This may be due to reduction in dissolved oxygen level in water (Soma Sekhar *et al.*, 1984). Inhibition in protein synthesis or due to accumulation of heavy metals, the major constituents of industrial effluents (Darjeetkaur, 1989). Sometimes impaired carbon flow due to distributed carbohydrate metabolism through the glycolysis pathway decrease the rate of sucrose formation (Downtol, 1977). Proline is considered to be involved in the adaptive mechanisms under stress condition (Palag and Aspinall, 1981). In our studies the accumulation of proline under effluent indicate the stress that the plants have been experiencing.

Table 20.3 : Effect of Treated Tannery Effluent on Metabolism in 30 Days Old Seedlings of *Hardwickia Binata* Roxb

S. No.	Parameters	Control A	Effluent B	SPIC Cytozyme C	% of Suppression B over A	% of enhancement after SC C over A
1.	Total Chlorophyll (mg.g/fw)	1.672 ± 0.02 (100.0)	0.976 ± 0.023 (58.0)	2.721 ± 0.02 (163)	42	63
2.	Total Sugar (mg.g/fw)	30.6 ± 0.65 (100.0)	25.7 ± 0.74 (84.0)	36.71 ± 0.095 (119.0)	16	19
3.	Total Soluble Protein (mg.g/fw)	35.7 ± 0.19 (100.0)	31.2 ± 0.12 (87.0)	42.6 ± 0.31 (119.0)	13	19
4.	Proline μ mole.g/fw	2.7 ± 0.002 (100.0)	3.01 ± 0.003 (111)	2.21 ± 0.002 (81.0)	11	19
5.	Catalase (mg.g/fw)	2.32 ± 0.004 (100.0)	3.01 ± 0.005 (130)	2.27 ± 0.004 (98.0)	30	2

Figures in braces indicate percentage;

± indicates standard error.

Retrieval effect of cytozyme on effluent suppressed growth and metabolism in 60 days on *H. binata* were shown in Tables 20.4 and 20.5. The growth and metabolism parameters (fresh weight, dry weight shoot length, root length and total chlorophyll, total sugar, soluble protein, proline, catalase) were high in SPIC cytozyme treated when compared to effluent treatment treated ones. Regarding the restoration of the chlorophyll level by cytozyme after stress Skene (1968) was of the opinion that it is possible because of high RUBP carboxylase activity and increase in the content of cytokinin a constituent of cytozyme. (Goswani and Srivastava, 1985). The development of chlorophyll repairing mechanism, prevention chlorophyll degradation, stability of pigment protein complex, regulation of plastid differentiation and chlorophyll synthesis by the combining activity and cytokinin are also the other factors responsible in bringing about normalcy in the affected plants (Telic and Bogdanovic, 1990). Dingra *et al.* (1995) revealed that the protein decreased during stress is restored after the application of cytokinin by the way of (a) incorporating amion acids into protein, (b) influencing translocational and transcriptional mechanism of protein synthesis, (c) inhibiting hydrolysis of protein, (d) synthesizing new protein like stress protein and shock protein. Here the proline content was found to be decreased with the application of SPIC cytozyme, a growth hormones.

Table 20.4 : Retrival Effect of Cytozyme on Treated Tannery Effluent on Growth in 60 Days Old Seedlings of *Hardwickia Binata* Roxb

S. No.	Parameters	Control D	Effluent + Cytozyme E	SPIC Cytozyme F	% of Suppression E over D	% of enhancement after SC F over E
1.	Fresh weight (mg)	4.350 ± 0.02 (100.0)	1.732 ± 0.071 (40.0)	6.716 ± 0.09 (154.0)	60	54
2.	Dry weight (mg)	1.792 ± 0.07 (100.0)	0.962 ± 0.021 (54.0)	2.002 ± 0.067 (112.0)	46	12
3.	Shoot length (cm)	26.7 ± 1.72 (100.0)	20.62 ± 1.21 (77.0)	32.6 ± 0.721 (122.0)	23	22
4.	Root length (cm)	12.90 ± 0.62 (100.0)	10.2 ± 0.91 (85.0)	24.7 ± 0.12 (206.0)	15	106

Figures in braces indicate percentage;

± indicates standard error.

Table 20.5 : Retrival Effect of Cytozyme on Treated Tannery Effluent on Metabolism in 60 Days Old Seedlings of *Hardwickia Binata* Roxb

S. No.	Parameters	Control D	Effluent + Cytozyme E	SPIC Cytozyme F	% of Suppression E over D	% of enhancement after SC F over E
1.	Total Chlorophyll (mg.g/fw)	2.267 ± 0.01 (100.0)	1.171 ± 0.071 (52.8)	4.76 ± 0.02 (210)	48	110
2.	Total Sugar (mg.g/fw)	39.72 ± 0.71 (100.0)	32.61 ± 0.69 (84.0)	45.72 ± 0.69 (115.0)	18	15
3.	Total Soluble Protein (mg.g/fw)	42.14 ± 0.91 (100.0)	36.71 ± 0.61 (87.0)	48.71 ± 0.312 (116.0)	13	16
4.	Proline μ mole.g/fw	2.92 ± 0.07 (100.0)	5.71 ± 0.09 (195)	2.67 ± 0.01 (91.0)	95	0.9
5.	Catalase (mg.g/fw)	2.97 ± 0.002 (100.0)	4.127 ± 0.04 (138)	2.56 ± 0.006 (0.86)	38	14

Figures in braces indicate percentage;

± indicates standard error.

There is a good number of literature available to support this concept. Kinetin showed significant decrease in proline content in mungbean subjected to stress (Singh *et al.*, 1994) It is justifiable and quite true that growth hormone like cytozyme can bring down the level accumulation of proline (Saha and Gupta, 1993) which may be due to (a) arrest to breakdown of protein (b) shift in metabolism and maintenance of osmotic balancer between cytoplasm and vacuole by some constituents of cytozyme (Flowers *et al.*, 1977). Catalase activity was very high in the treatment with SPIC cytozyme when compared to effluent treatment. Treated tannery effluent was toxic even at lower concentration. However, cytozyme applicated following effluent treatment promoted the plant growth. Incidental occurrence of inhibitory effect of treated tannery effluent on economically important crops could be overcome by applying SPIC cytozyme which is economically viable and very much useful to farmer.

References

APHA, AWWA, WPCF, 1990. Standard methods of examination of water and wastewater. 19th edition, American Public Health Association, Washington, DC.

Addy., K. and Goodman, R.K. 1978. Polyphenol oxidase and peroxidase activity in apple leaves inoculated with a virulant or avirulant of Erwinia amylovora. Indian Phytopathol. 25. : 575-579.

Appao Rao, B.V and Karthikeyan, G. 1990. Pollution tanneries of Dindigul a melady - Remedy analysis, Department of chemistry, Gandhigram Rural Institute, Gandhigram. Tamil Nadu. 45 - 52.

Arnon, D. I. 1949. Copper enzymes in isolated chloroplasts Polyphenol oxidase in *Beta vulgaris*. Plant. Physiol. 24 ; 1-15.

Bates L.S., Waldran R.P. and Teare I., 1973. Rapid determination of free proline in water stress studies. Plant and soil. 39 ; 205 - 208.

Bishop, A. W. 1983. Disposal of some industrial wastewater, I. P.W.P.C. Tech. Annual, X. p.149-156.

Daljeet Kaur, 1989. Toxic effect of heavy metals on flowering and pod development in *Pisum sativum* and their alleviation with some growth regulators. Thesis abstract vol. III.

Dingra, H.R., Chhaba, S., Kalal, N. and Varghese, T.M. 1995. Salinity and growth regulators induced changes in seed quality of chick pea. Indian J. Plant Physiol. XXXVIII. 4, 322-324.

Downtol, W.J.S. 1977. Photosynthesis in salt stressed grapesvines . Aust. J. Plant. Physiol. 4: 183-192.

Foda, H.A., E.T., Chobashy, A.S and Abu, Tabeikh, A.T., 1973. Egyptian. J. Botany, 16 : 191-203.

Flowers, T.J., Troke, P.F and Yeo, A.R., 1977. Salt tolerance in halophytes, Ann, Rev. Plant. Physiol. 28 : 89-96.

Geeta, Sand Vembu, B. 1998. Efect of tannery effluent on morphological and Biochemical features of *Eleucine corocana*, Gaetern. J. Environ. Monit. 8(3); 183. 186.

Goswani, B.K. and Srivastava, G.C. 1985. Effect of ABA and Zeatin on water stress induced photosynthesis and respiration in sunflower leaves. Indian. J. Plant. Physiol. 28(4): 350-357.

Jeyaraman, J. 1981. Laboratory manual of Biochemistry. Wiley Eastern Limited, Chennai. 234.

Khambatta, S. J and Kethker, C.M. 1960. Sewage sludge and effluent use in Agriculture. Land Appl. Waste materials, 4 : 138-153.

Lowry, O.H. Rosenburgh, N.J. Farr, A.L. and Randall, R.J. 1952. Protein measurement with, Folin phenol reagent. J. Biol. Chem 193. 265-275.

Mariappan, V., Balamurugan, T and Rajan, M.R. 2001. Irrigational utilization of treated tannery effluent on growth and biochemical characteristics of certain crop plants. Ecol. Envrio & Cons. 7(2); 103 -108,

Paleg L.G and Aspinall D, 1981. Proline accumulation physiological aspects pp. 206-140. In L. G. Palag and Aspinall (ed). The physiological and Biochemistry of drought resistance in plants. Academic Press, Syndey, Australia.

Rajan M.R. and Rukmani, S. 2000. Recycling of treated tannery effluent for growing selected tree species. Ecol. Env. Con. 6(3): 319-322.

Ryan, J., Sagair, A.R., Shafyuddin, M. and Barsuminan, A. 1982. Agronomic evaluation of cytozyme as a growth regulator. Argn. J. 74: 144-146.

Saha, K. and Gupta, K. 1993. Effect of Lab 150978. A plant growth retardant on sunflower and mung bean seedlings under salinity stress, Indian J. Plant. Physiol. 36 (3); 151-154.

Singh, R.V. 1982. Fodder trees of India, Oxford & IBH Public. Co., New Delhi, Bombay, Calcutta 663.

Singh, K., Yadav, J.S. and Sharma, S.K. 1990. Performance in shisham in salt affected soil. Indian. J. Forester. 8: 154 - 162.

Singh, S.P., Singh, B.B. and Singh, M. 1994. Effect of kinetin on chlorophyll nitrogen and proline content in mung bean under saline condition. Indian. J. Plant. Physiol. 37 (1) : 37-39.

Srivastava, R.K. 1991. Effect of papermill effluent on seed germination and early growth performance of Radish and onion. J. Ecotoxico. Environ. Monit. 1 (1).13 -18.

Seenivasan, A. and Sounderrajan, R. 1964. Effects of certain wastes on the water quality and fisheries of rivers Cauvery and Bhavani. J. Environ. Hlth. 9: 13-21.

Skene, K.G.M. 1968. Increase in cytokinin in bleeding sps. of *Vitis vinifera* after CCC (Cycocel Chloromequat Chloride) treatment science 159: 1477- 1478.

Somasekhar, P.K., Gowda, M.T. G., Shettinger, S.L.N and Srikanth, K.P. 1984. Effect of Industrial effluents on crop plants. Ind. J. Environ. HLTH. 25 : 136-146.

Strogonov, B.P. 1965. Physiological basis of salt tolerance of plants. English ed. Jerusalem, TPST.

Swaminathan, K., Arjunan, J. and Gurusamy, R. 1989. Effect of glucose factory effluents on seed germination and seedling development of groundnut (*Arachis hypogea*) Envriromental Impact. Biosystems. R.C. Dalelea, Muzzafarnagar, Environ. Bio. (18).187.

Telic, G., Bogdaniovc, M. 1990. The realtionship between chlorophyll accumulation and endogenous cytokinin the guening cotyledouous of *Pinus nigra*. Plant Science 71, 153-157.

Trivedy, R.K., Khatavkar, S.D., Kulkarni, A.Y and Shorti, A.C. 1990. Ecology and pollution of the river Krishna in Maharastra II. Physico-chemical characteristics, River pollution in India, Asian Publ. House. New Delhi.

Tiwari, T.N and Manzoor Ali, A. 1987. River pollution in Kathmandu valley - variation of WQI. Indian J. Environ, Protection 7 (5): 347- 352.

Wolestenholme, S. 1974, The tannery effluent problem, practical waste treatment and disposal. Applied Publishers Ltd., England, pp. 171 - 185.

Yadava, R.B.R and Sreenath, P.R. 1975. Indian. J. Plant. Physiol. 18 : 135-139.

ENVIRONMENTAL BIOTECHNOLOGY

Edited By :   Professor (Dr.) Arvind Kumar

Published By :   DAYA PUBLISHING HOUSE

21

IMPACT OF LINDANE (Υ-1, 2, 3, 4, 5, 6-HEXACHLOROCYCLOHEXANE) ON REPRODUCTIVE FUNCTION OF MALE ALBINO RAT

• *Suresh C. Joshi and Rekha Goyal*

Reproductive Physiology Section, Department of Zoology,
University of Rajasthan, Jaipur

Abstract

The present study was carried out to investigate the toxic effects of lindane (an organochlorine pesticide) on male rats at the dose level of 20mg/kg b.wt./day for 30 days. No significant change was observed in the body weight of the animals exposed to lindane. However, a significant decrease in the weights of testes and accessory sex organs were noticed. A severe impairment of sperm motility in caudaepididymis and sperm density in testes and caudaepididymis were observed. Fertility test was 90% negative. A significant reduction in sialic acid content of testes and sex accessory organs and testicular glycogen was observed. Total protein content in testes and sex accessory organs and testicular cholesterol were increased significantly. Haematological examination suggested that lindane administration leads to decrease in the number of RBC and increase in the number of WBC. These observations suggested that the lindane treatment not only altered the haematological and biochemical parameters but also showed toxic effects on male reproductive system of rats.

Key Words : Lindane, Organochlorine pesticide. Male rat, Reproductive toxicity

Introduction

The organochlorine pesticides are the most extensively used insecticides against a large variety of pests of household, agriculture and public health importance due to their excellent insecticidal properties. Organochlorine exposure has sub-chronic effect on plasma, gonadotrophins, testosterone and enzyme of androgen biosynthesis in rat. (Singh and Pandey, 1990).

Lindane, an organochlorine pesticide is used to control a wide variety of insect pests in agricultural, public health and medicinal applications. In pure form, lindane is a colourless crystalline solid. Acute toxicity of organochlorine pesticide has been observed in laboratory animals. (Mikhail, 1979, Gupta *et al.*, 1981, Labana *et al.*, 2001). Fausto *et al.*, 2001 reported that sperm quality and reproductive traits get affected in male offspring of female rabbits exposed to lindane during pregnancy and lactation.

In present work we examined the testicular sperm density, motility and density of caudaepididymis, biochemical and haematological parameters of reproductive organs of male rats treated with lindane.

Materials and Methods

Test animals - Healthy male albino rats (*Rattus norvegicus*, Wistar strain) weighing 150-180 gms were chosen for the experiment. The animals were housed in clean plastic cages covered with crome plate grills at room temperature (20°C ± 5°C) and uniform light (14 : 10 :: L : D). The animals were mostly maintained on standard rat feed procured from Ashirwad Industries Ltd., Chandigarh, India and occasionally on germinated/sprouted gram and wheat seeds as an alternative feed and fresh water *ad libitum*.

Pesticide - Organochlorine pesticide lindane (85% pure) obtained from Gupta Chemicals, Jaipur, India was used for experimental work.

Dose and Experimental design - Animals were divided into two groups. The first group animals served as vehicle treated control. The second group animals were administered lindane dissolved in olive oil by oral intubation at the dose level of 20 mg/kg b.wt./day for 30 days. At the end of experimental period the animals were weighed and autopsized under light ether anaesthesia. Blood was collected by cardiac puncture for haematological and serum biochemistry. Male reproductive organs viz. testis, epididymis, seminal veseicle, prostate gland and vas deferens were dissected out and weighed and processed for detailed biochemical studies.

Parameters studied

Haematological Parameters

The RBC, WBC (Lynch *et al.*, 1969), haemoglobin (Crossby *et al.*, 1954) and blood sugar (Astoor and King, 1954), were measured.

Sperm motility

Sperm motility was assayed by the method of Prasad *et al.* (1972). The epididymis was removed immediately after anaesthesia and known weights of caudaepididymis was gently teased in a specific volume of physiological saline (0.9% NaCl) to release the spermatozoa from the tubules. The sperm suspension was examined within 5 minutes after their isolation from epididymis. The results were determined by counting both motile and immotile sperms in at least ten separate and randomly selected fields. The results were finally expressed as per cent motility.

Sperm density

Sperm density was assayed by the method of Prasad *et al.* (1972). Briefly, total number of sperms were counted using haematocytometer after further dilution of the sperm suspension from caudaepididymis and testes. The sperm density was calculated in million/ml as per the dilution.

Fertility Test

The mating exposure test of all the animals was performed. They were cohabited with proestrous females in the ratio 1 : 3. The vaginal plug and presence of sperms in the vaginal smear was checked for positive mating. The mated females were separated to note the implantation sites on day 16th of pregnancy.

Biochemical Parameters

In biochemical parameters total protein (Lowry *et al.*, 1951), Sialic acid (Warren, 1959), Glycogen (Montgomery, 1957), Total cholesterol (Zlatkis *et al.*, 1953), Phospholipid (Zilversmit *et al.*, 1950) and Triglyceride (Gottfried and Rosenburg, 1973) were estimated.

Statistical analysis : Difference between the control and treated groups were evaluated statistically by using student's 't' test (Gad and Weil, 1982). The data are expressed as mean ± SEM. Significance was set at P ≤ 0.01 and P ≤ 0.001.

Results and Discussion

The present study showed that administration of lindane (20 mg/kg b.wt./day) orally for 30 days to male rats resulted in many degenerative changes in the male reproductive organs. The weight of testes (P ≤ 0.001), epididymis and seminal veseicle were decreased significantly (P ≤ 0.01) (Fig. 21.1). The reduction in the testicular weight reflects regressive changes in seminiferous tubules, which are associated with azoospermia and germinal aplasia. The change in testicular weight has also corresponded to the presence or absence of post meiotic germ cells. A reduction in the number of germinal cells leads to reduction in the weight of testes (Naqvi and Vaishnavi, 1993, Sinha *et al.*, 1995).

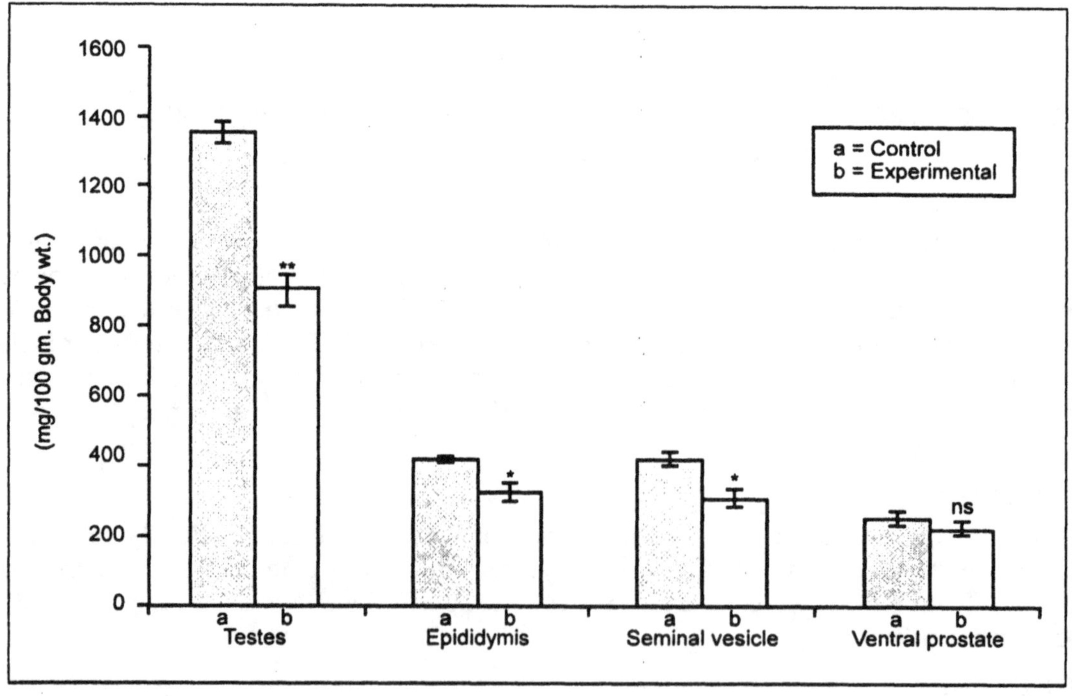

Fig. 21.1 : Changes in Organ Weights After Lindane Treatment.

Reduction in the weight of accessory reproductive organs directly support the reduced availability of androgens (Mukherjee, 1992, Patil *et al.*, 1998). Further decrease in the sperm reserve appears to be a reasonable cause for reduction in weight of epididymides (Akbarshah *et al.*, 1995, Sarkar *et al.*, 1997). Decrease in seminal veseicle and prostate gland weight after lindane treatment reflects an interference with testosterone output (Shaikh *et al.*, 1993, Malarvizhi and Mathur, 1995). Thus, the observed reduction in the weights of accessory sex organs may be due to reduced androgen availability by Lindane.

The sperm motility in caudaepididymis was declined significantly ($P \leq 0.001$) by 53.24% (Fig. 21.2). The sperm density in caudaepididymis and testes also decreased significantly ($P \leq 0.001$) by 24.16% and 78.19% respectively (Fig. 21.3). Suppression of gonadotrophins might have caused decrease in sperm density in testes (Reuber, 1981, Sinha *et al.*, 1995). Decline in sperm density in caudaepididymides may be due to alteration in androgen metabolism (Chitra *et al.*, 2001). Fertility test showed 90% negative fertility in comparison to control (Fig. 21.2). The 90% negative fertility may be due to lack of forward progression and reduction in density of spermatozoa and altered biochemical mileu of cuadaepididymis.

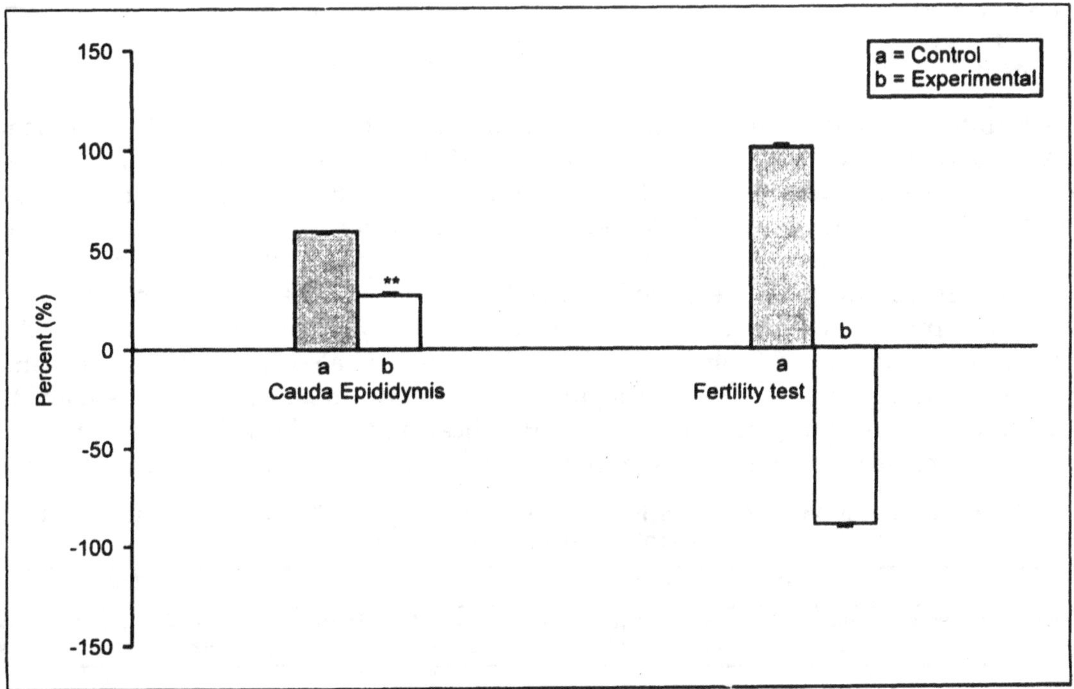

Values ± SEM 6 determinations
** = ($P \leq 0.001$)

Fig. 21.2 : Altered Sperm Motility in Cauda Epididymis and Fertility Test After Lindane Treatment.

There were no significant changes observed in blood and serum parameters except increase in total protein and W.B.C. counts and decrease in R.B.C. counts ($P \leq 0.001$) (Table 21.1). The decrease in erythrocyte count recorded in the present study may be due to either inhibition of erythrocyte production (erythropoiesis in bone marrow) or destruction of erythrocyte due to poisoning. The destruction of erythrocytes is indicated by haemolysis

(Reeves *et al.*, 1981, Qadri *et al.*, 1987). Pesticide have been reported to stimulate leukocyte production. Since pesticides act as chemical stressors, the stress causes a slight increase in adrenaline level and in consequence to neutrophil and lymphocytic leucocytosis (Thakur and Bais, 2000).

Table 21.1 : Blood and Serum Analysis of Lindane Treated Rats

| Treatment | Blood | | | | Serum (mg/dl) | | | |
	R.B.C. (million/mm^3)	W.B.C. (-/mm^3)	Haemo-globin (gm%)	Blood Sugar (mg/dl)	Total Protein	Total Cholesterol	Triglycerides	Phospholipids
Control (Vehicle Treated)	6.94 ± 0.07	6010 ± 93.54	14.32 ± 0.26	102.08 ± 2.68	12555.54 ± 143.44	115.0 ± 9.57	187.49 ± 8.86	150.0 ± 10.20
Experimental (20 mg/kg/b. wt./Day	5.90** ± 0.05	6987** ± 43.05	13.32[ns] ± 0.18	92.39[ns] ± 2.67	19055.53** ± 106.37	130[ns] ± 12.90	143.75[ns] ± 8.06	200[ns] ± 20.41

ns = Non-significant

* = (P ≤ 0.01)

** = (P ≤ 0.001).

A significant increase in the total protein content of testis, caudaepididymis, seminal vescicle and ventral prostate (P ≤ 0.001) by 21.62%, 11.36%, 7.35% and 13.85% was observed. Also, a significant reduction in sialic acid content of testes and accessory sex organs (P ≤ 0.001) by 7.31%, 10.55%, 10.59% and 13.29% was noticed. The reduction in testicular glycogen contents (P ≤ 0.001) and elevation in testicular cholesterol (P ≤ 0.01) by 32.5% and 25% were observed in lindane exposed rats (Table 21.2). Lindane also induces biochemical changes in reproductive tract. Reduction in testicular glycogen contents may be due to interference in glucose metabolism. Leydig cell function is also suppressed in absence of carbohydrates (Bedwal *et al.*, 1994). A significant depletion in sialic acid contents of testes and other sex accessory organs after treatment indicates reduced androgen supply to these organs and decrease in number of spermatozoa in lumen (Dixit and Gupta, 1987).

Table 21.2 : Biochemical Changes in Sialic Acid, Total Protein, Glycogen and Cholesterol Content of Reproductive Organs

| Treatment | Total Protein | | | | Sialic Acid | | | | Total | |
	Testes	Cauda epididymis	Seminal vesicle	Ventral prostate	Testes	Cauda epididymis	Seminal vesicle	Ventral prostate	Cholesterol (Testes)	Glycogen (Testes)
						(mg/gm)				
Control (Vehicle Treated)	251.09 ± 1.29	217.75 ± 1.81	240.7 ± 1.81	199.75 ± 1.15	4.24 ± 0.02	3.79 ± 0.05	4.06 ± 0.03	3.91 ± 0.02	8.00 ± 0.70	2.61 ± 0.13
Experimental (20 mg/kg/b. wt./Day)	305.4** ± 2.86	242.5** ± 2.87	258.4** ± 2.88	227.43** ± 2.77	3.93** ± 0.03	3.39** ± 0.03	3.63** ± 0.38	3.39** ± 0.03	10.0* ± 1.11	1.76** ± 0.20

* = (P ≤ 0.01)

** = (P ≤ 0.001).

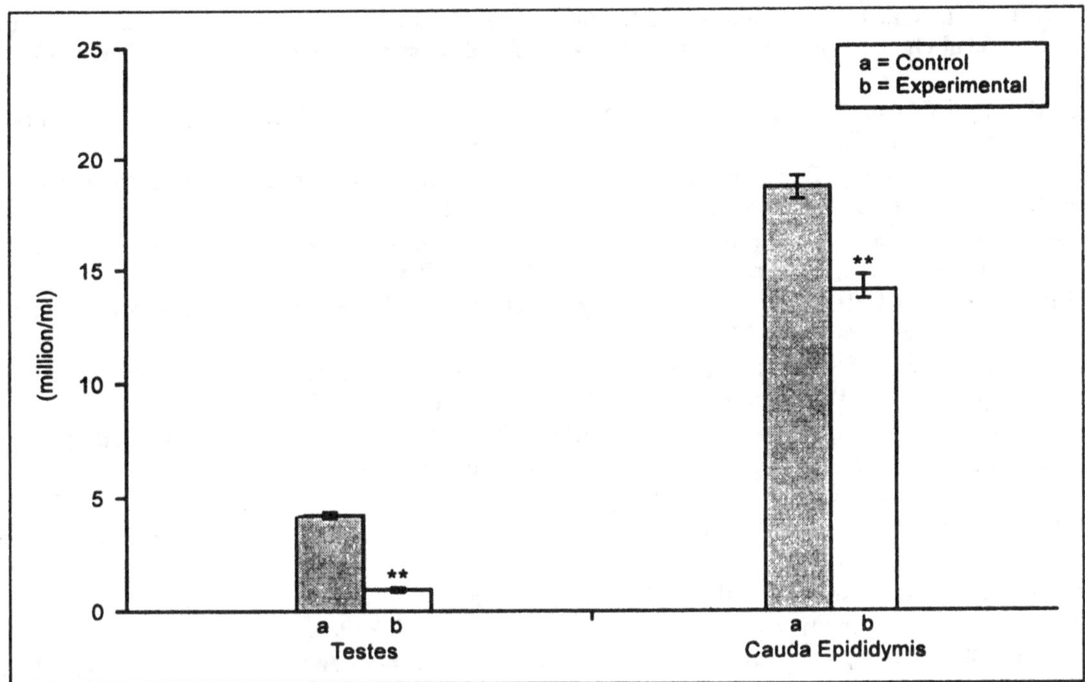

Values ± SEM 6 determinations
** = (P ≤ 0.001)

Fig. 21.3 : Altered Sperm Density in Testes and Cauda Epididymis After Lindane Treatment.

The elevation in the testicular protein may be due to the hepatic detoxification activities cause by lindane which resulted in the inhibitory effects on the activities of enzyme involved in the androgen bio-transformation (Dikshit and Dutta, 1972).

Administration of lindane also increases the total serum cholesterol. The changes in the serum cholesterol levels can be attributed to alteration in hepatic synthesis or rate of tissue utilization and degradation. Increased level of cholesterol in testes is attributed to decreased androgen concentration (Bedwal *et al.*, 1994). It also indicates non-availability of pituitary gonadotrophins essential for steroidogenesis (Reddy *et al.*, 1997)

Thus, these observations revealed that lindane produces alteration not only in haematological and biochemical parameters but also produces toxic effects on reproductive function of male rats.

Acknowledgement

The authors are grateful to SAP (Special Assistance Programme), Department of Zoology, University of Rajasthan, Jaipur for financial assistance.

References

Akbarshah, M.A., Stanley, A. and Averal, H.I. 1995. Effects of Vincristine on Leydig cell and accessory reproductive organs. Current Science. 68(10): 1053-57.

Astoor, A.M. and King, E.J. 1954. Simplified colorimetric blood sugar method. Biochem. J. 56,44.

Bedwal, R.S., Edwards, M.S., Katoch, M., Bahuguna, A. and Dewan, R. 1994. Histological and biochemical changes in testes of zinc deficient BAL B/C strain mice. Ind. J. Exp. Bio. 32 (4) : 243-47.

Boyd, G.S. and Oliver, M.F. 1958. The physiology of circulating cholesterol and lipoproteins. In : cholesterol. Cook, R.P. (ed.) Academic Press, New York.

Chitra, K.C., Latchoumy Candane, C. and Mathur, P.P. 2001. Chronic effect of endosulfan on the testicular functions of rat. Asian. J. Androl. 1 : 203-06.

Crossby, W.H., Munn, J.I. and Furth, F.W. 1954. Standardizing a method for clinical haemoglobinometry, US. Armed. Force. Med. J. 5, 695-703.

Dikshith, T.S.S. and Datta, K.K. 1972. Pathological changes induced by pesticides in the testes and liver of rats. Exp. Path. 7 : 309-16.

Dixit, V.P. and Gupta, R.S. 1987. Antispermatogenic and antiandrogenic activity of *sapindus trifoliatus* extract in intact and castrated male gerbils. Planta Medica 46 : 242-46.

Fausto, A.M., Morera, P., Margarit, R. and Taddei, A.R. 2001. Sperm quality and reproductive traits in male offspring of female rabbits exposed to lindane (gamma-HCH) during pregnancy and lactation. Reprod. Nutr. Dev. May-June, 41(3) : 217-25.

Gad, S. and Weil, C.S. 1982. Statistics for toxicologist. In Principles and Methods of Toxicology (Wallance H.A., Ed.), pp. 273-320.

Gottfried, S.P. and Rosenburg, B. 1973. Improved manual Spectrophotometric procedure for determination of serum triglycerides. Clin. Chem., 19, 1077-1078.

Gupta, P.K., Shrivastava, S.C. and Ansari, R.A. 1981. Toxic effects of endosulfan on male reproductive organs in rats. Ind. J. Biochem. Biophys. 18 (Suppl.) : 159.

Labana, S., Bansal, R.C and Mahmood, A. 2001. Age related effects of organochlorine insecticide lindane on intestinal brush border membrane in rats. Indian J. Biochem. Biophys. Aug 38(4) : 249-52.

Lowry, O.H., Rosenburg, N.J., Far, A.L. and Roudall, R. 1951. Protein measurement with Folin-Pheno reagent. J. Biol. Chem., 193, 265-275.

Lynch J.M., Raphel, S.S., Melier, L.D., Spare, P.D. and Inwood, M.J.H. 1969. Collection of blood sample and haemocytometry red cell count, white cell count. In : Medicinal laboratory Technology and clinical pathology Pub. W.B. Saunders Company Igakar Sohim Ltd. Tokyo, 626-647.

Malarvizhi, D. and Mathur, P.P. 1995. Effect of cisplastin on physiological status of normal rat testis. Ind. J. Exp. Bio. 33 : 281-83.

Mikhail, T.H. 1979. Acute toxicity of organophosphorous and organochlorine insecticides in laboratory animals. Z. Emahrung. Swiss. 18 : 258-68.

Montgommery, R. 1957. Detemination of glycogen. Arch. Biochem. Biophys. 67, 612-614.

Mukherjee, M., Chattopadhyay, S. and Mathur, P.P. 1992. Effects of flutamide on the physiological status of epididymis and epididymal sperms. Andrologia. 24: 113-16.

Naqvi, S.M. and Vaishnavi, C. 1993. Bioaccumulative potential and toxicity of endosulfan insecticide to non-target animals. Comp. Bio. Chem. Physiol. 105: 347-61.

Patil, S., Patil, S.R., Londonkar, R. and Patil, S.B. 1998. Effect of pathidine on spermatogenesis in albino rats. Ind. J. Pharmacol. 30: 249-53.

Prasad, M.R.N., Chinoy, N.J. and Kadam, K.M. 1972. Changes in succinic dehydrogenase levels in the rat epididymis under normal and altered physiological conditions. Fertil. Steril. 23, 186-190.

Qadri, S.S., Usha, G., Jabben, K., Rahman, M.F. and Mustafa, M. 1987. Effect of dermal application of phosphamidon on different tissue and haematobiochemical parameters in albino rat. J. Toxicol. Environ. Hlth. 20(1553): 273-86.

Reddy, C.M., Murthy, D.R.H. and Patil, S.B. 1997. Antispermatogenic and antiandrogenic activities of various extracts of Hibiscus rosa sinensis in albino mice. Ind. J. Exp. Bio. 35: 1170-74.

Reeves, J.D., Driggers, D.A. and Vincent, A.K. 1981. Child leukemia and Aplastic anemia after malathion exposure. The Lancet. 300.

Reuber, M.D. 1981. The role of toxicity in the carcinogenicity of endosulfan Sci. Total. Environ. 20: 23-47.

Sarkar, S.N., Majumbar, A.C. and Chattopadhyay, S.K. 1997. Effects of isoproturon on male reproductive system : Clinical histological and histoenzymological studies in rats. Ind. J. Exp Bio. 35: 133-38.

Shaikh, P.D., Manivannan, B., Pathan, K.M., Kasturi, M. and Ahmed, R.N. 1993. Antispermatic activity of *Azardirarchta Indica* leaves in albino rats. Current Science, 64(9): 688-89.

Singh, S.K. and Pandey, R.S. 1990. Effect of sub-chronic endosulfan exposure on plasma gondadotrophins, testosterone, testicular testosterone and enzyme of androgen biosynthesis in rat. Ind. J. Exp. Biol. 28: 953-56.

Sinha, N., Narayan, R., Shanker, R. and Saxena, D.K. 1995. Endosulfan induced biochemical changes in the testes of rats. Vet. Hum. Toxicol. 37: 547-49.

Thakur, P.B. and Bais, V.S. 2000. Toxic effects of aldrin and fanvalerate on certin hamatological parameters of fresh water teleost *Heteropneustes fosilis*. J. Environ. Biol. 21(2): 161-63.

Warren, L. 1959. Thio-barbituric acid assay of silaic acid. J. Biol. Chem. 234, 1971-1975.

Zilversmit, D.B., Davis, A.K. and Memphis, B.S. 1950. Microdetermination of plasma phospholipid by trichloracetic acid precipitation. J. Lab. clin. Med. 35: 155-160.

Zlatkis, A., Zak, B. and Boyle, A.J. 1955. A new method for direct determination of cholesterol. J Lab. Clin. Med. 41: 486-492.

ENVIRONMENTAL BIOTECHNOLOGY

Edited By :   Professor (Dr.) Arvind Kumar

Published By : DAYA PUBLISHING HOUSE

22

BIOACCUMULATION OF MERCURY IN UPPANAR ESTUARY, CUDDALORE, SOUTHEAST COAST OF INDIA

• *R. Rajaram and M. Srinivasan*

CAS in Marine Biology, Annamalai University,
Parangipettai-(Tamil Nadu)

Abstract

Estuarine ecosystem are highly complex, dynamic and subject to many internal and external relationship that are subject to change over time. Uppanar estuary is considered to be one of the highly polluted estuary in south east coast of India due to industrialization. SIPCOT (Small Industrial Promotion Corporation of Tamil Nadu, covering an area of about 520 acres with 44 industries) is located on the bank of Uppanar estuary at Cuddalore. It was established for chemical, petrochemical, pharmaceutical, biocides, fertilizer, fungicides, chlor-alkali and metal processing industries etc. Indiscriminate discharges of wastewater from SIPCOT industrial complex into coastal environment affect both biotic and abiotic system and finally cause some ill effects to human beings through food-chain. A detailed study was made on the bioaccumulation of mercury in the food-chain and the results were discussed in this paper.

Key Words : Mercury, Food-chain, Uppanar Estuary.

Introduction

Coastal zone comprises varied biotopes such as estuaries, backwaters, mangroves, salt marshes, coral reefs, lagoons and near shores. Estuaries are highly protective, sensitive, feeding and breeding ground for fin and shell fishes with rich biodiversity. Realizing the importance of the estuaries, the ever exploding human population exploit not only the biological resources but also interferes and modifies the basic coastal processes. This results in many environmental episodes such as "Minamatta" and "Itai Itai" diseases.

The coastal zone receives the pollutants principally from three major routes - viz. atmosphere, riverine and glaciers. Of course man also serve as a "Geological Agent" by the way of discharging the effluents through piped outfalls, direct pimping, operation of ships etc. When the pollutants exceed the threshold limits coupled with environmental variables such as salinity, temperature, pH, DO, H_2S etc give stress to organisms of the environment. So monitoring of estuarine environment needs be carried out periodically to estimate the concentration and distribution of varied pollutant levels to observe the changes in the ecosystem. Discharges of wastewater from SIPCOT industrial complex estimated to be 56.92 gallons in lakhs/day into the coastal environment. So the present study were, therefore; initiated to understand the mercury distribution pattern in Uppanar estuarine food-chain with regard to effluents from SIPCOT industries.

Sources of mercury

1. Natural occurrences

2700 - 6000 tonnes/year due to earth's crust, emission from volcanoes and evaporation from natural water bodies (Food and Drug Administration).

2. Anthropogenic sources

3000 tonnes are released annually into the atmosphere by human activities (Judith, 2000). About 180 tonnes of mercury and its compounds are introduced into Indian coastal environment per annum (Patel and Chandy, 1988) by the following industries :

(a) Pharmaceutical and cosmetic industries :
  (i)   Phenyl mercury acetate used in contraceptive vaginal jellies
  (ii) Various organo-mercury compounds used as diuretics
  (iii) Organo-mercurial are used as antiseptic products

(b) Electrical apparatus industries
  Mercury was used in the manufacture of mercury batteries, mercury pool rectifiers and power tubes, varieties of lamps includes fluorescent, germicidal, photocopying and high intensity arc discharge lamps.

(c) Industrial control instrument industries
  It is used in the manufacture of switches, relays, gauges, pump seals and valves.

(d) General laboratory
  Mercury used in experimental equipments such as diffusion pumps, barameters, monometers, thermometers and vibration dampers

(e) Chlor-alkali industries uses continuous flow of mercury cathode cell to produce chlorine and caustic soda

(f) Paint industries
  (i)   Organo-mercurial compounds: Bactericide-fungicide agents to protect water based paints from bacterial fermentation
  (ii) Phenyl mercury derivatives are used as paint preservatives

(g) Amalgam alloy is made of silver, tin and copper mixed with mercury to form plastic mass used to fill the tooth cavities

(h) Mercury is used to prepare various catalytic salts such as chloride, oxide, sulfate, acetate etc.

(i) Various mercury compounds are used as slimicides to control microbial growth in paper and pulp industries.

Materials and Methods

Uppanar estuary (Lat. 11° 42'N: Long. 79° 46'E) is formed by the confluence of Gadilam and Paravanar river, and opens into the Bay of Bengal near Cuddalore old town on the south east coast of India (Fig. 22.1). It forms a potential fishing ground with an annual average landing of about 2000 tonnes. The tidal effect extends up to a distance of about 6 km and width of the estuary is about 30m near the mouth and 20m in the upstream. SIPCOT industrial complex is located on the bank of Uppanar estuary, where the effluents are discharging, mixing and diluting. Apart from the industrial effluents, drainage of municipal and domestic sewage from Cuddalore old and new towns and waste from coconut husk retting ground are discharging.

The bottom soil is characterized by blackish clay in nature, salinity ranges from 5 to 36 ppt. Sediment, phytoplankton, zooplankton, bivalve, fish, water and effluent were collected from the study areas for the analysis of mercury.

Water and Effluents

Effluent from SIPCOT industrial discharging site taken carefully and surface water samples were collected in pre-cleaned, acid washed polypropylene bottles and the samples were filtered in Millipore filter paper (Pore size 0.45μ). The samples were pre-concentrated with APDC-MIBK extraction procedure (Brooks *et al.*, 1967).

Sediment

Samples were collected in pre-cleaned, acid washed PVC corers and washed with metal free double distilled water and dried in an oven at 40°C (EPA, 1979) for 5 to 6 hours and ground to powder and redrying the samples from which 500 mg was taken and digested with a mixture of 1 ml of Con. H_2SO_4, 5 ml of Con. HNO_3 and 2 ml of $HClO_4$. A few drops of hydrofluoric acid add to achieve complete digestion and filter the sample to make tip 25 ml with metal free double distilled water for mercury analysis (Chester and Hughes, 1967).

Phytoplankton and Zooplankton

Phytoplankton samples were collected from surface water by towing plankton net having 0.35 mouth diameter and 48 μm mesh size (300 μm for zooplankton) for half an hour and dried the sample in an oven at 40°C till the constant weight was obtained. A known weight of sample (500 mg) was digested in Con. NHO_3 and H_2O_2. After centrifugation to remove silica fractions, the solutions were diluted to 25 ml and analyzed in standard mercury analyzer (Knauer and Martin, 1973).

Bivalve and fish

The bivalve *Meretrix casta* and fish *Mugil cephalus* were freshly collected from mouth of Uppanar estuary. A stainless steel scalpel steel was used to remove foot and muscle from bivalve and gill, liver, kidney and tissue from fish and the dissected portions were dried in hot air oven at 40°C. The dried samples were powdered and digested in 3:1 ratio of HNO_3 and $HClO_4$. The digested samples were diluted to make up 25 ml and stored for mercury analysis (Topping, 1973).

Total mercury was quantified by adopting cold vapour technique in mercury analyzer, model No. 5800D, Serial No. 209, fabricated by Electronic Corporation of India, Bangalore.

Results

Distribution and accumulation of mercury in water, effluent, sediment, phytoplankton, zooplankton and various organs in bivalve and fish are shown in Table 22.1.

Discussion

Natural sources of mercury in water are through land and river runoff and mechanical and chemical weathering of rocks. The high value of mercury in water due to rainfall causing flooding the stagnant effluent and facilitating higher influence of land derived materials including agricultural runoff and washing of waste coupled with river discharges and direct discharge of effluents. The estuarine sediments may be exchanging part of the exchangeable phase of the metals in water. The high value of mercury in sediment may be due to land drainage, microbial activity, sediment particle type and size and organic content which influence the absorption and accumulation of mercury in sediments.

Many environmental factors influence and affect the uptake of mercury by phytoplankton. This includes physico-chemical, biological and bio-chemical parameters and also the interaction between the mercury by environment. Phytoplankton should have concentrated more heavy metal by absorption and absorption mechanisms from the surface waters (Knauer and Martin, 1973, Bryan, 1976). Mercury accumulation by zooplankton is mainly by two pathways i.e., direct uptake from the water through and the assimilation from ingested food and detritus

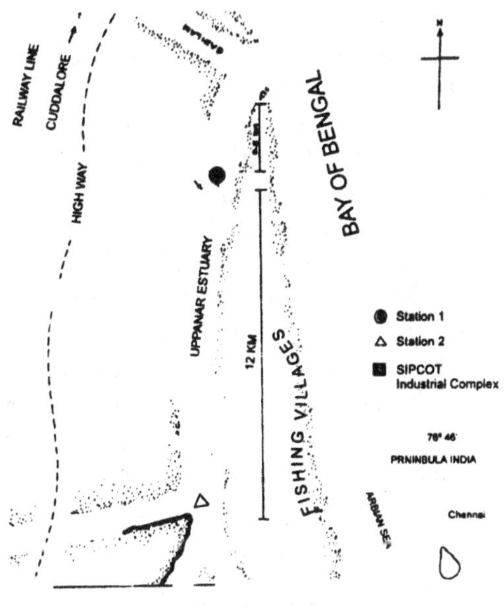

Fig. 22.1

(Davis, 1978). High concentration of mercury in zooplankton due to land runoff, heavy fresh water flux, industrial effluents and domestic sewage. Knauer and Martin (1972) stated that zooplankton ingested food (phytoplankton) with high mercury content seems to increase the concentration of mercury in zooplankton.

Accumulation of metals, by mussel is influenced by salinity, temperature and concentration of metal in water (Boyden and Romeril, 1974). The size, sex, reproductive condition and season are also factors for the variation. In the present study the accumulation of mercury in *Meretrix casta* showed N.D to 6.42 μg g⁻¹ and more amount of mercury accumulation due to filter feeding habit (Paul Raj, 1987).

Fishes are known to concentrate metals in their body tissues in varying proportions depending upon the species, environmental conditions and feeding habits and accumulates

high values then their basal environment due to the process of bioamplification (Cooper, 1983). The bioaccumulation of mercury in fish *Mugil cephalus* showed N.D - 3.92 μg g⁻¹ may be due to behaviour of the animal with respect to their environment i.e. muddy bottom feeder. The effluent containing organic wastes, exudates, dead and decayed matters are termed as detritus and it is not lost to the ecosystem, rather it serves as the source of mercury flow through detrivores like *Mugil cephalus*. The high level of mercury accumulation (3.92 μg g⁻¹) is mainly due to intake of detritus food, absorption by gill and intestine of skin. The excretory materials

Table 22.1 : Bioaccumulation of Mercury in Uppanar Estuary

No.	Samples		Value
1.	Effluent		0.17 – 11.49 μg l⁻¹
2.	Water		N.D. – 1.92 μg l⁻¹
3.	Sediment		0.17 – 6.710 μg g⁻¹
4.	Phytoplankton		0.12 – 6.87 μg g⁻¹
5.	Zooplankton		0.49 – 8.87 μg g⁻¹
6.	*Meretrix casta*	– Foot	N.D – 6.42 μg g⁻¹
		– Body muscle	N.D. – 5.48 μg g⁻¹
7.	*Mugil cephalus*	– Gill	0.08 – 3.1 μg g⁻¹
		– Liver	0.11 – 3.92 μg g⁻¹
		– Kidney	N.D – 2.1 μg g⁻¹
		– Tissue	N.D – 0.23 μg g⁻¹

N.D. – Non-detectable value.

present in the bottom sediments are the route of recycling of mercury. Hence, *Mugil cephalus* forms an important item of food for man and it can become an indirect source of mercury entering the body, leading to toxic effects. The mercury accumulation in different food-chain organisms was found to be in the decreasing order from the source to water (*Effluent > Zooplankton> Phytoplankton> Sediment > Bivalve> Fish*). In the present study mullet showed 0.23 $\mu g\ g^{-1}$ (ppm) in the edible portion (tissue) which is lesser than the maximum permissible limit of mercury for human consumption prescribed as 0.5 ppm by USSR, 0.7 ppm by Italy and 0.4 ppm by Japan (FAO, 1983). In the present study the bio-accumulation of mercury found to be safer level but due course the level may increase and cause several problems to the environment. Already there were reports on flourosis in the rural areas near by the industries. Further ground water turns useless for drinking and bathing, smoke released by the industries cause bad odour and affects the respiratory system also. During monsoon season, the effluents washed from all the industries and discharge in to the Uppanar and hence mass mortality of fin fish and shell fish was noted. Therefore, Government of India should take necessary steps to control the pollutants released from all the industries in this area.

Acknowledgements

The authors are thankful to the Director, Centre of Advanced Study in Marine Biology, Annamalai University, Parangipettai for providing the facilities and encouragements during the study.

References

Boyden, C.R and Romeril, M.G. 1974. A trace metal problem in pond oyster culture. Mar. Pollut. Bull., 5: 74-78.

Brooks, R.R., Presley, B.J. and Kalplan, I.R. 1967. APDC-MIBK extraction system for the determination of trace metals in saline water by atomic absorption spectroscopy, Talanta, 14: 809 - 816.

Bryan, G.W., 1976. Some aspects of heavy metal tolerance in aquatic organisms. In: Lockwood, A.P.M. (Ed.). Effects of pollutants on aquatic organism. Cambridge University Press, Cambridge, pp. 7-34.

Chester, R. and Hughes, M.J. 1967. A chemical technique for the separation of Ferro-manganese minerals, carbamate and absorbed trace elements from pelagic sediments. Chem. Geol., 2: 249-263.

Cooper, J.J. 1983. Total mercury in fishes and selected biota in Lohotan reservoir, Navada. Bull. Environ. Contom. Toxicol., 31: 9 - 17.

Davis, A.G. 1978. Pollution status with marine plankton. Part. II. Heavy metals. Adv. Mar. Biol., 15: 381-508.

FAO. 1983. Compilation of legal limits for hazardous substance in fish and fishery products. pp. 102.

Foulke Judith E. 2000. Mercury in fish; cause for concern. Sea Food Export Journal, Vol. XXXI, 1: 123-134.

Knauer, G.A. and Martin J.H. 1973. Seasonal variation of cadmium, manganese, lead and zinc in water and phytoplankton in Monterey Bay, California. Limnol. Oceanogr., 18(4): 597 - 604.

Patel, B and Chandy, J.P. 1988. Mercury in the biotic and abiotic matrices along Bombay coast. Indian J. Mar. Sci., 17: 56 - 58.

Topping, G. 1973. Heavy metals in fish from Scottish water. Aquaculture, 1: 373 - 377.

Paul Raj, S. 1986 -1987. Pollution Research 5(2): 39 - 43 and J. Environ. Biol., 8(2): 151 - 155.

ENVIRONMENTAL BIOTECHNOLOGY

Edited By :   **Professor (Dr.) Arvind Kumar**

Published By : **DAYA PUBLISHING HOUSE**

23

Electrochemical Study of Propoxur

• *G. Suresh, A. Hemasundaram and N.V.S. Naidu**
Department of Chemistry, S.V. University, Tirupati

Abstract

The Electrochemical reduction of propoxur has been studied employing d.c polarographic, cyclic voltammetric and coulometric techniques in 25% DMF in the pH range 3.5 - 8.0. Two well defined, irreversible and diffusion controlled, 12 electron cathodic waves/peaks are observed, due to the simultaneous reduction of the two nitro groups, to give diamine. The number of electrons involved in the reduction are interpreted based on cyclic voltammetric and coulometric techniques. The method developed is extended for the analysis of water samples.

Introduction

Propoxur (2-isopropoxyphenyl-N-methyl Carbamate) belongs to carbamate class of pesticide. It provides control of insects which have become resistant to chlorinated hydrocarbons and organophosphates.

Literature survey reveals that several techniques used for the determination of propoxur there are spectrophotometric (Agarwal *et al.*, 1993, Das *et al.*, 1994, Appaiah *et al.*, 1985 Raju and Abraham, 1992, Naidu and Naidu, 1990) and chromatographic techniques (Sianley *et al.*, 1972, Blagg and Rawis, 1972, Erust *et al.*, 1975, Lawrene and Frei, 1972, Ishii and Otake, 1973). In continuation of our earlier work (Sarswathi and Padanja, 1998) in developing simple and rapid methods of various pesticides/insecticides, propoxur is studied using d.c. polarography, cyclic voltammetry and coulometry. It may be mentioned here that there are no reports available so far on its electrochemical behaviour.

Experimental

D.C. polarograph model CL-357 coupled with model LR-101 P strip chart recorder manufactured by Elico Pvt. Ltd., Voltammograph model CV-27 coupled with R-XY recorder manufactured by BAS, USA and Millicoulometer setup was supplied by Radelkis (Budapest) consisting of integrator unit No. 322 and Radelkis potentiostat No. 318 are used in the present study.

---

* Author for Correspondence

pH measurements are made using pH meter model LI-120 manufactured by Elico Pvt. Ltd. All experiments were performed at 25±1°C. Dissolved oxygen was removed by purging nitrogen gas for 10 minutes.

The compound, propoxur is supplied by M/s. Bayer India Ltd. Mumbai. The purity of the compound was tested by melting point determination. The chemicals, sodiumacetate, acetic acid, dimethyl formamide and Triton X-100 (to suppress the maxima) are supplied by M/s. Glaxo.

Results and Discussion

Effect of pH

At 0.15 M concentration of sodium acetate supporting electrolyte and 0.2 mM propoxur, the pH is varied from 3.5 to 8.0. With increasing pH the reduction potentials are found to shift towards more negative side indicating the proton involvement in the electrode process. A well defined wave is seen in case of pH ~4.0. So pH ~4.0 is selected as optimum pH for all other studies.

Effect of supporting electrolyte

At pH~4.0 keeping the concentration of propoxur at 0.2 mM the concentration of the supporting electrolyte is varied from 0.05 M to 1.0 M and found that nature and shape of the polarogram remains the same. A well defined wave is seen in 0.15 M sodium acetate. So 0.15 M is selected concentration for all other studies.

The polarographic behaviour of propoxur has been examined over the pH range 3.5 to 8.0 in 0.15 M sodium acetate in 30% DMF. Typical polarogram of propoxur is shown in Fig. 23.1.

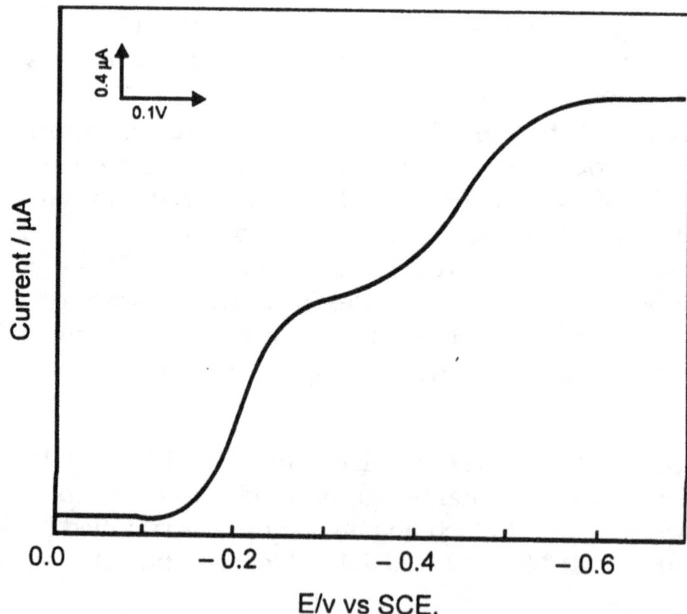

Fig. 23.1 : Typical Polarogram of Nitrated Propoxur in 0.15 M Sodium Acetate
pH : 4.0; Nitrated propoxur : 0.2 mM; DMF : 30%.

Effect of propoxur

The reduction process of the propoxur is diffusion controlled and irreversible based on the usual criteria.

The kinetic parameters like diffusion coefficient, D and forward rate constant, $K^o_{f.h}$ values are evaluated and reported in Table 23.1. The diffusion coefficient values are seen to gradually decrease which account for decrease in diffusion current with increase in pH due to less availability of protons at higher pH. The $K^o_{f.h}$ values are also found to decrease with increase in pH indicating that the electrode reduction becomes more irriversible with pH.

Table 23.1 : Diffusion Coefficient and Forward Rate Constants of Nitrated Propoxur

Sodium acetate : 0.15   DMF : 30%   Nitrated proposur : 0.2 mM

| pH | First wave | | | | Second Wave | | | |
	$E_{1/2}$ (I) -v vs SCE	i_d (I) μA	$D \times 10^6$ $Cm^2 S^{-1}$	K^o_{fin} $Cm\ S^{-1}$	$E_{1/2}$ (II) -v vs SCE	i_d (II) μA	$D \times 10^6$ $Cm^2 S^{-1}$	K^o_{fin} $Cm\ S^{-1}$
3.5	0.18	1.50	0.2304	1.49×10^{-3}	0.42	1.38	0.195	5.141×10^{-7}
4.0	0.20	1.45	0.2020	8.4×10^{-4}	0.44	1.32	0.180	4.21×10^{-7}
5.0	0.30	1.30	0.1747	3.60×10^{-4}	0.51	1.23	0.155	1.37×10^{-7}
6.0	0.34	1.20	0.1489	3.43×10^{-6}	0.53	1.00	0.103	9.33×10^{-9}
7.0	0.40	1.15	0.1360	8.30×10^{-7}	0.63	0.98	0.099	5.04×10^{-9}
8.0	0.42	1.00	0.1034	2.80×10^{-7}	0.68	0.90	0.083	1.0×10^{-9}

Cyclic Voltammetry

Cyclic voltammogram of nirated propoxur showed two peaks in 0.15M sodium acetate in 30% DMF. Peak potentials of both peaks shifted to negetive values with increasing pH indicating proton involvement in the reduction process. Both peaks also shifted to negetive potentials with increasing concentration and scan rate indicating the irreversible nature of the electrode process. Further, the absence of anodic peak in the reverse scan confirms the irreversible nature. The current function, in both the cases is found to be constant showing that the electrode process is diffusion controlled. The results shown in Tables 23.2 and 23.3. A typical voltammogram of propoxur is shown in Fig. 23.2.

Table 23.2 : Effect Concentration of nitrated propoxur

Sodium cetate : 0.25 M   DMF : 30%

pH : ~ 4.0   Scan rate : 0.04 Vs^{-1}

Concentration, 10^{-4}M	E_p (I) -V vs SCE	i_p (I) μA	E_p (II) -V vs SCE	i_p (II) μA
1.0	0.205	0.800	0.440	0.68
2.0	0.210	1.575	0.445	1.45
3.0	0.212	2.250	0.452	1.98
4.0	0.214	3.025	0.455	2.80

Coulometry

From coulometric results number of electrons involved in the reduction process is found to be twelve due to the reduction of two nitro groups to diamine stage.

Mechanism: From the data obtain in d.c polarography, voltammetry, cyclic Voltammetry and coulometry. The mechanism of the electrode reaction is suggested.

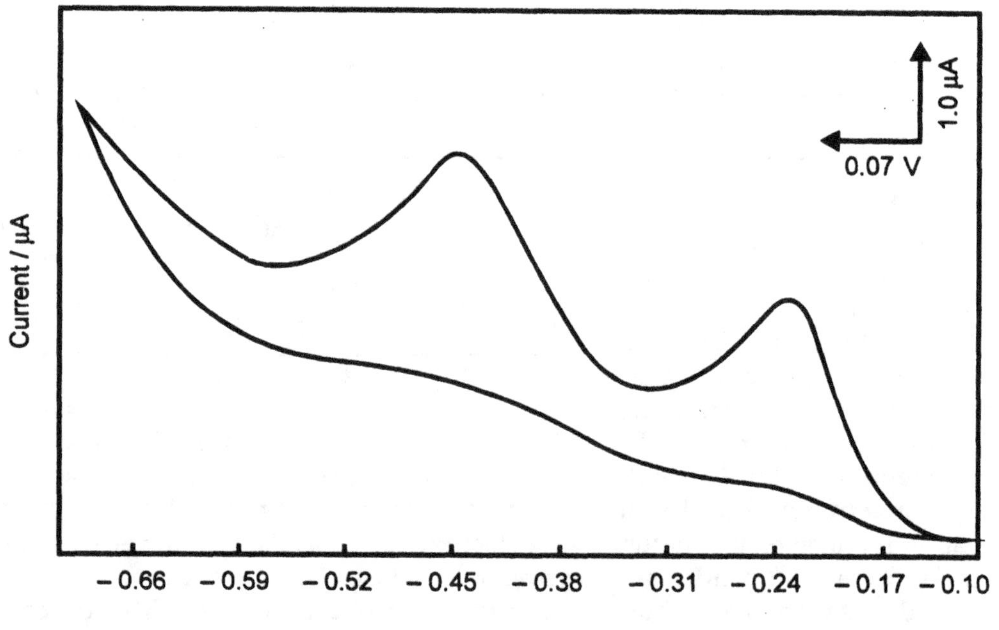

Table 23.3 : Effect Sacen Rate

Sodium cetate : 0.15 M		DMF : 30%		Nitrated propoxur : 0.2 mM		pH : ~4.0
Scan rate	$E_p(I)$	$i_p(I)$	$\dfrac{i_p(I)}{V^{\frac{1}{2}}}$	$E_p(II)$	$i_p(II)$	$\dfrac{i_p(I)}{V^{\frac{1}{2}}}$
VS^{-1}	-V vs SCE	μA		-V vs SCE	μA	
0.02	0.200	1.100	7.780	0.442	1.025	7.24
0.04	0.210	1.575	7.875	0.445	1.450	7.25
0.08	0.215	2.185	7.730	0.448	2.050	7.24
0.10	0.216	2.480	7.840	0.450	2.280	7.21
0.20	0.18	3.500	7.820	0.455	3.250	7.27

Fig. 23.2 : Typical Cyclic Voltammogram of Nitrated Propoxur in 0.15 M Sodium Acetate
pH : ~4.0; DMF : 30%;   Nitrated propoxur : 0.2 mM; Scan rate : 0.1 Vs⁻¹

Analysis of Propoxur (on nitrition) in spiked water samples

100 ml of water is spiked with known concentration of the pesticide in DMF. The sample is extracted with 20 ml portions of trichloromethane. The extracted insecticide is washed with potassium carbonate dried over sodium sulphate. The organic phase is after evaporation nitrated and the residue is dissolved in 30% DMF in 50 ml flask. Known aliquots

of the solution is taken in polarographic cell and polorographed maintaining the experimental conditions developed. The amount of the pesticide present in the water samples is determined from calibration graph already drawn and the results are shown in Table 23.4.

Table 23.4 : Determination of the Propoxur Concentration in Spiked Water

S. No.	Spileked level µg/ml	Amount found*, µg/ml	Recovery %	Standard deviation
1	2.90	2.80	96.5	± 0.052
2	6.78	6.54	96.5	± 0.053
3	8.70	8.50	97.0	± 0.040
4	11.60	11.40	98.3	± 0.060
5	29.0	28.5	98.3	± 0.040

* Average of three determinations.

References

Agarwal, V. and Gupta, V.K. 1993. Chemia Analityczna, 38. 197-200.

Appaiah, K.M., Sreenivasa, M.A. and Kapur, O.P. 1985. Bull. Env. Cont. Tax., 35. 296.

Blagg, A.A. and Rawis, J.L. 1972. Amer. Lab., 4. 17.

Das, J.V., Ramachandran, K.N. and Gupta, V.K. 1994. Fres. J. Anal. Chem., 348. 840.

Ernst, G.F., Rocdar. S.J., Tian, G.R. and Tansen, T.T.A. 1975. J. Assoc. Off. Anal. Chem., 58. 1015.

Ishii, Y. and Otake, T. 1973. Bull. Chem. Inst. Sta., 13. 32.

K. Saraswathi and K. Padmaja, 1998. Bull. Elect. Chem. 14. 276.

Lawrene, J.F. and Frei, R. W. 1972. Int. J. Environ.Anal. Chem., 1. 317.

Naidu D.V. and P.R. Naidu., 1990. Talanta, 37, 629.

Raju, J. and Abraham, R. 1992. Analyst, 20. 489.

Stanley, C.W., Thornton, J.S. and Katague, D.B. 1972. J. Agric. Food Chem., 20. 1265.

ENVIRONMENTAL BIOTECHNOLOGY

Edited By :   **Professor (Dr.) Arvind Kumar**

Published By :   **DAYA PUBLISHING HOUSE**

24

Bulb Yield, Water Use Efficiency and Moisture Extraction Pattern in Onion as Influenced by Levels of Irrigation and Nitrogen

• *J.M.L. Gulati, M.M. Mishra, A.K. Patra and D.N. Singh*
Regional Research and Technology Transfer Stations (OUAT) Chiplima,
Sambalpur - 768 025, Orissa

Abstract

A field experiment was conducted during rabi season of 1993-94 to 1995-96 to study the influence of irrigation and nitrogen on onion (*Allium cepa*) cultivar Nasik Red on sandy loam soil at Regional Research and Technology Transfer station, Chiplima. Pooled data revealed a significant increase in bulb yield (106.6 q/ha) under frequently irrigated plots of 1.2 I.W./C.P.E. requiring 9 irrigations of 5 cm depth each. The corresponding water requirement and evapotranspiration was 46.7 and 3.4 cm, respectively with crop water use efficiency of 350.47 kg/ha. Crop extracted maximum moisture from 0-20 cm (12.8 cm) accounting to 62.1% of the total moisture used by the crop. Mid season ET/Epan ratio ranged between 0.89-0.91 in February-March. Bulb yield increased with increase in N levels upto 150 kg N/ha though the difference among the levels was statistically non significant.

Key Words : Bulb yield, Water use efficiency, Moisture extraction pattern, IW/CPE ratio.

Introduction

Onion is an important commercially valuable crop. Its cultivation promises has increased to tail end areas of different canal commands owing to its ability to withstand under lesser water suply conditions. However, its shallow and fibrous root systems requires precise application of water along with optimal use of nitrogen which is by far the most important nutrient influencing yeild potential of the crop. Keeping this in view, present study was planned under the agro-climatic condition of Hirakud irrigation command in western Orissa.

Materials and Methods

Field experiments were carried out during *rabi* season from 1993-94 to 1995-96 at Regional Research and Technology Transfer Station, Chiplima, Sambalpur. Soil of the experimental site was sandy loam in texture having 17.0% and 8.0% moisture at field capacity and permanent wilting point, respectively. Soil was low in nitrogen (212 kg N/ha) medium in phosphorus (16.0 kg P_2O_5/ha) and potash (312 kg K_2O/ha). Crop received 19.3 mm, 6.9 mm and 117.9 mm through rainfall and lost 394.9 mm, 461.5 mm and 456.6 mm moisture through evaporation during first, second and third year of study respectively. About 45 days old seedlings of variety Nasik red were transplanted in 3rd week of December with a crop geometry of 15 cm row to row and 10 cm plant to plant and fertilized with a common level of phosphorus (h) 60 kg P_2O_5/ha (h) and potash (100 kg K_2O/ha). The experiment was laid out in split plot design where four irrigation levels (0.6, 0.8, 1.0 and 1.2 IW/CPE) in main plot and three nitrogen levels (90, 120 and 150 kg N/ha) in sub plots were replicated three times. Phosphorus was applied as basal while nitrogen and potash were applied in two equal splits i.e. 50% as basal and 50% at the time of first intercultural operation. Irrigation treatments were imposed after the first top dressing and measured amount of water (IW-5.0 cm) was applied through irrigation modules having a discharge capacity of 3 litres/second. Soil moisture study was carried out by gravimetric method and data on yield and growth characters were analysed statistically using standard method (Panse and Sukhatme, 1980). Crop matured in 117, 126 and 124 days during first, second and third year respectively.

Results and Discussion

Effect of irrigation

Bulb yeild of onion variety Nasik red increased with increasing IW:CPE ratios and was significantly the highest (121.1 q/ha) under most frequently irrigated lots of 1.2 IW/CPE in the first year of experimentation. Similar trend was also observed during second and third year with declined trend over the years. Yield reduction was the minimum (12.3%) under 1.0 IW/CPE and was the maximum (21.8%) under 1.2 IW/CPE. Data on pooled over 3 seasons revealed a significantly higher bulb yield (106.6 q/ha) under 1.2 IW/CPE ratio which reduced by 14.5, 16.4 and 23.3% under 1.0, 0.8 and 0.6 IW/CPE respectively. Similar higher bulb yield at 1.2 IW/CPE has also been reported from Memari, West Bengal (Annonymus, 1995). In the present study the increase in bulb yield due to increased levels of irrigation was associated with increase in single bulb weight and its diameter (Table 24.1).

Table 24.1 : Bulb Yield and Growth of Onion as Influenced by Levels of Irrigation and Nitrogen

Particulars	Bulb yield (q/ha)				Plant height (cm)	Stem girth (cm)	Fresh bulb weight (g)	Bulb diameter (cm)	% moisture loss 40 days after harvest	% rotting 40 days after harvest
	S_1	S_2	S_3	Pooled						
Irrigation levels (IW:CPE)										
0.6	92.4	72.9	80.2	81.8	35.2	2.6	20.7	10.9	17.7	2.5
0.9	98.1	84.8	84.4	89.1	35.7	2.8	22.1	11.0	19.0	7.4
1.0	99.0	87.7	86.8	91.2	36.8	3.1	23.7	11.5	19.1	7.5
1.2	121.1	104.0	94.7	106.6	36.9	3.1	25.0	11.8	19.3	10.3
C.D (0.05)	7.69	4.48	9.08	3.71	NS	0.20	0.72	NS		
N level (kg/ha)										
90	100.1	84.9	82.4	89.1	32.4	2.7	20.4	10.6	17.2	4.2
120	103.5	87.9	87.0	92.8	37.2	2.9	22.7	11.4	19.5	6.6
150	104.3	90.5	90.2	95.0	38.9	3.1	25.5	11.8	19.5	10.1
CD (0.05)	NS	5.32	3.40	NS	1.45	0.14	1.45	0.52		

Effect of nitrogen

No significant difference in yield of Bulb was observed during 1st year though yeild increased marginally with increase in nitrogen levels. However, during 2nd and 3rd year, significantly the highest yeild was observed at 150 kg N/ha with a bulb yield of 90.5 and 90.2 q/ha respectively. The superiority was however, only over the lowest level of Nitrogen (90 kg/ha). Pooled data also did not show any significant difference in bulb yield due to levels of nitrogen (Table 24.1). Response of onion to applied nitrogen was not consistent and the response to a wide range of N-levels has been reported by various workers. Its response to a lowest N level of 40 kg N/ha (Shulka *et al.*, 1992), to a moderate level of 90 kg N/ha (Patel and Patel, 1990) and to a higher level of 150 kg N/ha (Vacchori and Patel, 1990) are also under report. Economically, 120 kg N/ha gave more profit than the 150 kg N/ha (Table 24.3).

Table 24.2 : Bulb Yield and Water use Efficiency of Onion Under Level of Irrigation (Mean of three seasons)

IW:CPE	Bulb yield (q/ha)	Irrigation requirement (cm)	Water requirement (cm)	Consumtive use (cm)	Water use efficiency		Crop susceptibility factor (cf)
					Field (kg/ha-cm)	Crop (kg/ha-cm)	
0.6	81.8	30 (6)	37.8	21.5	216.40	380.47	0.23
0.8	89.1	35 (7)	41.9	24.4	212.65	365.16	0.16
1.0	91.2	40 (8)	42.0	26.1	217.14	349.43	0.14
1.2	106.6	45 (9)	46.7	30.4	228.27	350.66	–

Table 24.3 : Economics of Nitrogen Use in Onion

Nitrogen levels (kg/ha)	Bulb yield (q/ha)	Cost per kg. N applied Rs/ha	Total cost of Nitrogen Rs/ha	Cost of produce Rs/ha	Return from N Rs/ha	Profit over 90 kg N/ha Rs/ha	Profit over each N level Rs/ha
90	89.1	9.78	888.20	35640	34759.8	-	-
120	92.8	9.78	1173.60	37120	35946.4	1186.6	1186.6
150	95.0	9.78	1467.00	38000	36533.0	1773.2	586.6

Cost of urea = Rs 450/q;
Cost of onion = Rs 400/q.

Keeping quality of onion

Onion bulbs are subjected to losses due to desiccation, sprouting, rotting, respiration and shrinkage during storage which is due to presence of high moisture content in scales and physiologically active growing points (Stow, 1975). In the present investigation the percent moisture in onion stabilized after 40 days of storage and, the loss of moisture was comparatively less under stress regime of 0.6 IW:CPE and at a lower N levels of 90 kg/ha (17 to 17.2%). Percent rotting increased markedly with increase in irrigation and N levels. A minimum of 2.5% rotting, observed at 0.6 IW/CPE, increased by 3 and 4 times at moderate level of 0.8 and 1.0, and frequent level of 1.2 IW/CPE respectively. Similarly, lowest rotting percentage of 4.2 was observed at 90 kg N/ha which increased by 57 and 140% at 120 and 150 kg N/ha respectively (Table 24.1). Singh and Singh (2000) has also reported a minimum loss in onion at moderate dose of nitrogen.

Water requirement and water use efficiency

Pooled study over 3 seasons indicated that water requirement of onion varied from 37.8 cm (ET 21.5 cm) at 0.6 IW:CPE to 46.7 cm (ET 30.4 cm) at 1.2 IW/CPE ratio requiring 6 and 9 irrigation respectively. An increase in irrigation by 3 numbers and in WR by 10 cm resulted in an increase of 30.3% in bulb yield. However, the efficiency of water use by the crop was the maximum at lowest irrigation level (380.47 kg/ha-cm) which declined with increase in IW:CPE with lowest of about 350 kg/ha-cm at 1.0 and 1.2 IW/CPE ratio (Table 24.2).

Moisture extraction pattern

Crop utilized more water from 0-10 cm profile accounting to 34.4% of the total water used. Moistrue utilization increased from stress regime of 0.6 IW/CPE (17.2 cm) to 24.5 cm at 1.2 IW/CPE upto 50 cm profile. It decreased with increasing profile depth in all the irrigation treatments (Table 24.4). An average reduction in utilization by 83% from surface (0-10 cm) to lowest depth of 40-50 cm indicated more root configuration at the surface in onion.

Table 24.4 : Moisture Extraction Pattern (cm) in Onion Under Levels of Irrigation (Mean of three seasons)

Profile depth (cm)	Irrigation levels, IW/CPE				Mean
	0.6	0.8	1.0	1.2	
0-10	5.60 (32.6)	7.00 (36.5)	7.60 (35.5)	8.10 (33.1)	7.08 (34.4)
10-20	4.80 (27.9)	5.20 (27.1)	5.90 (27.6)	6.90 (28.2)	5.70 (27.7)
20-30	3.50 (20.3)	3.90 (20.3)	4.20 (19.6)	5.00 (20.4)	4.15 (20.2)
30-40	2.00 (11.6)	2.30 (12.0)	2.40 (11.2)	3.00 (12.3)	2.43 (11.8)
40-50	1.30 (7.6)	0.80 (4.1)	1.30 (6.1)	1.50 (6.0)	1.23 (5.9)
Total	17.20	19.20	21.40	24.50	20.60

ET/Epan ratio

Crop coefficient (ET/Epan ratio) increased with increasing levels of IW/CPE ratio (Table 24.5). Monthly crop coefficient under frequently irrigated plots (1.2 IW/CPE) increased from 0.43 in December (early stage) to 0.89-0.91 in February-March (mid season) and, which declined to 0.29 at maturity in April. Mid season crop coefficient values varying from 0.85-0.95 has also been reported by Doorenbos and Pruitt (1977).

Table 24.5 : Monthly ET/Epan Ratio in Onion Under Different Levels of Irrigation (Mean of three seasons)

IW:CPE	Months				
	December	January	February	March	April
0.6	0.43	0.51	0.61	0.50	0.31
0.8	0.43	0.51	0.71	0.66	0.31
1.0	0.43	0.51	0.77	0.71	0.24
1.2	0.43	0.51	0.91	0.89	0.29

References

Annonymous, 1995. Annual report of Directorate of Water Management Research Patna, 1994-95, pp. 30-31.

Doorenbos, J. and Pruitt, W.D. 1997. Guidelines for predicting crop water requirements. Food and Agriculture Organisation, Rome 24 p. 41.

Panse, V.G. and Sukhatme, P.V. 1985. Statistical method for Agricultural Workers. Indian Council of Agricultural Research, New Delhi.

Patel, J.J. and Patel, A.T. 1990. Effect of Nitrogen and phosphorus levels on growth and yield of onion (*Allium Cepa L.*) cultivar, Pusa Red. Research Journal. Gujarat Agricultural University. 15(2) : 1-5.

Shukla, V., Rao, K.P.G.K. and Prabhar, B.S. 1992. Effect of nitrogen on bulb yield and keeping quality of onion cultivars. Progressive Horticulture. 21(3-4) : 244-245.

Singh, Sanjay and Singh, T. 2000. Quality and post harvest response of rainy season onion to nitrogen, sulphur and zinc application. National Symposium-Onion and Garlic production and post harvest technology : challenges and strategies held at Nasik : 19-21 November p. 222.

Stow, J.R. 1975. Effect of humidity on losses of bulb stand at high temperature. Experimental Agriculture 11 : 85-87.

Vachhani, M.V. and Patel, Z.G. 1993. Effect of nitrogen, phosphorus and potash on bulb yield and quality of onion (*Allium cepa*). Indian Journal of Agronomy. 38(2) : 333-334.

ENVIRONMENTAL BIOTECHNOLOGY

Edited By :   **Professor (Dr.) Arvind Kumar**

Published By : **DAYA PUBLISHING HOUSE**

25

A Study of Biochemical Parameters, Chromosomal Aberration and Sister Chromatid Exchange in Tannery Workers and People Living Near The Tanner Units

• *Radha Munagala, Parvatham, R. and Anitha Subash*

Department of Biochemistry, Avinashilingam Deemed University,
Coimbatore - 641 043

Abstract

Tannery workers are exposed to hazardous chemical and physical environment in their work place. Study of various biochemical parameters reflected the extent of damage caused to the health status of the workers and the people living near the tannery units when compared to the control group. Chromosomal aberration and sister chromatid exchange analysis also revealed a slight rise in the frequency of occurance of the abnormalities in the test group when compared to the control group.

Introduction

India produces about 100 million pieces of hides and skins equivalent to 500,000 tons, which constitutes 12 per cent of the output of the whole world. The minimum amount of effluent produced is 3000 litres per 100 kg hide which is rich in polluting constitutents such as high pH, high sodium with chloride, high sulphate, high chromium and tannins, high BOD and COD. Nearly 90 per cent of leather in the world is produced through chrome tanning. Trivalent chromium when it enters the environment is doubted to be oxidized to the hexavalent form, which is toxic. In addition, about 400 tons of tannery solid wastes are being generated per day. The present traditional way of handling solid wastes, unhygienic procedures adopted in transportation and drying, constrains in disposing them off during monsoon etc., cause serious environmental problems and health hazards (Jogand, 1995).

The health status of the tannery workers was shown to be poor when compared to the control group (Munagala *et. al.*, 2002). Nagaya *et. al.*, 1989 had reported on increase in chromosomal aberration of peripheral blood leucocytes in workers handling chromium compounds. Lai *et. al.*, 1984 conducted a study to understand the mutagenic effects of Cr(VI) in humans and described SCE in lymphocytes as an useful method for evaluation of the biological effects of environment mutagen.

The present study was an attempt aimed to analyze the health status of tannery workers and the people living near the tannery units through biochemical analysis. Chromosomal aberrations and sister chromatid exchange studies were also carried out to determine the environmentally induced chromosomal damage in these subjects, if any.

Materials and Methods

Twelve subjects working in the tannery industry (Group I) and thirteen subjects living near the industry (Group II) were selected for the study from Erode. Ten healthy volunteers whose age and sex matched served as control group.

The venous blood samples of the test and control subjects were collected under aseptic precautions with and without anticoagulant. Serum was separated and used for biochemical analysis, namely serum chromium, tocopherol and enzymatic estimations like amylase, aspartate transaminase, alanine transaminase, lactate dehydrogenase and antioxidant enzymes peroxidase. For lipids, fasting blood samples were collected. Blood sample containing heparin was used for lymphocyte culture to study the chromosomal aberration and sister chromatid exchange (*in vitro*) was also done to determine the environmentally - induced chromosome damage in the selected subjects. The data obtained from various estimations were subjected to Analysis of Variance (ANOVA).

Results and Discussion

Biochemical Estimations

Serum Chromium

The serum samples of the selected subjects were analyzed using atomic absorption spectroscopy (AAS) through direct aspiration into air acetylene flame (sensitivity 0.1 ng/L) and found to contain undetectable levels of chromium indicating that chromium levels were less than 1 µg/ml. In a study conducted with tannery workers, Randall and Gibson, 1994 had reported a mean serum chromium level of 0.49 ng/ml with a range from 0.37-0.81 ng/ml and in control subjects as 0.15 ng/ml having a range from 0.12-0.20 ng/ml using electro-thermal AAS (sensitivity < 50 µg/L).

Lipid Profile

Table 25.1 represents the mean levels of total cholesterol, HDL-C, LDL-C, VLDL-C, and triglycerides of tannery workers, people living near the tannery units and control group.

The values of the above were within their respective normal ranges. The results of the lipid profile in the present study reflect that all the three groups did not suffer from any abnormality of lipid metabolism.

Enzyme Analysis

The mean serum levels of amylase of Group I, Group II and Control Group were 107.0, 90 and 84.5 U/dl respectively (Table 25.2). These levels were found to be within the normal range of 80 - 180 U/dl (Murray *et. al.*, 2000).

The mean serum levels of ALT and AST (IU/L) showed significant increase (P<0.01) in Group I and Group II when compared to the Control Group (Table 25.2). The higher serum aminotransferase levels in Group I and Group II reflect liver damage. The liver damage in these groups could have resulted from the exposure to various chemicals in the work environment and/or from alcohol consumption.

Table 25.1 : Mean Levels of Lipid Profile of the Selected Subjects

Parameter	Group (I)	Group (II)	Control (C)	Groups compared	CD 0.01	CD 0.05	Mean difference
TC (mg/dl)	163.1	152.0	151.8	I Vs II	42.3	32.2	11.1[NS]
				II Vs C	44.4	33.8	0.2[NS]
				I Vs C	45.3	34.4	11.3[NS]
HDL-C (mg/dl)	51.2	46.5	48.3	I Vs II	8.3	6.3	4.7[NS]
				II Vs C	8.8	6.7	1.8[NS]
				I Vs C	8.9	6.8	2.9[NS]
LDL-C (mg/dl)	93.3	88.2	85.4	I Vs II	44.8	34.1	5.1[NS]
				II Vs C	47.0	35.8	2.8[NS]
				I Vs C	48.0	36.5	7.9[NS]
VLDL-C (mg/dl)	18.6	17.2	17.0	I Vs II	5.7	4.4	1.4[NS]
				II Vs C	6.0	4.6	0.2[NS]
				I Vs C	6.1	4.7	1.6[NS]
TG (mg/dl)	93.2	86.3	85.5	I Vs II	29.0	22.0	6.9[NS]
				II Vs C	30.4	23.1	0.8[NS]
				I Vs C	31.0	23.6	7.7[NS]

NS = Not significant.

Table 25.2 : Mean Levels of Enzyme Activities and Tocopherol of the Selected Subjects

Parameter	Group (I)	Group (II)	Control (C)	Groups compared	CD 0.01	CD 0.05	Mean difference
Amylase (U/dl)	107.0	90	84.5	I Vs II	23.7	18.0	17[NS]
				II Vs C	24.9	18.9	0.5[NS]
				I Vs C	25.4	19.3	17.5[NS]
ALT (IU/L)	28.5	26.0	11.3	I Vs II	12.4	9.4	2.5[NS]
				II Vs C	13.0	9.4	14.7**
				I Vs C	13.2	10.1	17.2**
AST (IU/L)	23.8	19.3	10.7	I Vs II	9.6	7.3	4.5[NS]
				II Vs C	10.1	7.6	8.6**
				I Vs C	10.3	7.8	13.1**
LDH (U/L)	352.5	279.2	231.0	I Vs II	84.6	64.4	73.3*
				II Vs C	88.7	67.5	121.5**
				I Vs C	90.5	38.8	48.2*
Catalase[a] (U/ml)	4.0	4.8	5.5	I Vs II	0.83	0.63	0.8[NS]
				II Vs C	0.87	0.66	0.7*
				I Vs C	0.89	0.68	1.5**
SOD[b] (U/ml)	3.2	3.5	4.3	I Vs II	0.84	0.64	0.3[NS]
				II Vs C	0.87	0.67	0.8*
				I Vs C	0.90	0.68	1.1**
Glutathione Peroxidase (U/ml)	18.2	19.8	22.4	I Vs II	2.5	1.9	1.6[NS]
				II Vs C	2.7	2.0	2.6*
				I Vs C	2.7	2.1	4.2**
Tocopherol (mg/dl)	0.37	0.46	0.58	I Vs II	0.35	0.26	0.09[NS]
				II Vs C	0.36	0.28	0.21[NS]
				I Vs C	0.37	0.28	0.21[NS]

a: U = μ moles of H_2O_2 consumed / min/ml,

b: U = amount of anzyme required to causes 50% reduction of NBT,

** - (P<0.01), * - (P<0.05), NS - Not significant.

Serum LDH levels (U/L) in Group I, Group II and Control Group were found to be 353.5, 279.2 and 231.0 respectively (Table 25.2). Significant increase (P<0.05) in serum LDH levels in Group I were noticed when compared to Group II and Control Group. The increase in LDH levels is indicative of either hepatocelluar, renal or chronic liver damage.

A significant decrease (P<0.05) in the mean serum levels of the antioxidant enzymes, namely, catalase, SOD and glutathione peroxidase, were observed in Group I and Group II subjects when compared to the Control Group (Table 25.2).

The significant decrease in the activity of catalase, superoxide dismutase and glutathione peroxidase could be due to the exposure of hazardous chemicals in the work environment, which cause lipid peroxidation (Kikugawa *et. al.*, 1984). During reduction of chromium (VI) to chromium (III), highly reactive radical species and paramagnetic chromium such as chromium (V) are formed which are believed to induce cellular damage (Sugiyama *et. al.*, 1994). Habits like alcohol consumption, smoking and tobacco are also found to decrease serum peroxidase levels while increasing formation of free radicals (Reddy *et al.*, 1994).

The mean serum tocopherol levels in tannery workers, people living near the tannery units and Control Groups were 0.37, 0.46 and 0.58 mg/dl respectively. The levels of tocopherol were below the normal range of 0.5 - 1.8mg/dl (Bauer,1990), but the decrease was not statistically significant.

Chromosomal aberration and sister chromatid exchange

The mean levels of chromosomal aberration (% of metaphase with aberrations) of Group I, Group II and control group were 5.3, 5.2 and 5.0% respectively.

Table 25.3 : Mean Levels of CA and SCE of the Selected Subjects

Parameter	Group (I)	Group (II)	Control (C)	Groups compared	CD		Mean difference
					0.01	0.05	
CA	5.3	5.2	5.0	I Vs II	1.9	1.5	0.2[NS]
				II Vs C	2.0	1.6	0.4[NS]
				I Vs C	2.1	1.6	0.6[NS]
SCE	3.89	3.4	3.82	I Vs II	1.52	1.14	0.99[NS]
				II Vs C	1.59	1.14	0.99[NS]
				I Vs C	1.63	1.24	0.15[NS]

NS - Not Significant.

Analysis revealed a slight rise in the frequency of occurrence of the abnormalities in the Group I and Group II subjects when compared to the Control Groups. However, there were no significant differences among the three groups.

The work environment, additions like alcohol consumption, smoking and tobacco chewing could be the possible factors (Munagala *et. al.*, 2002) responsible for the rise in the appearance chromosomal exchange and sister chromatid exchange in Group I and Group II subjects.

The results were in accordance with the study conducted by Gonzaleze *et. al.*, 1994 with the tannery workers, where he concluded that the exposure to chemicals during leather tanning did not produce genotoxic effects measured by chromosomal aberrations in peripheral lymphocytes and micronuclei in urine of this group of workers.

The mean sister chromatid exchange (SCE) levels were 3.89, 3.4 and **3.82%** in Group I, Group II and Control Group respectively. The present study showed that the percentage of SCE levels was not significantly different among the three groups. In a study carried out with workers exposed to chromium by Wu *et al.*, 2000, higher frequency of SCEs were shown when compared to the Control Group. Smokers showed a significantly higher frequency of SCE than non-smokers in both groups. Duration of exposure was also an important factor (Mandal *et al.*, 1992).

Summary and Conclusion

It was found that the mean levels of lipid variables, namely, total cholesterol, HDL-C, LDL-C, VLDL-C, and triglycerides were within the normal range in all the selected groups suggesting no lipid metabolism abnormality.

Serum amylase levels were found to be within the normal range in all the three groups in the study. Aminotransferases, namely, alanine aminotransferase and aspartate aminotransferase showed a significant rise ($P<0.01$) in Group I and Group II when compared to the Control Group, indicating hepatocellular damage in these individuals. The rise was higher in Group I than in Group II subjects. This could be due to exposure to hazardous chemicals during their work and also due to alcohol consumption. Enzymatic antioxidants, namely, catalase, superoxide dismutase and glutathione peroxidase were found to be significantly low ($P<0.01$) in Group I and Group II when compared to the Control Group. The decrease in antioxidant levels signifies the higher rates of lipid peroxidation and oxidation stress in these groups. The mean levels of non-enzymatic antioxidants-tocopherol in Group I and Group II were found to be below the normal range.

References

Bauer, D. John. Clinical Laboratory Methods. 9th edition, B.I. Publication Ltd, New Delhi, India, 472-473, 1990.

Gonzaleze, H., Dora Loria, Marta, Vilensky, Jose Luis Miotti and Elena Motos. Leather tanning workers: Chromosomal Aberration in peripheral lymphocytes and micronuclei exfoliated cells in urine. *Mutat. Res.*, 1994, 259(2) , 197-201.

Jogand, S.N., Environmental Biotechnology. 15th edition, Himalaya Publishing House, Bombay, India,160-167, 1996.

Kikugawa, K., Kougi, H. and Asa Kura, J. Effects of MDA, a product of lipid peroxidation on the function and stability of haemoglobins, *Arch .Biochem. Biophys.*, 1984, 229, 7.

Lai, J., S., Kuo, H.W., Liao, F.C. and Lien, C.H. Sister chromatid exchange induced by chromium compounds in human, *Int. Arch Occup. Environ. Hlth.*, 1984, 71 (8), 550-3.

Mandal and Drabir Kumar. Sister chromatid exchange observed in individuals addicted to processed betal - nut, quid and zarda, *J. Environ. Biol.*, 1992, 13(3), 247-251.

Murray. K. Robert., Doryl, K. Gamner, Peter, A. Mayes and Victor, W. Rodwell. Harper's Biochemistry. 25th edition, Appleton and Lange, 869-870, 2000.

Nagaya, T., N. Ishikawa and H. Hate. Sister chromatid exchange analysis in Lymphocytes of workers exposed to hexavalent chromium, *Br., J. Ind. Med.*, 1989, 46(1), 48-51.

Radha Munagala., Parvatham, M.and Anitha Subash. Economic, Nutritional and Health Status of the Tannery Workers and the People Living near the Tannery units, Indian. j. Environ and Ecoplan, (2002), 6(3) : 545-550.

Randall, J.A. and Gibson, R.S. Serum and Urinary Chromium as indices of chromium status in tannery workers, *Proc. Soc. Exp. Biol. Med.*, 1994, 185(1), 16-23.

Reddy, K.K., Ramachandraiah, T., Reddanna, P. and Thyagaraju, K. Serum lipids and lipid peroxidation pattern in urban and rural Indian men, *Arch. Environ. Health.*, 1994, 49, 123-127.

Sugiyama, Masayasu. Role of paramagnetic chromium in chromium (VI) induced damages in mammalian cells, *Environ. Hlth. Perspet.*, 1994, 102(3), 31-33.

Wu, T.Y., Tsai, F.J., Kuo, H.W., Tsai, C.H., Wu, W.Y., Wang, R.Y. and Lai, J.S. Cytogenetic study of workers exposed chromium compounds, *Muta. Res.*, 2000, 464(2), 289-96.

ENVIRONMENTAL BIOTECHNOLOGY

Edited By : **Professor (Dr.) Arvind Kumar**

Published By : **DAYA PUBLISHING HOUSE**

26

VAM Fungi Association of Biomass Timber Yielding Trees of Alkali Affected Soil in Mainpuri Region (U.P.)

• *Versha Rani and Seema Bhadauria*

Microbiological Research Laboratory,
Department of Botany, R.B.S. College, Agra - 282 002 (U.P.)

Abstract

An assessment of vesicular-arbuscular mycorrhizal association in 15 timber-yielding trees collected from alkali soil was performed. All the trees showed the VAM infection, but with varying incidence. The VAM fungi associated with these tree species mainly belonged to the species *Glomus*, *Gigaspora* and *Sclerocystis* were recorded.

Key Words: *Endomycorrhizae, Alkaline soil, Timber yielding.*

Introduction

Mycorrhiza plays an important role in establishment and survival of plant communities in hostile soils of salt affected land (Smith and Read, 1997). *VAM* association is geographically ubiquitous and occurs over a broad ecological range (Gerdeman, 1968). In natural communities, approximately 70% of the higher plants are obligatorily dependent on fungal associates, 12% facultatively dependent and 18% typically non mycorrhizal (Mohan and Subashini 1995). A lot of interest has been developed in the study of *VAM* in the recent years because of their role in improving plant growth, controlling root pathogens, increasing in biological nitrogen fixation, and improving the ability of plants to withstand water stress, salinity, and heavy metal toxicity (Verma and Jamaluddin, 1994). Elucidation of a predominant *VAM* fungi, if any, associated with these timber yielding tree might be helpful in devising an indigenous reclamation strategy based on microbes (mycorrical); till recently reclamation strategy have been solely chemical (gypsum/Pyrites) based. Experiments have shown that the endotrophic association can greatly increase phosphorous uptake of plants in the phosphorus deficient soils, and can help in survival, growth performance and yield (Rachel *et al.*, 1993).

Hence a survey of the salt affected districts of Mainpuri (U.P) was carried out, and the degree of mycorrhization, occurrence, and spore density of *VAM* fungi is reported here.

	Table 26.1 : Physio-chemical Characterisitcs of Alkali Soil Mainpuri

Parameters	Value
pH	9.0 - 10.4
EC	20.1 - 10.4 mmhos/cm
Texture (Sand : Silt : Clay)	62 : 24 : 14%
Exhangeable Sodium	80 - 88%
SAR	830 - 950%
$CaCO_3$	3%
Total Porosity	35.9 - 40.6%
Non capillary porosity	2.4 - 5.9%
Intrinsic Permeabilty	0.01 - 0.08%

Methods and Materials

Soil and root samples were collected from 15 commonly occurring timber-yielding trees in Mainpuri district. Roots were properly washed in running tap water and cleared in a near -boiling 10% aqueous solution of KOH for 48 hr. Roots were stained in Trypan blue following several washings in distilled water to drain out the KOH (Phillips and Hayman, 1970). The stained roots were cut into 1-cm segments, 80-100 such segments were picked up and examined under the microscope. *VAM* colonization was determined by Nicoloson's formula (1995) as follow.

$$\text{Percent colonization} = \frac{\text{Number of root segments colonized}}{\text{Total No. of segments examined}} \times 100$$

The spore density of *VAM* fungi was quantities in 1 kg of soil following the wet sieving and decanting technique of Geredemann and Nicolson (1963) and the frequency of occurrence was calculated by Norton's formula (1978) as follow. The identification of *VAM* fungi was based on taxonomic keys prepared by Hall (1984), Schenck and Perez (1990). The physio-chemical characteristics of soil in Mainpuri district is shown in Table 26.1.

$$\text{Absolute frequency} = \frac{\text{Number of samples containing a species}}{\text{Total number of segments examined}} \times 100$$

$$\text{Relative frequency} = \frac{\text{Frequency of a specis}}{\text{Sum of frequencies of all the species}} \times 100$$

Results

Fifteen species (as listed in Table 26.2) covering twelve genera of timber yielding trees from different places of the experimental site were surveyed periodically for the association of *VAM* fungi.

Table 26.2, it is clear that *VAM* fungi affected all the timber-yielding trees under investigation. However, degree of infection varied with the plant species and place of collection. Highly colonized roots were seen in plants of *Syzygium cumini* (100%), *Dalbergia sissoo* (100%), *Tectona grandis* (80%) and *Casuarina equsilifolia* (90%). *Acacia catechu* (60%) *Diterocarpus turbinatus* (35.66 %), *Hardwickia pinnata* (51%), *Terminalia arjuna* (47) and *Zizyphus jujuba* (59) were next in the degree of infection. On the other hand, *Azadirachta indica* (28%) followed by *Morus alba* (25%), *Tamarindus indica* (24.3%), *Albizzia Procera* (16%) and *Aegle marmelous* (12%) were with low percentage of infection. Thaper *et al.* (1992) have also recorded varying degree of infection among different plants within a family.

Table 26.2 : Occurrence of VAM Fugnal Spores in Rhizosphoric Soils of Alkali Tolerance

Sl. No.	Plant Species	% Infection	Vesicles	Arbuscies	No. of spores per 100 gm of soil	VAM fungi in rhizospheric soils
1.	*Acacia catechu*	60	+	−	114	*Glomus mossae, Gl. fasciclulatum*
2.	*Acacia nilotica*	100	+	+	170	*Gl. macrocarpum, Gl. mossae*
3.	*Albizza procera*	16	+	−	60	*Gl. fasciculatum, Gl. geosporum*
4.	*Azadirachta indica*	28	−	−	87	*Gigaspora calospora, Gl. mossae*
5.	*Aegle marmelous*	12	+	+	92	*Gl. caladonius, Gl. macrocarpum*
6.	*Casuarina equsilifolia*	90	+	−	59	*Gl. albidium, sclerocystis coremoiJes*
7.	*Dalbergia sissoo*	100	+	+	55	*Gl. monosporum, Gl. mossae*
8.	*Dietrocarpus tubinatus*	36.66	−	−	40	*Sclerocystis sinusa*
9.	*Hardwickia pinnata*	51	+	−	104	*Gl. multicaulis*
10.	*Morus alba*	25	−	−	88	*Gl. mossae, Gl. macrocarpum*
11.	*Tectona grandis*	80	+	+	125	*Sclerocystis coremoides, Gl. geosporum*
12.	*Terminalia arjuna*	47	+	+	116	*Gl. reticulates*
13.	*Tamarindus indica*	24.3	+	−	45	*Gl. mosseae, Gl. fasciculatum*
14.	*Syzgium cumini*	100	+	−	201	*Gl. margarita*
15.	*Zizyphus jujuba*	59	−	−	124	*Gl. mossae, Gl. macrocarpum*

Acacia nilotica, Aegle marmelous, Dalbergia sissoo, Tectona grandis, Terminalia arjuna supported the formation of both vesicles and Arbusculus. While *Acacia catchu, Albizzia procera, Casuarina equsilifolia, Hardwickia pinnata, Tamarindus indica* and *syzygium cumini* supported only vesicles formation. On the other hand, roots of *Azadirchta indica, Diterocarpus turbinatus, Morus alba* and *Zizyphus jujuba* could show presence of only infection of mycorrhizal threads. The number of propagule of *VAM* mycorrhizal fungi varied with the tree species. High number of *VAM* propagules in rhizosphere soil of *Syzygium cumini, Acacia nilotica, Tectona grandis, Zizyphus jujuba, Terminalia arjuna, Hardwickia pinnata* and *Acacia Catechu* contained significantly high number of *VAM* proagules, while *Dipterocarpus turbinatus* and *Tamarindus indica* supported the least number of *VAM* propagules. Rest of the timber tree species supported intermediate number of fungal propagules.

In all the 15 species of endogonaceous fungi were isolated out of which ten belonged to Glomus and two to sclerocystis (Table 26.2). *Glomus mossae* was most abundant, next followed by *Gl. fasciculatum, Gl. macrocarpum, Sclerocystis coremoides,* where as *Gl. margarita, Gl. multicaulis, Sclerocystis sinusa, Gigaspora calospora, Gl. Geosporum* occurred rarely. *VAM* fungal propagules also differed qualitatively. For instance, the rhizosphere of *Ditercarpus turbinatus, Hardwickia pinnata, Terminalia arjuna* and *Syzygium cumini* recorded the dominance of one species each. Such host specific *VAM* fungi has been reported earlier by Bloss and Walker (1987). Rest of the plant species under study supported the growth of more than one *VAM* fungal species. Variation in the percentage *VAM* fungi colonization indicated that colonization might be regulated at the species level. Variation of the spore population in rhizosphere soil indicated that the multiplication of spores depends on species to species level.

Discussion

It is well known that mycorrhizal infection is heaviest on infertile soil. Low levels of nitrogen or phosphorus can themselves lead to an increased intensity of infection (Mosse, 1973). Most plant species are typically mycorrhizal with approximately four fifths of all land plants forming *VAM* (Malloch *et al.*, 1980). Altogether 15 plants species were positive in response towards *VAM* fungi infection. Highly responsive species were *Cassia fistula* (94%), *Cassia alata* (69%), *Leucaena leucophala* (100%) and *Convoluvlus arvensis* (100%). High level of colonization indicates its deep susceptibility towards *VAM* fungi as well as the ineffectiveness of *VAM* fungi. All the plants observed in the young stage, effect of *VAM* fungi cannot be spell out in the form of growth responses of the host, but the efficiency of *VAM* fungi through a high rate of colonization was indicated. Therefore, the study brings out that effective VAM fungi is not enough to generalise the endophyte species composition an alkali soils, because the endophyte species composition known to change with time and crop (Schenck and Kinloch, 1980). Thus, extensive sampling over a long period is required to determine the species diversity (Walker *et al.*, 1982). It is apparent from the present results that in alkali soil it is possible that *VAM* infection increases the capacity of plants to accumulate increase the capacity of plants to accumulate phosphate as it is released by other micro-organisms.

According to Mosse (1973) *Gigaspora* and *Acaulospora* are more tolerant to acidity whereas *Glomus* sp. favour neutral and alkaline soils. This difference is attributed to edaphic factors and host plant interactions at the particular site. Variation in spore frequency and incidence of external hyphae may be due to high crop density, pure composition of specific crop, age and host species compatability with *VAM* fungi. Comparatively few arbuscules or may be due to poor soil quality and age of plants and reportedly arbuscular formation is most prominent to maximum growth during early stage whereas vesicles may be abundant in older tissues (Saif *et al.*, 1975).

It is apparent from the present results that some *VAM* fungi isolates tolerate high alkalinity or pH, low nutrient and irrigation regimes, hence they might be establishment under survey conditions and thus be available to aid plant growth or survival when tailored nursery stock is out painted in distributed or deficient sites of alkaline soils.

References

Bloss, H.E. and Walker, C. (1987). Some endogonaceous fungi of Santa Caltalina mountains in Arizona *Mycologia*, 79(a) : 649-654.

Gerdeman, J.W. (1968). Vesicular arbuscular Mycorrhiza and plant growth. Annual review of Phytopathology. 60 : 397-418.

Gerdemann, J.W. and Nicolsan, T.H. (1963). Spores of mycorrhizal *Endogone* species extracted from soil by wet sieving and decanting. Transactions of the British Mycological society. 46 : 235-244.

Hall, I.R. (1984). Taxanomy of *VAM* mycorrhizal fungi. In *VAM* Mycorrhiza, pp. 58-94, edited by CLL Powell and DL Bagyaraj Florida; CRC Press. p. 234.

Malloch, D.W., Pirozynski, K.A. and Raven, P.H. (1980). Ecological and evolutionary significance of mycorrhizal symbiosis in Vesicular plants. Proc. Nat. Acad. Sci. USA 75 : 2113-2118.

Mohan, M.S.S. and Subashini , H.D. 1995. : Mycorrhizas as bioindicators of ecosystem health. In Bioindicators Identification and use in Assessing Environmental Disturbances, pp. 11-20, edited by MSS Mohan, S. Nair , R.J. Ranjit Daniels, Madras: Reliance Printers.

Mosse, B. (1974). Ann Rev. Phytopathology 11: 71.

Nicolson, T.H. 1995. The mycotrophic habit in grass (Thesis, pp. 227-286) Nottingham, U.K: University of Nottingham.

Phillips, J.M. and Hayman, D.S. (1970). Improved procedures for clearing roots and staining Parasilic and Vesicular - arbuscular mycorrhizal for rapid assessment of infection. *Trans. Brit. Mycol soc.* 55 : 158-161.

Rachel, E.K. Reddy, S.R. and Reddy, S.M. (1993). Effect of different mycorrhizal treatments on the growth and yield of sunflower. *Indian J. of Microbial Ecol.,* 3(2) : 99-103.

Saif, S.R. and Khan, A.G. (1975). Can. J. Microbial. 21 : 10120.

Schenk, N.C. and Kinloch, R.A. (1980). Incidence of mycorrhizal fungi on six field crops in monoculture on a newly cleared wood land site Mycologia, 72 : 445-456.

Schenk, N.C. and Perez, Y. (1990). Manual for Identification of *VAM* mycorrhizal fungi. Gainesville, USA : University of Florida.

Smith, S.E. and Read, D.J. (1987). Mycorrhizal symbiosis. London : Academic press p. 605.

Verma, R.K. and Jamaluddin. (1994). Effect of *VAM* fungi on growth and survival of *Acacia nilotica* seedlings under different moisture regime, Proceedings of the National Academy of Sciences 64(B) : 205-210.

Walker, C., Mize, C.W. and Mcnabb, H.S. Jr. (1982). Population of endogonaceous fungi at two locations in central Iowa Can. Bot. 60 : 2518-2529.

ENVIRONMENTAL BIOTECHNOLOGY

Edited By :   **Professor (Dr.) Arvind Kumar**

Published By :   **DAYA PUBLISHING HOUSE**

27

PATHOGENIC VARIATION WITHIN *PHYTOPHTHORA CACTORUM* ISOLATES FROM APPLE

• *Bhupesh Gupta, L.N. Bhardwaj and Anil Handa*

Department of Mycology and Plant Pathology,
Dr. Y.S. Parmar University of Horticulture and Forestry.
Nauni, Solan - 173 230 (H.P.)

Abstract

Five isolates of collar rot pathogen collected from different apple growing areas of Himachal Pradesh showed considerable variation in their pathogenic behaviour to different commericial cultivars and susceptible rootstock *Malus prunifolia* Shaishie. Excised twig method was used to study the virulence of the isolates. P_1 isolate of *Phytophthora cactorum* was found to be highly virulent.

Key Words : Pathogenic variation, Apple, Collar rot.

Introduction

Phytophthora cactorum (Leb and Cohn.) Schroet., the causal agent of collar rot of apple causes a considerable loss to the crop. The extensive cultivation of apple with varied genetic background and non judicious use of fungicides are expected to influence virulence(s) of pathogen population through directional selection pressure. Understanding of pathogen population is necessary for successful execution of a disease management programme. Therefore, the present study was undertaken to determine the variation in five isolates of pathogen representing major apple growing localities of Himachal Pradesh.

Materials and Methods

To study the virulence in different isolates of *Phytophthora cactorum*, artificial inoculations were carried out on susceptible rootstock (*Malus prunifolia* Shaishie) by excised twig method (Borecki and Millikan, 1969). The excised twigs after inoculation with test isolates were incubated at 25±1°C in BOD incubator. The observations on lesion length and lesion colour were recorded 3, 6 and 9 days after inoculation. For testing the virulence of different isolates of *P. cactorum* on different apple cultivars excised twig method was followed. Twenty cultivars of apple grown in the orchard of Dr. Y.S. Parmar University of Horticulture and Forestry, Solan were used in these studies. One year old twigs of each cultivar was taken in the month of July and inoculations were done with each of the five isolates of *P. cactorum*. The length of lesion on each cultivar was measured after 9 days of pathogen inoculation.

Results and Discussion

It is clear from the data presented in Table 27.1 that the isolates of *P. cactorum* were able to infect the susceptible rootstock, *M. prunifolia* Shaishie within 3 to 9 days after pathogen inoculation. However, isolate P_1 was found highly virulent as it produced the largest lesion of 58.33 mm, 9 days after pathogen inoculation, P_2 was next virulent isolate producing 42.67 mm average lesion length on susceptible rootstock. Isolate P_5 was observed to be the least virulent isolate producing 28.00 mm length of lesion 9 days after pathogen inoculation. In addition to this, virulent isolate produced dark brown lesions whereas, less virulent produced light brown lesions on susceptible rootstock. Sewell and Wilson (1959) and Aldwinckle *et al.* (1975) also reported existence of variability in *P. cactorum* isolates of different areas which gave different reaction on the same rootstock. These differences in susceptibility have been attributed to the existence of pathotypes in the fungus (McIntosh, 1975). In 1987, Grag also reported difference in virulence in four isolates of *P. cactorum* from apple.

Table 27.1 : Virulence of *Phytophthora cactorum* isolates to apple rootstock *Malus prunifolia* Shaishie by excised twig method

Isolate	Average lession size (mm)			Lesion colour
	3 days	6 days	9 days	
P_1	13.67	28.33	58.33	Dark brown
P_2	10.33	21.67	42.67	Dark brown
P_3	4.33	12.00	30.67	Light brown
P_4	6.33	15.33	36.33	Slightly dark brown
P_5	5.67	10.67	28.00	Light brown

It is evident from the Table 27.2 that all the isolates of *P. cactorum* were able to infect the apple cultivars, however, the degree of infectivity varied with the isolate. Apple cv. Well Spur was found highly susceptible to all isolates of *P. cactorum* showing a mean lesion length of 27.93 mm. Red Free, Starking Delicious, Red Delicious, Mollie's Delicious, Red Gold and Gold Spur were the next in order of their susceptibility. The lesion length in all these cultivars ranged between 13.67-28.67 mm. Nevertheless cv. Vance Delicious was found least susceptible against all the fungal isolates showing a mean lesion length of 5.33 mm. Isolate P_5 was found most virulent on Vance Delicious and P_3 produced minimum lesion length (4.33 mm). Royal Delicious (6.80 mm), Red Spur Delicious (8.20 mm) and Golden Delicious (9.13 mm) showed resistant type of reaction. Isolate P_1 was most virulent on Royal Delicious followed by P_4 and P_3, though both are statistically at par with P_1. Similar results were observed with Red Spur Delicious. In case of Golden Delicious, P_4 was found to be the most virulent followed by P_1 an P_3, while P_2 and P_5 were least virulent on this cultivar. Of the five *P. cactorum* isolates, isolate P_1 was found highly virulent irrespective of cultivars with mean lesion length of 20.02 mm when assessed days after pathogen inoculation, isolate P_5 was least virulent with mean lesion length of 12.67 mm. Various workers have reported differential reactions of apple cultivars to *P. cactorum* isolates in different parts of the world (Stevens, 1938; Welsh, 1942; Braun and Krober, 1958; Aldwinckle *et al.*, 1975; Gupta *et. al.*, 1985; Harris and Tobutt, 1986; Garg, 1987; Boneti and Katsurayama, 1993). Schmidle (1957) found Golden Delicious to be resistant to *P. cactorum* but his findings were contrary to the findings of Rana (1981) who observed Red Delicious, Red Gold and Grany Smith to be resistant to *P. cactorum*. Similar results were also obtained by Garg (1987). These variations in the results could be due to strainal variations in *P. cactorum* isolates.

Table 27.2 : Response of Apple Cultivars to *Phytophthora cactorum* Isolates

S. No.	Cultivars	Lession length (MM)					Overall Mean
		P_1	P_2	P_3	P_4	P_5	
1.	Royal Delicious	8.33	5.67	7.00	7.67	5.33	6.80
2.	Red Delicious	24.67	24.33	21.33	17.67	15.33	20.67
3.	Rich-a-Red	24.33	16.00	18.33	19.67	13.00	18.27
4.	Vance Delicious	5.33	5.00	4.33	5.33	6.67	5.33
5.	Top Red	14.00	17.33	12.67	9.00	11.67	12.93
6.	Red Chief	19.00	14.67	14.00	12.33	9.33	13.87
7.	Red Spur Delicious	9.67	7.67	8.00	8.33	7.33	8.20
8.	Golden Delicious	9.33	8.33	9.00	10.67	8.33	9.13
9.	Granny Smith	26.67	18.33	21.00	10.33	8.67	17.00
10.	Starking Delicious	26.33	18.33	23.67	16.00	23.33	20.73
11.	Mollie's Delicious	23.67	26.33	14.00	19.33	15.33	19.73
12.	Gold Spur	25.67	20.33	19.00	16.33	13.67	19.00
13.	Red Free	28.67	26.67	16.67	18.67	14.67	21.07
14.	Red Gold	17.67	18.33	15.67	22.67	21.33	19.13
15.	Tydeman's Early Worcester	20.67	13.67	18.00	17.33	12.33	16.40
16.	Oregon spur	19.33	17.33	15.33	13.67	10.67	15.27
17.	Well Spur	33.00	29.33	30.00	26.67	20.67	27.93
18.	Silver Spur	19.00	17.33	13.00	14.67	13.00	15.40
19.	Scarlet Gala	25.67	18.33	16.67	19.67	15.33	19.13
20.	Red Fuji	19.33	16.33	13.67	13.33	7.33	14.00
	Overall Mean	20.02	16.98	15.37	14.97	12.67	

$CD_{0.05}$

Cultivar = 0.71

Isolate = 0.36

Cultivar × Isolate = 1.59.

References

Aldwinckle, H.S., Polach, F.J., Molin, W.T. and Pearson, R.C. 1975. Pathogenicity of *Phytophthora cactorum* isolates from New York apple trees and other sources. Phytopathology 65 : 989-998.

Boneti, J.I., Da.S. and Katsurayama, Y. 1993. [*Phytophthora* species associated with apple collar rot in Santa Catarina State.] Fitopatologia Brasileria 18 : 206-212.

Borecki, Z. and Millikan, D.F. 1969. A rapid method for determining pathogenicity of *Phytophthora cactorum*. Phytopathology 59 : 247-248.

Braun, H. and Krober, H. 1958. Studies of collar rot of apple caused by *Phytophthora cactorum* (Leb. and Cohn.) Schroet. Phytopathol. Z. 35 : 35-94.

Garg, R.C. 1987. Variability and biocontrol of *Phytophthora cactorum* on Apple. Ph. D. Thesis, UHF, Nauni, Solan. p. 110.

Gupta, V.K., Rana, K.S. and Mir, N.M. 1985. Variability in *Phytophthora cactorum* in India. In: Ecology and Management of soil borne plant pathogens ed. C.A. Parker, A.D. Rovira, K.J. Moore, P.T.W. Wong and J.F. Koilmongen. American Phytopathological Society, USA. pp. 169-171.

Harris, D.C. and Tobutt, K.R. 1986. Factors affecting the mortality of apple seedlings inoculated with zoospores of *Phytophthora cactorum*. J. Hortic. Sci. 61 : 457-464.

McIntosh, D.L. 1975. Proceedings of the 1974 APDW Workshop on crown rot caused by *Phytophthora cactorum* in planting of apple trees aged 1 to 7 years. Plant Dis. Reptr. 59 : 539-541.

Rana, K.S. 1981. Population Dynamics of *Phytophthora cactorum* (Leb. and Cohn.) Schroet. In relation to soil factors and fungicides. Ph. D. thesis, Himachal Pradesh Krishi Vishva Vidyalaya, College of Agriculture, Solan, p. 137.

Schmidle, A. 1957. On inoculation experiments of apple trees with *Phytophthora cactorum* (Leb. and Cohn.) Schroet. the agent of collar rot. Phytopathol. Z. 28 : 329-342.

Sewell, G.W.F. and Wilson, J.F. 1959. Resistance trials of some apple rootstocks to *Phytophthora cactorum* (Leb. and Cohn.) Schroet. J. Hortic. Sci. 34 : 51-58.

Stevens, N.E. 1938. Departure from ordinary methods in controlling plant diseases. Suppl. Note Bot. Rev. 12: 677-678.

Welsh, M.F. 1942. Studies on crown rot of apple trees. Can. J. Res. 20 : 457-490.

ENVIRONMENTAL BIOTECHNOLOGY

Edited By :   **Professor (Dr.) Arvind Kumar**

Published By :   **DAYA PUBLISHING HOUSE**

28

COMPARATIVE STUDY OF SOILS WITH REFERENCE TO THE EFFECT OF AGRO-CHEMICALS, USING PHYSICO-CHEMICAL AND MICROBIAL PARAMETERS – A CASE REPORT

• *Mrs. Jaya Vikas Kurhekar*

Department of Microbiology, Dr. Patangrao Kadam College,
Sangli–416 416, Maharashtra, India

Abstract

The present study was aimed at studying various physico-chemical and microbial parameters of acidic soil and sugarcane field soil as comapred to a control soil sample in Sangliwadi area. In all the three seasons, electrical conductivity of acidic soil sample was found to be more, as compared to the sugarcane field and control soils, indicating increased salt concentration and water-logging. Nitrogen, organic Carbon and Manganese contents were found to be maximum in sugarcane field while Zn content was found to be low. In acidic soils, Ca, Mg and Fe contents were maximum and organic matter contents were found to be minimum. All these parameters were seen to influence micro flora and enzyme activity of sugarcane field soils and acidic soils, as compared to the control samples.

Key Words : Physico-chemical, Microbial, Acidic soil, Electrical conductivity, Organic matter, Micro-flora, Enzyme activity.

Introduction

Uncontrolled and excessive use of chemical fertilizers and pesticides for increasing yield in agriculture has become a major problem in India. (Russell, E.W., 1973). Soil is a dynamic system because of the presence of micro-organisms and their biochemcial activities in soil. Their activities liberate a lot of enzymes in soil, which may be coming from roots of plants or may be excretion products of micro-organisms or their secretions, their degradation products or their synthesized products. These enzymes become stabilized in soil by binding to soil components (Alexander Martin, 1985).

India is basically an agricultural country. 80% of its population depends on agriculture as its primary occupation. The need for a good economy is based on good crop yield, becasue of which the soil is being uncontrollably exposed to a variety of agro-chemicals in order to increase crop yield, decrease crop infections and infestations, increase soil fertility etc. (Deshpande M.S., 1996). This has led to decreased fertility and acidic soil formation in various region of Maharashtra. About 0.534 million hectare of land is affected by saline, acidic and alkaline conditions in Maharashtra alone, while India, as a whole, shows 7 million hectare of land under salinity. In Maharashtra, an area of about 33,200 hectare has been affected by acidity/alkalinity/both. In addition, 34,000 hectare land has been converted to

"Khar" land because of ingress of sea-water (Russell, E.W., 1973). Salinity and arises because of key factors like; nature of carbonates, sulfates and chlorides of Na, K, Ca, Mg and their solubility behaviour. A number of geological, climate-related and hydrological factors combine to give the formation of acidic/alkaline/saline soils. Restricted drainage of water in high ground water table or low permeability is one of the major factors contributing towards salinity in Western Maharashtra (Vijaya, T., B.V. Prasada Reddy, 1998). Bringing such soil under cultivation, is one of the major issues of concern. Similarly, for increasing the yield of a cash-crop like sugarcane, agro-chemicals are being used heavily. This may also have certain effects on future condition of soil.

The present study was carried out at a sugarcane field and a four-acre field in Sangliwadi, which was rendered infertile, according to the owner, for the last six years. Before that, the same field showed maximum yield. An area close to this field, which was totally untreated and to which no agro-chemicals had been added was taken as a control soil sample.

Material and Methods

Selection of Samples

For this study, the samples selected were a sugarcane field and a richly irrigated area, a four acre acidic field, in Sangliwadi, Sangli. At the same time, soil samples which were totally free of any agro-chemcials, were selected as a control sample from a close-by area.

Method of Collection

The samples were collected as per the Instruction Leaflet, published by the Agriculture Information Department, Director of Agriculture, Pune. The site was roughly categorized into four parts. Sites where cattle sit or debris is dumped or is close to human habitat or is close to water canal or is water-logged, is avoided. The area from where the sample is collected is marked. A 20 cm. deep "V" shaped trench is dug from which the soil is discarded. From the walls of the trench, soil is scraped and collected in a clean vessel and spread on a jute cloth. Soil samples collected in this manner from all the sites is mixed thoroughly and divided equally in four parts. If moist, it is dried in the shade and strained. Fine soil is then marked appropriately, dated and labelled with the name of the site, village, district, name of the owner, water drainage of the field, nature of the field, last crop taken and next crop to be taken (Instruction leaflet, Pune, 1997).

Season of Collection

Climatic conditions are known to affect soil (Russell, E.W., 1973). Hence, the samples were collected in three different seasons. The months in which the sugarcane field and acidic soil samples were collected, were as follows:

(1) February 2001, 2002 (2) June 2001, 2002 (3) October 2001, 2002. At the same time, control soil sample were also collected in these months.

Selection of Physico-chemical and Microbial Parameters

Different physico-chemical and microbial parameters were decided upon, in order to get a reasonably correct picture of the soil sample, as given in the Table 28.1 & Table 28.2.

Temperature of the soil sample was checked with the help of a Thermometer, inserted in the soil, at the site of sample collection. pH was measured with the help of a portable pH-meter while electrical conductivity was measured with an electrical conductivity meter. Nitrogen content was estimated by Kjehldahl method while Phosphorous content with Olsen-Blue method and Potassium content with a Plame photo-meter. Organic Carbon content was estimated by Walkley and Black method. Calcium and Magnesium contents were measured with the help of EDTA method. Iron, Zinc, Copper and Manganese contents were checked by Atomic Absorption Spectrophotometer. Organo-Phosphorous and Organo-chlorine contents were measured with the help of AOAC-multi-residue method. Enzymes Nitrate reductase, Dehydrogenase and Phosphatase were analysed by High-Pressure Liquid Chromatography (Bear Firman, E., 1964). Bacterial count was taken on Nutrient agar, Fungal count on Sabouraud's agar while Actinomycetal count on Actinomycetes medium, using Standard Plate Count method.

Results and Discussion

The three soil samples, collected in three different seasons, were subjected to analysis (Bear Firman, 1964 and Burns, R.G., 1978) and the data obtained is tabulated as in the Table 28.1 and Table 28.2. The soil samples were analysed and the sample soils compared with the respective control soil samples. It was observed from the tables that; as compared to control and sugarcane field soil samples, pH of the acidic soil is remarkably low in all the seasons. EC of this soil is markedly elevated indicating presence of salts and water-logging. N, P, K, Cu and organic matter contents were found to be quite low as compared to control soils, while Ca, Mg, Fe, Zn and Mn contents were markedly elevated in all three months. Enzyme Nitrate Reductase and Dehydrogenase levels are observed to be minimum in acidic sample soils indicating dry or stored soils, low temperatures and aerobic conditions. Control soils show optimum analytical results; pH around 7.66, maximum P, K, enzyme nitratase & dehydrogenase levels. Actinomycetal count and minimum EC, Mn, Enzyme Phosphatase and fungal count and minimum in acidic sample soils. Organo-Phosphrous and Organo-chlorine contents indicating presence or absence of agro-chemcials is below detection levels in control soils, in all the seasons (Gupta, V.K., 1993). High nitratase activity is shown in 40-60% water-holding capacity of soil, at pH 3.6-8.2, in increased nitrate content in the atmosphere, at temperatures between 5-40 degree C, large soil particles and under poor aeration. Maximum dehydrogenase activity is shown at 37 degree C under anaerobic conditions (Burns, R.G., 1978).

Sugarcane field soils show maximum N contents, probably coming from nitrogenous fertilizers, high organic carbon content, Cu, Mn, organo-phosphorus, organo-chlorine and enzyme Phosphatase levels as compared to others. They show low levels of Ca, Mg, Fe and Zn. Enzyme phosphatase level is found to be higher (D. Leo, P. and Sacher, J.A., 1970) in sugarcane field soils. It indicates neutral pH, higher substrate levels, temperature between 37-60 degree C, flooded conditions and low moisture content.

Enzyme activity in any environment is mainly dependent on its micro flora (Alexander, M., 1971). In considering the interpretations of enzymatic activity of soil, it is necessary to remember that much or all of the activity measured is that of extra-cellular enzymes and that many of the enzymes may not exist apart from the cell because they are intracellular. Only those enzymes which are excreted by the members of the community, and those that may be released to the surroundings when the cells are lysed are found free in the soil. A number of environmental factors affect the rate of reactions brought about by these enzymes like pH, temperature, soil type, season of the year, kind of vegetation, size of microbial community etc. (Burns, R.G., A. H. Pukite and A.D. McLaren, 1972). Bacterial count is found to be reduced in acidic soils as compared to the control and sugarcane field soils, while fungal count is maximum in acidic soils owing to low pH and minimum in the control soils. Actinomycetal count is found to be highest in the month of October in control soils and minimum in June.

Table : 28.1

Sl. No.	Parameter	February 2001			June 2001			October 2001		
		Control soil	Acid soil	Sugar-cane soil	Control soil	Acid soil	Sugar-cane soil	Control soil	Acid soil	Sugar-cane soil
1.	Temp.(deg.C)	30	30	30	20	20	20	22	22	22
2.	pH	7.66	6.93	7.41	7.23	6.99	7.00	6.98	6.48	7.03
3.	EC(mmhos/c)	0.24	3.57	0.66	0.215	3.5	0.990	1.285	2.98	1.628
4.	N(kg/hec)	9.66	4.2	25.01	9.5	4.08	27.18	88.72	4.82	110
5.	P(kg/hec)	25.85	0.5	7.09	26.00	0.43	8.00	48.66	0.51	40.76
6.	K(kg/hec)	313.6	2.1	148	308.41	2.00	150	372.78	2.03	370
7.	Organic C(%)	0.27	0.18	0.37	0.31	0.10	0.40	0.90	0.39	1.03
8.	Calcium(%)	1.02	2.99	0.52	1.50	2.90	0.55	1.42	1.89	1.33
9.	Magnesium(%)	0.33	0.92	0.13	0.35	0.85	0.16	0.75	0.77	0.77
10.	Iron(ppm)	3.48	8.16	3.68	3.40	8.01	3.73	4.8	8.88	2.98
11.	Zinc(ppm)	3.12	3.34	0.42	3.15	3.20	0.44	5.16	6.82	3.16
12.	Copper (ppm)	0.39	0.20	1.13	0.40	0.15	1.16	0.98	0.38	2.22
13.	Mn (ppm)	0.26	0.39	2.12	0.25	0.31	2.16	2.16	0.44	0.32
14.	Organo P(ppb)	BDL	0.00	0.04	BDL	0.02	0.16	0.001	0.32	0.002
15.	Organo Cl(ppb)	BDL	0.016	0.082	BDL	0.01	0.22	0.02	0.124	0.013
16.	Enz.NO$_3$ red.(μmol/gm)	0.46	0.21	0.30	0.42	0.12	0.31	0.321	0.06	0.16
17.	Enz.DeH$_2$ase (umol/gm)	0.22	0.18	0.16	0.18	0.16	0.22	0.112	0.08	0.09
18.	Enz.PO$_4$ase (unol/gm)	0.36	0.52	0.92	0.36	0.48	0.83	0.106	0.21	0.77
19.	Bacterial count (colonies/ml)	2.2×10^6	9×10^5	2.5×10^6	4×10^6	7×10^5	5×10^6	2×10^8	1.68×10^8	1×10^8
20.	Fungal count (colonies/ml)	3×10^6	5×10^7	4×10^5	7×10^7	9×10^8	3×10^7	2×10^5	8×10^6	4×10^5
21.	Actinomycetes (col./ml.)	1×10^7	3×10^7	3.8×10^7	5×10^5	3.5×10^6	3×10^6	1.7×10^6	1.2×10^6	5×10^5

Conclusion

In conclusion, acidic soils show a definite tendency towards high EC, (the normal range being 0.5-3.0 optimum being 1.0 mmhos/cm. for good growth of crop); increased Ca (normal range being 0.1-3.2%), Mg (normal being 0.1-0.3%), Fe (normal being 2-7 ppm) and Mn (Normal being 3-29 ppm) contents; as compared to the control soil samples. The levels of N(>100 kg/hec. good for crop growth), P (30-40 kg/hec. normal range), K (10-280 kg/hec.

Table : 28.2

Sl. No.	Parameter	February 2002			June 2002			October 2002		
		Control soil	Acid soil	Sugar-cane soil	Control soil	Acid soil	Sugar-cane soil	Control soil	Acid soil	Sugar-cane soil
1.	Temp.(deg.C)	30	30	37	20	20	25	22	22	28
2.	pH	7.45	6.50	7.12	7.19	6.85	6.89	6.85	6.37	7.01
3.	EC(mmhos/cm)	0.35	3.45	0.59	0.198	3.35	0.968	1.199	2.89	1.598
4.	N(kg/hec)	8.95	3.95	24.96	9.28	3.97	26.28	87.88	4.75	109.5
5.	P(kg/hec)	24.67	0.68	6.89	24.85	0.37	7.85	47.96	0.48	39.68
6.	K(kg/hec)	336.6	3.2	139	306.83	1.98	148	358.88	4.99	368
7.	Organic C(%)	0.28	0.22	0.41	0.28	0.09	0.38	0.88	0.29	0.98
8.	Calcium(%)	1.015	2.85	0.48	1.45	2.75	0.48	1.39	1.78	1.25
9.	Magnesium(%)	0.36	0.98	0.19	0.29	0.78	0.13	0.67	0.69	0.70
10.	Iron(ppm)	3.23	7.95	4.02	3.29	7.95	3.55	3.99	8.08	2.85
11.	Zinc(ppm)	2.98	2.96	0.39	3.09	3.16	0.38	5.01	6.75	2.99
12.	Copper (ppm)	0.385	0.195	1.096	0.35	0.11	1.13	0.95	0.36	2.09
13.	Manganese (ppm)	0.32	0.35	2.52	0.18	0.29	2.11	2.01	0.37	0.25
14.	Organo P(ppb)	BDL	0.00	0.035	BDL	0.03	0.15	0.001	0.29	0.0015
15.	Organo Cl(ppb)	BDL	0.009	0.085	BDL	0.01	0.19	0.019	0.119	0.009
16.	Enz.NO$_3$ red.(umol/gm)	0.52	0.32	0.45	0.38	0.09	0.25	0.299	0.059	0.145
17.	Enz.DeH$_2$ase (umol/gm)	0.31	0.21	0.20	0.17	0.15	0.19	0.111	0.075	0.088
18.	Enz.PO$_4$ase (umol/gm)	0.41	0.66	0.98	0.33	0.46	0.79	0.099	0.198	0.69
19.	Bacterial count (colonies/ml)	5.5×10^6	3×10^4	9×10^6	3×10^6	1.5×10^4	3.3×10^6	5×10^8	7×10^8	1×10^8
20.	Fungal count (colonies/ml)	8×10^6	1×10^7	4×10^5	5×10^6	6×10^7	2×10^6	1×10^6	3×10^7	2.5×10^6
21.	Actinomycetal Count (col./ml.)	6×10^6	2×10^7	3.5×10^7	1×10^5	2.2×10^6	4×10^6	5×10^7	3×10^7	1×10^6

normal range), Cu (3-22 ppm normal range), Organic C (0.6-1.0% opt. for crop) are seen to have decreased in acidic soil samples and levels of Organo Phosphorous and Organo Chlorine compounds (Gupta, P.K., 1999) slightly elevated as compared to control samples. Enzyme Nitrate reductase, Enzyme Dehydrogenase levels, which are some of the factors responsible for decresed soil fertility and crop yields, if in below normal amounts (Kiss, S.M., Dragan Bularda and D. Radulesco, 1975), are found to be lower (Gupta, P.K., 1999) in acidic soil samples. Bacterial count is also lowered in acidic soils indicating decreased biochemical activities while Fungal and Actinomycetal count has slightly increased, owing to acidic conditions.

As per the analytical picture of sugarcane field soils, highest Nitrogen content is seen to be present in it (110 kg/hec) which comes from left-over fertilisers, maximum organic carbon content (1.03%), highest copper content (2.22 ppm) and maximum manganese (2.52 ppm). The levels of organo-chlorines and organo-phosphorus compounds are seen to be highest in sugarcane soils (0.16 and 0.22 ppb respectively), probably indicating excessive use of agro-chemicals.

The level of enzyme Phosphatase is highest in sugarcane field soils, indicating senescence and stress conditions (D. Leo P. and J.A. Sacher, 1970), while bacterial, fungal and Actinomycetal counts are moderate as compared to the acidic and control soil samples. Thus, acidic soils depict a gloomy picture and some methods of remedy should be applied in order to decrease water logging, acidity and increase soil fertility for a better crop-yield.

Comparatively higher levels of organo-chlorine and organo-phosphorus compounds and lower enzyme contents as compared to the control samples also indicate ensuing problems in the sugarcane field samples.

Indiscriminate use of agro-chemcials for increased crop yield, in the 80's & 90's, has proved to be a very short-sighted method, dangerous in the long run, as it has led to vast expanses of land lying absolutely idle. Getting a moderate yield throughout, with sensible and appropriate use of agro-chemicals, seems to be a better idea.

Acknowledgement

The author is thankful to the University Grants Commission, New Delhi, for the financial support to carry out this work.

The author is also thankful to Dr. M.S. Sagare, Principal and Dr. M.G. Bodhankar, Head, Dept. of Microbiology, Dr. Patangrao Kadam Mahavidyalaya, Sangli, for granting permission to carry out this work in the laboratory.

References

Alexander Martin, (April 1985). Introduction to soil microbiology, second edn., Wiley Eastern Limited, New Delhi.

Alexander M. (1971). Biochemcial ecology of micro-organisms, Annu. Rev. Microbiol. 25 : 361-392.

Bear Firman, E.. (1964). Chemistry of the soil, Reinhold publishing corporation, New York.

Burns, R.G. (1978). Soil Enzymes, Academic press, London.

Burns, R.G. A.H. Pukite and A.D. McLaren, 1972. Soil Sc., Soc. Amer. Proc., 36 : 308-311.

Deshpande, M.S. (1996). Bhumi Labh, Bio-organic Fertilearth, Kolhapur.

Gupta. P.K. (1999). Soil, Plant, Water and Fertilizer analysis, Agro Botanica publishers and distributors, Bikaner.

Gupta, V.K. (1993). Methods of analysis of soils, waters and fertilizers, HLS Tandon, FDCO, New Delhi.

Instruction Leaflet (1977). Published by Agriculture information dept., Director of Agriculture, Pune.

Kiss, S.M., Dragan Bularda and D. Radulescu, (1975). Biological significance of enzymes accumulated in soil, Advan. Agron., 27 : 25-87.

Leo, D.P. and Sacher J.A. (1970). Plant physiology, 46 : 208-211.

Russell, E.W. (1973). "Soil conditions and plant growth", Longman group Ltd., London.

Vijaya, T. and B.V. Prasada Reddy, "Saline soil tolerance of Sapindus emarginatus (Vahl) seedlings with established Glomus fasciculatum infection", pp. 99-104, in Biofertilisers and Biopesticides (1998), editor Dr. A.M. Deshmukh, Technoscience publication, Jaipur.

ENVIRONMENTAL BIOTECHNOLOGY

Edited By :   **Professor (Dr.) Arvind Kumar**

Published By :   **DAYA PUBLISHING HOUSE**

29

EFFECT OF ARBUSCULAR MYCORRHIZAL (AM) FUNGI AND SALINE WATER WITH AND WITHOUT WATER ADDITIONAL PHOSPHATE ON *ELEUSINE CORACANA* GAERTN (FINGER MILLET)

• *Geeta, B. Patil and H.C. Lakshman*

Post Graduate Department of Botany, (Microbiology Laboratory)
Karnataka University, Dharwad - 580003 (India)

Abstract

Arbuscular Mycorrhizal (AM) fungi (*Glomus fasiculatum*) was isolated from coastal zones of Karvar. Pot cultured arbuscular mycorrhizal fungi were inoculated with and without additional phosphate levels, to evaluate the growth of finger millet plant under artificial Salinization. There was significant plant growth, percent root colonization, spore population and P uptake. The results indicated the increased supply of P nutrition for AMF plants growing under saline stress.

Key Words : Arbuscular - mycorrhizal Fungi (AMF), Eleusine coracana, Saline soil, Coastal zone, Percent root colonization, Spore population.

Introduction

Eleusine coracana (finger millet) is third millet of India. About 75% of the area under the crop lies in south India, especially in Karnataka. In India, arid and semi-arid regions where sufficient good quality irrigation water is not available, Finger millets are grown. When the plants become saline tolerant then it reflects a biomass productivity, adversely, and the magnitude depends upon the quality of soluble salts that get incorporated through irrigation water. In many important crops salinity increases and plant P concentration decreases (Ross *et. al.*, 1985), and these plants may be most beneficial to inoculate with efficient stains of arbuscular *mycorrhizal fungi*.

In recent years many experiments have been documented. Many crops; such as tomato (*Lycopersion escuentum* mill.), onion (*Allium cepa* L.), and bell pepper (*Capsium annum* L.) have increased growth of biomass production, when these plants grown under saline conditions with AM fungal colonization (Ojala *et al.*, 1983; Mass *et al.*, 1986). Most of these studies have been carried out under low soil P conditions. Studies on annual herbaceous perennial plants are very meagre.

The main objective of the present study is to distinguish between enhancement of phosphate nutrition by AMF with and without additional phosphate treatment of finger millet. This information will help in formlulating on annual plants for the best use of saline water with AM fungi to improve plant growth and biomass productivity in sand and semi-arid regions.

Materials and Methods

Finger millet grains were grown in pots contained sterilized sand and garden soil in the ratio of (1: 1). Uniformly grown 25 days old seedlings were transplanted to experimental pots measuring 25 cm. diameter contained 4 kg of soil. 2 kg red Laterite soil mixed with 2 kg pure sand 2.6 ppm of available (NH_4) phosphorus was determined. Experiments were conducted by using different levels of super phosphate and rock phosphate fertilization. The following treatment were given to all the experimental pots.

1. NM - (Nonmycorrhizal / control).
2. M - (MYcorrhizal).
3. M + SP_1 (Mycorrhizal inoculum + 2 mg superphosphate / kg soil)
4. M + SP_2 (Mycorrhizal inoculum + 4 mg superphosphate / kg soil)
5. M + SP_3 (Mycorrhizal inoculum + 6 mg superphosphate / kg soil)
6. M + RP_1 (Mycorrhizal inoculum + 2 mg rockphosphate / kg soil)
7. M + RP_2 (Mycorrhizal inoculum + 4 mg rockphosphate / kg soil)
8. M + RP_3 (Mycorrhizal inoculum + 6 mg rockphosphate / kg soil)

Mycorrhizal mixed inoculum of (25 g/pot) was provided by placing a thin layer of inoculum 2 cm. below the seeds at the time of sowing. The mycorrhizal inoculum consists of roots and soil from the pot culture of sudangrass [*Sorghym bicolour* (L.) moench. Var. Sudanese] wluch was infected with *Glomus fasiculatum*. The inoculum contained hyphae, vesicles, Chlamydospores (146 per pot). The pots were arranged in green house in a randomized block design with four replicates for each treatment. Artificial saline water was given on alternate days. Hoagland solution of minus P of 5 ml/pot was given once in fifteen days. Plants were harvested at two intervals, i.e., 40 and 80 days after sowing. Observation was made on plant height, shoot dry weight, percent root colonization, spore number per 50 g soil and P uptake in shoots. The dry weight of shoot was taken after constant drying at 70° C in oven for 12 hrs. Phosphorus content of the shoot was detemined colorimetrically by vanadomolybedate phosphoric-yellow colour method outlined by Jackson (1973). Percentage of mycorrhizal colonization of the roots were determined after clearing in 10% KOH and stained 0.05% trypan blue (Phillips and Hayman, 1970). Number of arbuscular mycorrhizal fungal spore were isolated following wet-sieving and decanting technique of (Gerdemann and Nicolson, 1963).

Results and Discussion

Soils contain low salt concentration may interfere with the growth of many crop plants. This statement is applicable to a annual plant like *Eleusine coracana*, (finger millet). When the experimental plants given fresh water with mycorrhizal inoculation showed better growth, biomass production, mycorrhizal colonization and spore population and P uptake, compared to the plants given saline water with mycorrhizal inoculation (Table 29.1). However, the percent of mycorrhizal colonization and spore number decreases in *Eluesine corecana* plants, which were given saline water with mycorrhizal inoculation. But there was sequential increased percent of mycorrhizal colonization and spore number per 50 g soil was recorded in plants which were treated with saline water with higher level 6 mg Rock Phosphate/kg soil with *Glomus fasciculatum* inoculation. On the other hand the sequential decreased percent of (AM) fungal colonization and spore number per 50 g soil was recorded, in those plants treated with higher super phosphate (Tabe 29.2). *Eluesina corecana* levels of plants treated with super phosphate or rock phosphate level of (2 mg/pot) does not responded significantly with *G. fasciculatum*. The percent AM fungal colonization in the roots was greater after *G. fasciculatum* inoculation at 80 days harvest. When phosphorus was not given to experimental plants, mycorrhizal inoculated plants grew poorly. But *E. coracana* showed significant increase in plant height, dry matter of shoot, percent root colonization, spore population and P uptake in shoots. Any change or build up insoluble salt content of soil may influence on crop production in several through changes in the proportions of

Table 29.1 : Plant Height, Shoot Dry Weight, percent Root Colonization, Spore Number and P Uptake in Finger Millet as Influenced by *Glomus fasciculatum*, With and Without Saline Water (Treatment for 40 days)

Treatment	Plant height (cm)	Shoot dry weight (g)	Percent AMF root colonization	AMF spore number/50 g soil	% P uptake in shoot
N.M	8.7 ± 3.2	1.4 ± 1.0	–	–	0.05
M+F-W	22.3 ± 5.3	4.1 ± 3.2	51.3 ± 5.4	168.5 ± 4.3	0.16
M+W-SW	19.6 ± 4.1	3.9 ± 0.0	49.4 ± 4.0	141.5 ± 4.3	0.10
L.S.D. (5%)	8.0 ± 2.1	0.2 ± 0.0	11.0 ± 2.0	14.3 ± 0.0	0.12

Table 29.2 : Plant Height, Shoot Dry Weight, % AMF Colonization, AMF Spore Number and P Uptake in Shoots in Finger millet as Influenced by *Glomus fasciculatum*, in Fresh Water and Saline Water and Without Super and Rock Phosphate (Treatment for 40 days)

Treatment	Plant height (cm)	Shoot dry weight (g)	Percent AMF colonization	AMF spore number/50 g soil	% P uptake in shoot
N.M	12.4 ± 5.2	1.8 ± 2.3	–	–	0.10
M+F-W+SP$_1$	21.3 ± 5.2	3.8 ± 1.2	46.2 ± 2.2	152.2 ± 5.2	0.19
M+F-W+SP$_2$	23.1 ± 1.1	4.0 ± 0.0	44.4 ± 3.1	152.3 ± 2.4	0.22
M+F-W+SP$_3$	25.4 ± 1.0	4.6 ± 1.1	43.1 ± 1.1	143.4 ± 1.2	0.24
M+F-WRP$_1$	32.3 ± 2.4	8.2 ± 4.2	47.4 ± 4.3	179.5 ± 1.1	0.31
M+F-WRP$_2$	43.2 ± 3.2	11.4 ± 3.1	49.5 ± 5.2	193.3 ± 2.1	0.32
M+F-WRP$_3$	48.5 ± 4.1	13.2 ± 4.3	51.3 ± 7.1	186.2 ± 0.0	0.32
M+W-SW+SP$_1$	27.4 ± 1.1	5.7 ± 5.1	44.4 ± 3.5	48.6 ± 4.6	0.18
M+W-SW+SP$_2$	24.6 ± 4.5	4.9 ± 2.2	44.2 ± 2.2	48.7 ± 1.1	0.20
M+W-SW+SP$_3$	28.7 ± 7.5	5.8 ± 4.4	40.5 ± 1.0	39.4 ± 2.4	0.24
M+W-SW+RP$_1$	36.4 ± 6.1	10.5 ± 6.3	45.2 ± 7.2	76.3 ± 3.6	0.34
M+W-SW+RP$_2$	44.6 ± 4.3	12.7 ± 4.6	48.4 ± 3.4	88.2 ± 4.2	0.35
M+W-SW+RP$_3$	76.1 ± 5.8	15.8 ± 5.6	67.3 ± 5.1	81.5 ± 5.0	0.38
L.S.D. (5%)	10.5 ± 0.0	0.5 ± 1.0	9.2 ± 1.0	19.3 ± 0.0	0.04

NM = Non mycorrhizal, M = Mycorrhizal, F-W = Fresh Water, W-SW = With Saline Water, SP = Super Phosphate, RP = Rock Phosphate.

—

exchangeable cations. This may be due to, whenever soil have problems of excessive soil moisture then there will be least secondary salanization in the root zone. Plant growth is restricted or entirely prevented even mycorrhizal association (Jackson *et al.*, 1972; Massana Hoffman, 1986). However, in the present study, *E. coracana* plants inoculated with AM fungi *Glomus fasciculatum* which were grown in P deficient soil, with additional rock phosphate (6 mg RP$_3$/kg soil) and saline water treatment showered a significant increased plant growth percent root colonization, spore number and P content in shoots, compare to non-mycorrhizal plants or plants treated 2 or 4 mg SP$_2$ and SP$_3$ with mycorrhizal inoculation. These findings are supported to early workers contribution of (Alien and Cunnigham, 1983; Ross *et al.*, 1985). The difference in phosphorus concentrations between mycorrhiza inoculated and non-mycorrhizal plants are well marked. Further the increase of P content in shoots is due to added rock-phosphate for the plants at the required level. Considering the plants height and dry weights of shoots the treatments were clearly comparable, and it is suggestive that mycorrhiza (*G. fasciculatum*) was able to extract adequate or near optimum levels of phosphorus for growth of *E. coracana* even from a phosphorus deficient soils. These findings are in consistent with (Pfeiffer and Bloss Lakshman, 1994).

The main role of the AM fungus *Glomus fasiculatum* in improving growth in finger millet saline soil appeared to be increasing P accumulation and concentration when soil P was low. When sufficient P is added to the saline soil a parasitic relationship may exist in that fungal respiration presents a carbon drain on the experimental plants (Hirrel and Gerdmann, 1980; Lakshman, 1999).

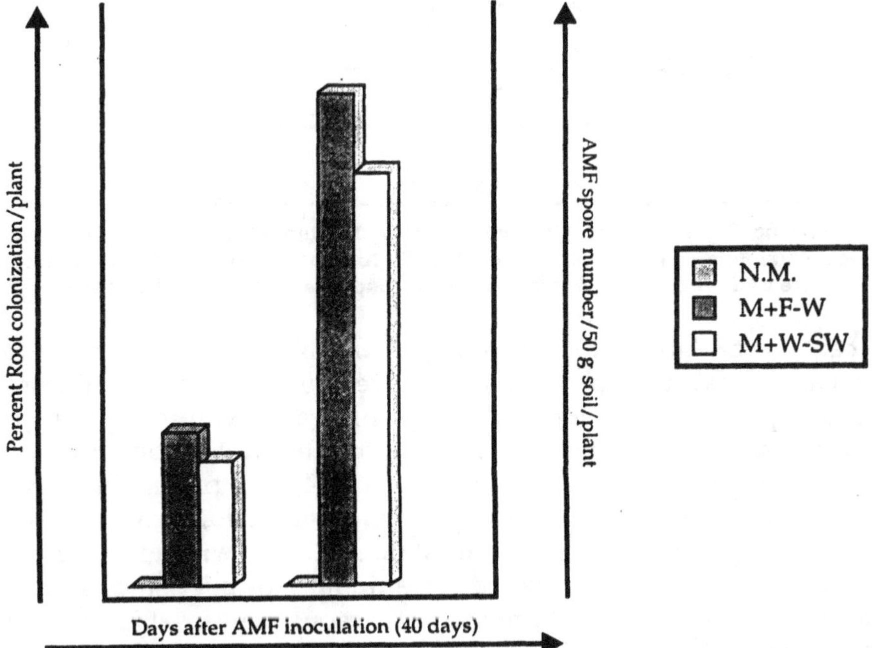

Fig. 29.1 : Finger Millet Influenced by *Glomus fasciculatum* and With and Without Saline Water Showing Percent Root Colonization and Spore Number for 50 g soil.

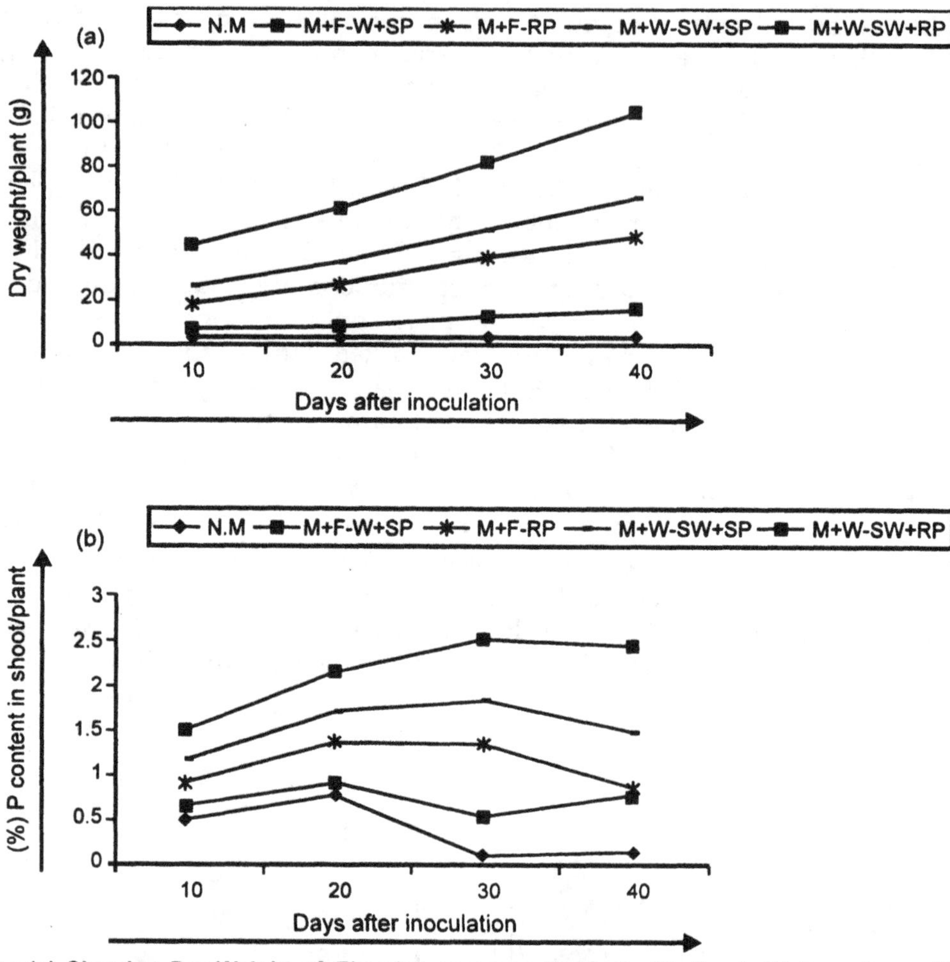

Fig. 29.2 : (a) Showing Dry Weight of *Eleusina coracana* Treated with Fresh Water + Super Phosphate and Rock Phosphate. (b) Showing Percent of P Content in Shoots of *Eleusina coracana* Treated with Fresh Water + Super Phosphate and Rock Phosphate.

The percentage of AMF colonization, spore number in root system was gradually reduced, when additional super-phosphate was given, but, there was significantly increased. spore number and percent root colonization in plant treated with rock phosphate. Increased height of mycorhizal plants can be partially be attributed to improve P nutrition (Gemma and Kosk G., 1977; Puppi and Riess, 1987; Bellgrad, 1994). The present investigation brought out clearly that AMF fungi *G. fasciculatum* greatly assisted fold, increased growth of finger millet plants. These plants were grown in stalinized soil with and without additional phosphate. At the same time it emphasise more number of pilot studies are warranted in both laboratory and field, on other annuals/herbaceous plants, which are colonized with arbuscular mycorrhizal fungi of saline treatment with other beneficial microorganisms.

Acknowledgements

Authors are thankful to SAP, DRS II COSIST, UGC New Delhi for the financial support.

References

Alien, F.B. and Cunningham, G.L. 1983. Effects of Vescicular - arbuscular mycorrhizae as Danish Chils spicata under three levels. New Phytol. 93 : 227-236.

Bellgard, S.E. 1994. The bi-functional nature of the mycelial network associated with plants colonized by VA-mycorrhizal fungi. Mycorrhiza News 6(1) : 1-5.

Gemma, J.N. and Koske, K. 1977. Arbuscular mycorrhizae in sand dune plane of the north Atlantic coast of the US. Field and green house inoculation and presence of mycorrhizae in planting stock. J. Environ. Management. 50(3) : 1251-1264.

Gerdmann, J.W. and Nicolson, T.I.T. 1963. Spores of mycorrhizal Endogene species extracted from soil by wet-sieving and decanting Trans. Brit. Mycol. Soc. 46(2) : 235-244.

Hirrel, M.C. and Gerdmann, J.W. 1980. Improved growth of onion and bell pepper in saline soils by two vescicular-arbuscular fungi Soil. Sci. Amer. 44 : 654-655.

Hougland, D.R. and Arnon, D.L. 1950. The water culture method of growing plants without soil. University of California, College of Agriculture, Circular No. 347. University of California, Berkley.

Jackson, M.L., 1973. Soil chemical analysis prentice Hall, New Delhi.

Jackson, N.E., Frankline, R.E. and Miller, R.H. 1972. Effects of Vescicular - arbuscular mycorrhizae on growth and phosphorus content of three agronomic crop. Proc. Soil Sci. SoC. America. 36 : 64-72.

Lakshman, H.C. 1994. Occuring of VA mycorrhizal in some tropical Hydro plastic xerophytic plants of Dharwad District. J. conservation 105(9), 79-86.

Lakshman, H.C. 1999. VA mycorrhizal survey of plants species colonizing beach ponds of Karwar in south India. Asian. J. Microbial. Bitech and Environ 191(2) : 15-21.

Logan, V.S., Clark, P.J. and Allway, W.G. 1989. Mycorrhizae and root attributes of plants if coastal sand dunes of New South Wales. Aust. J. Plant Physiol. 16 : 141-146.

Mass, E.V. and Hoffman, G.J. 1986. Salinity of Sorghum at three 1977. Crop alt tolerance current assessment. J. Irrigation and Drainage Division A3CE. No. IR2, Proc. Pap. 12993.03 : 7-14.

Nicolson, T.H. 1960. Determination of Nitrogen, in soil by Kjeldhal method. J. Agri. Sci. 55 : 11-17.

Ojala, J.C. Jarrell, W.M. Menge, J.A. and Johnson, E.L.V. 1983. Influence of mycorrhizal fungi on mineral nutrition and yield of onion in saline soil. Agron. J. 75 : 255-259.

Pfeiffer, C.M. and Bloss, H.E. 1988. Growth and nutrition of guayule (Partenium aggentatm) in saline soil as influenced by arbuscular mycorrhiza and phosphorus fertilization. New Phytol. 108(3) : 321-351.

Phillips, J.M. and Hayman, D.S. 1970. Improved procedure for clearing roots and staining parasitics and Vesicicular - arbuscular mycorrhizal fungi for rapid assessment of infection. Trans. Brit. Mycol. Soc. 55 : 158-161.

Puppi, G. and Riess, D.J. 1987. Role of Ecology of VA-mycorrhizae in sand dunes Ange. Dot. 67 : 115-126.

Ross, S.A. Pond, E.C. Menge, J.A. and Jarrell, W.M. 1985. Effect of salinity on mycorrhizal onion tomato in soil with and without additional phosphate. Plant and Soil. 88 : 309-319.

Walkely, A. and Black, T.A. 1934. Estimation of organic matter and proposed modification of the chromic titration method. Soil Sci. 37 : 29-38.

ENVIRONMENTAL BIOTECHNOLOGY

Edited By :   **Professor (Dr.) Arvind Kumar**

Published By :   **DAYA PUBLISHING HOUSE**

30

GENETIC VARIABILITY AND CHARACTER ASSOCIATION FOR YIELD AND PROCESSING TRAITS IN TOMATO (*LYCOPERSICON ESCULENTUM* MILL.)

• *Hemandeep Kaur, D.S. Cheema, Avtar Singh* and Jasjit Singh**

Department of Vegetable and *Department of Agronomy,
Punjab Agricultural University, Ludhiana-141 004 (Punjab)

Abstract

Twenty tomato genotypes were evaluated along with two check varieties, over two years, for isolating superior lines. Genetic variability, heritability, genetic advance, correlation coefficients, were estimated. Phenotypic as well as genotypic coefficients of variation were high for carotene, lycopene and average fruit weight. Heritability and genetic advance were also high for carotene, lycopene and average fruit weight. Marketable fruit yield and average fruit weight were positively correlated with total yield per plant. Total soluble solids had negative correlation with total yield per plant. So emphasis should be given on traits like average fruit weight, carotene and lycopene.

Key Words : Heritability, Genetic advance, Correlation, Quality.

Introduction

Tomato (*Lycopersicon esculentum* Mill.) is one of the most important summer vegetable grown throughout the world for fresh consumption as well as for processing. It is used in various culinary and processed preparations. Tomato is rich in vitamin A, C, sugars, organic acids, carotenoids, lycopene and due to its high nutritive value, it is referred as "Poor Man's Orange".

In India tomato production is 6.8×10^5 MT (Anon., 2000). Among vegetables, it ranks second in area and production after potato, however, it ranks first in processed vegetables. India contributes only 7 to 10 per cent of the total production which is processed (Anon., 1996). One of the probable reasons could be the lack of suitable varieties for processing. Along with increasing tomato productivity, there is need to extensively collect, evaluate and exploit germplasm of tomato in order to develop genotypes possessing superior quality attributes.

Among the quality factors total soluble solids, pH, lycopene, carotene, acidity, vitamin C, and dry matter are very important for processing purposes. Most of the traits of economic importance in crop plants are dependent on one or the other trait. Improved processing qualities along with high yield have been the major objectives of processing tomato breeding programmes. To achieve these targets, information on variability parameters like coefficient of variation, heritability, genetic advance and degree of association between the various characters are of paramount significance. Therefore, the present investigation was undertaken to study above listed parameters with twenty genotypes of tomato.

Materials and Methods

The present investigations were carried out at the Vegetable Research Farm of the Punjab Agricultural University (PAU), Ludhiana during the years 1999-2000 and 2000-2001. The field trials were conducted with twenty genotypes, which were sown in October during both years, and seedlings were transplanted in end of December. The twenty genotypes were tested in randomized block design (RBD) with three replications. Each entry consisted of single row of ten plants. The other cultural practices were same as recommended by the PAU (Anon, 1999.).

The data was recorded on five plants selected at random from each variety. The character under study were total yield per plant, marketable yield per plant, average fruit weight, total soluble solids, vitamin C, lycopene and carotene content. Pooled data over two years was statistically analyzed for genetic variability, heritability, genetic advance and genotypic and phenotypic coefficients of correlation.

The genotypic and phenotypic coefficients of correlation were calculated following Burton and Devane (1952) method. Heritability and genetic advance were calculated according to the formula suggested by Allard (1960). The correlation coefficients at phenotypic and genotypic levels were calculated following Al-Jibouri *et al.* (1958) method.

Results and Discussion

Results pertaining to the pooled analysis of variance for the characters under study are reported in Table 30.1. The analysis revealed that there was significant difference among the genotypes for all the traits studied, suggesting presence of sufficient variability among the genotypes.

In the present study, carotene showed the highest genotypic coefficient of variation (46.74%), followed by lycopene (41.31%), average fruit weight (37.35%), marketable yield per plant (26.80%). Quite low value of genotypic coefficient of variation was observed for vitamin C (7.83%) and total soluble solids (9.84%). More or less similar trend of variation was observed for phenotypic coefficients of variation for the traits under study. Jaya and Seemanthini (1994) have also reported the highest value of genotypic and phenotypic coefficients of variation for carotene.

In all the traits studied, the genotypic coefficients of variation were less than the phenotypic coefficients of variation. In case of total yield per plant and marketable yield per plant, the difference between genotypic and phenotypic coefficients of variation is more than other traits. Similar trend of genotypic and phenotypic coefficients of variation was observed by Nandpuri *et al.* (1973) and Pankaj *et al.* (1996). It is due to greater effect of environment on the expression of these traits. According to Reddy and Reddy (1992) and Pankaj *et al.* (1996), the less difference between phenotypic and genotypic coefficient of variation was due to less influence of environment. The genotypes having high genotypic coefficients of variation for traits possessed better potential for further gain and improvement through selection.

Table 30.1 : Mean, Genotypic and Phenotypic Coefficient of Variation, Heritability in Broad Sense and Genteic Advance as Percent Mean for Different Characters in Tomato as per Pooled Data Analysis.

Character	Mean	Genotypic coefficient of variation	Phenotypic coefficient of variation	Heritability	Genetic Advance
Total Yield Per Plant	1.75	26.80	37.47	51.15	39.48
Marketable Yield Per Plant	1.25	30.68	43.59	49.53	44.47
Average Fruit Weight	68.47	37.35	38.60	93.65	74.46
T.S.S.	4.25	9.84	10.51	87.65	18.97
Lycopene	1.08	41.31	42.20	95.79	83.28
Vitamin C	51.02	7.83	9.51	66.95	13.19
Carotene	0.64	46.74	47.20	98.04	95.34

Table 30.2 : Phenotypic and Genotypic Correlation Coefficient among various Characters in Tomato on the Basis of Data Pooled over 1999-2000 and 2000-2001

Character		Total Yield	Marketable Yield	Averae Fruit Weight	T.S.S.	Lycopene	Vitamin C	Carotene
Marketable Yield	(P)	0.9491[xx]						
	(G)	0.9697[xx]						
Average Fruit Weight	(P)	0.1531	0.1991[x]					
	(G)	0.1835[x]	0.2828[x]					
T.S.S.	(P)	−0.0157	−0.1219	−.2850[xx]				
	(G)	−0.0497	−0.1886[x]	−.3209[xx]				
Lycopene	(P)	−0.3044[xx]	−0.254[x]	−.1547	.0000			
	(G)	−0.3985[xx]	−0.3237[xx]	−.1597	−.0101			
Vitamin C	(P)	.007	−0.448	−.0098	−.0889	.0073		
	(G)	−.0966	−.1719	−.0516	−.0952	.0423		
Carotene	(P)	−.1326	−.1661	−.0386	−.2750[x]	.2429[x]	.3479[xx]	
	(G)	−.1924	−.2480	−.0325	−.2956[xx]	.2420[*]	.4547[xx]	

x, xx significant at both 5% and 1% levels respectively.

Knowledge of heritability coupled with expected genetic advance of a trait is necessary for assessing the scope of its improvement through selection. High values of heritability coupled with genetic advance were observed for carotene, lycopene and average fruit weight. Similar results of high estimates of heritability along with genetic advance were reported by Reddy and Reddy (1992), Jaya and Seemanthini (1994), Mahajan *et al.* (1996) and Anita Kumari *et al.* (1998). The high estimates of heritability along with genetic advance may be attributed to the effect of additive genes.

Hence, based on the results of coefficients of variation, heritability and genetic advance, selection can be based on the traits viz. carotene, lycopene and average fruit weight. Heritability and genetic advance were low for both total yield per plant and marketable yield per plant, so direct selection of yield is not sufficiently effective.

The genotypic, phenotypic and environmental correlations among seven traits were worked out in the present study. Results pertaining to the coefficients of correlation are reported in Table 30.2. The analysis revealed that genotypic coefficient of correlation, in general, were higher in magnitude than the corresponding phenotypic coefficients of correlation. This indicates that there was inherent association among various characters studied and phenotypic expression of correlation was lessened under the influence of environment. The results corroborated the views of Dudi and Kalloo (1982).

Nandpuri *et al.* (1973) reported that total yield per plant had positive and highly significant correlation with marketable yield per plant as is also indicated in Table 30.2. Total yield per plant also showed positive and significant correlation with average fruit weight. These results were also obtained by Dudi and Kalloo (1982) and Aruna and Veeraragavathatham (1997). According to them as the fruit weight increases, the yield also increases.

Total soluble solids is an important biochemical constituent and lycopene is important for colour of tomato varieties. These both had negative correlation with total yield and marketable yield per plant. These results are in accordance with the findings of Thamburaj (1995) and Verma *et al.* (1997). These both can be used as indirect traits for selection against total yield per plant. From this, it is evident that with an increase in total yield, there is decrease in both total soluble solids and lycopene. The best way to improve quantitative and qualitative traits would be to lay emphasis on average fruit weight, carotene and lycopene as is evidenced by various parameters like correlation's and variability studies.

References

Al-Jibouri, H.R., Niller, P.A. and Robinson, H.F. 1958. Genotypic and environment variations in upland cotton crosses of inter specific origin. *Agron. J.,* 50 : 633-637.

Allard, R.W. 1960. Principles of Plant Breeding. *John Wiley and Sons, New York,* London, p. 584.

Anita Kumari, Banerjee, M.K. and Grewal, R.B. 1998. Assessment of physico-chemical characteristics of different tomato genotypes. *Veg. Sci.,* 85 (2) : 127-130.

Anonymous, 1996. Horticulture Information Service, National Horticulture Board, Gurgaon, India, p. 27.

Anonymous, 1999. Package of Practices for Fruits and Vegetable Crops. Punjab Agricultural University, Ludhiana.

Anonymous, 2000. Horticulture Information Service, National Horticulture Board, Gurgaon, India, p. 54.

Aruna, S. and Veeragavathatham, O. 1997. Correlation among yield and component traints in tomato. South Indian Hort., 45 : 1-2

Burton, G.W. and Devane, E.N. 1952. Estimating heritability in all Fischue (*Festhica arundinacea*) from replicated clonal material. *Agron. J.,* 45 : 478-481.

Dudi, B.S. and Kalloo, 1982. Correlation and path analysis studies in tomato (*Lycopersicon esculentum* Mill.). *Haryana J. Hort. Sci.,* 11(2) : 122-126.

Jaya Jasmine, A.P. and Seemanthini Ramdass, 1994. Qualitative evaluation of tomato hybrids and varieties. *South Indian Hart.,* 42 (1) : 26-28.

Mahajan, R., Sucheta Sharma and Bajaj, K.L. 1996. Biochemical evaluation of some tomato varieties. *Veg. Sci.,* 23(1) : 42-47.

Nandpuri, K.S., Singh Surjan and Lal Tarsem, 1973. Studies on the genetic variability and correlation of some economic character in tomatoes. *J. Res. Pb. Agri. Univ.,* 10 (3) : 316-321.

Pankaj Mittal, Sant Parkash and SINGH, A.K. 1996. Variability studies in tomato (*Lycopersicon esculentum* Mill.) under sub-humid conditions of Himachal Pradesh. *South Indian Hort.*, 44 : 132-134.

Reddy, V.V.P. and Reddy, K.V. 1992. Studies on variability in tomato. *South Indian Hort.*, 40 (5) : 257-260.

Thamburaj, S. and Nair Indu, P., 1995. Correlation studies in certain exotic collections of Tomato. *South Indian Hort.*, 43 : 35-37.

Verma, S.K., Sarnaik, D.A. and Verma, D.P. 1997. Studies on phenotypic and genotypic correlations on yield and quality attributes in tomato (*Lycopersicon esculentum* Mill.). *Orissa J., Hort.*, 25(1) : 55-57.

ENVIRONMENTAL BIOTECHNOLOGY

Edited By :   **Professor (Dr.) Arvind Kumar**

Published By : **DAYA PUBLISHING HOUSE**

31

EFFECT OF *GLOUMUS FASICULATUM* AND *RHIZOBIUM LEGMINOSARUM* AND ROCK PHOSPHATE APPLICATION ON GROWTH NUTRIENT DYNAMICS OF *PTEROCARPUS MARSUPIUM*

• *H.C. Lakshaman and Geeta B. Patil*

Post Graduate Department of Botany (Microbiology Lab.)
Karnataka University, Dharwad-580 003

Abstract

At nursery stage, experiments were conducted on *pterocarpus marsupium* to evaluate the effect of three levels of rock phosphate with *G. fasiculatum* and *Rizobium leguminosarum* inoculation. Out of three level of rock phosphate levels (5, 10, 15 kg ha^{-1}), the results indicated that the application of p/15 ha^{-1} had the significantly improved the growth parameter with mycorrhizal and *Rhizobium* inoculation. On contrary there was a reduced number of mycorrhizal spores and root/shoot ratio of plants, when the plants received all the three inoculants. Similarilty, N and P of enhanced up take in shoots of inoculated plants at p/15 kg ha treatment over the control seedlings. The importance of rock phosphate, vesicular arbuscular mycorrhiza and *Rhizobium* inoculation to young legume nursery plants has been discussed.

Key Words : *Pterocarpus marsupium, Vesicular arbuscular mycorrhizal (VAM) fungi, Phosphate, Rhizobium leguminosarum, Glomus fasiculatum.*

Introduction

The problems of degradation of natural resources have received increasing attention in recent days. Total of 177 million ha of land resources in India are subjected to various types of degradation. This has many direct and indirect ill effects. Mainly it causes imperative to revegetate the vast areas with high efficient and valuable forest species to cater to the bonafide needs of the growing population. *Pterocarpus marsupium* is one of the promising tree species, which could be taken as an alternative to achieve the goal of sustainability. *Pterocarpus marsupium* is a rare and good timber, legume species and it is usually associated with *Rhizobium* to fix the atmospheric nitrogen and make it available to the plants. In natural conditions, these plants cannot form the health symbiosis because of lack of specificity of the host. Therefore, there is exigency either to inoculate the species with specific *Rhizobium* strain or to fertilize with phosphorus to get the better nursery stock.

The role of *Vesicular arbuscular mycorrhizal* (VAM) fungi in improving the quality and survival of forest nursery seedlings and their growth after planting has been well recognised (Trappe, 1977). Plant roots increase the efficiency of absorption of nutrients and thus enhance the growth of plants and trees (Munns and Mouse, 1980; Tilak, 1993). The existence of host preference has been suggested by Singh, 1994. This preferential association between certain plants and fungal species can be evaluated with respect to combined inoculation with two organisms which provide the greatest plant growth stimulations and the higher root colonization or maximum sporulation (Jasper, 1994). Combined inoculation of *Glomus fasiculatum* and *Rhizobium* species with additional rock phosphate was found to more superior than un-inoculated control plants (Mosse *et al.*, 1976). Keeping in view the importance of nitrogen fixer (*Rhizobium*) *Vesicular arbuscular mycorrhizal* (VAM) fungi (*G. fasiculatum*) and phosphate, the present study carried out to evaluate the influence of *Rhizobium*, VAM and P fertilization on growth and nutrient dynamics of *pterocarpus marsupium* seedlings.

Materials and Methods

The experiments were carried out in randomized block design with three replication at green house, Karnataka University Department of Botany Dharwad during 2001-02. Earthen pots with 5 kg of garden soil was sterilized in 5% methyl bromide for one hour and stored at room temperature for one week. *Glomus fasciulatum* (VAM) maintained in host plant *Panicum maxium* in pot cultures was used as inoculum. 50 ml of inoculum containing spores 215 and infected roots (81%) was uniformly distributed below 2 cm soil in all mycorrhizal pots. There phosphorus levels (5, 10, 15 kg P/ha) was used when the seedling reach 2 weeks old. Jagary based *Rhizobium legumnosarum* culture was obtained from university of agricultural science Dharwad. *Rhizobium inoculum* was uniformly pasted on each *pterocarpus seed* to obtain 10^9 cells/mg of final product having 30% moisture content before sowing. For control treatment the sieved soil suspension was added to maintain the rhzosphere of soil. Pots were watered on alternate day. Hoagland (-p) solution was added once in a week. Plants were harvested after every 60 days up to 12 days.

Dry weight of shoots and roots was recorded after drying at 80° C for one day. Total P of the plants were analysed by method of (Olsen *et al.*, 1954). N content of plants and root nodules were analysed following the procedure of (Bermner, 1960). The percentage of mycorrhizal colonization was calculated after clearing the roots with 10% KOH and stained with 0.05% trypan blue (Phillips and Hayman, 1970). Spore count in 50 g soil was determined by wet-sieving and decanting technique (Gerdmann and Nicolson, 1963).

Results and Discussion

Pot experiments were conducted in sandyloam with low phosphorus (33. 87/kg). The data regarding the growth characters of *pterocarpus marsupium* seedling represented in (Table 31.2). The rock phosphate application with *G. fasciculatum* and *Rhizobium* had significant effect on seedling height, root length, dry weight of shoot and roots, nodule numbers, dry weight, percent root colonization and spore population in compare to control. These parameters increased with an increasing phosphorus (30 mg/kg soil) *G. fasiculatum* and *Rhizobium* similar results were reported by (Cruz *et al.*, 1988; Koide and Schreiner, 1992) . the number of nodules increased with an increased P level up to (30 mg/kg soil) with *Rhizobium* inoculation. However, further increase in the levels of P had suppressing effect even the presence of *Rhizobium*. This may be due to the reason that when the P is freely available in the soil, the nodule may or may not take place and even if the nodules are formed, they are in sterile state. At the same time higher concentration may act as inhibitor

Table 31.1 : Physico-chemical Analysis of Garden Soil Used for Pot Experiments

Parameters	Test Value	Method employed	Inference
Texture	Sandy loam	International pipette method (Jackson, 1973)	Silt dominated solid
pH	6.8	1 : 2 : 5 water suspension	Acidic
Organic Carbon (%)	0.47	Walkley and Black method, 1934	Low/Medium
Available N (kg ha^{-1})	242.91	Microjkeldhal procedure (Brener, 1960)	Low/Medium
Available P (kg ha^{-1})	33.87	Olsen *et. al.* (1954)	Low

* Each value is the mean of 12 samples

Table 31.2 : Effect of AMF (*G. fasciculatum*), *Rhizobium* and Phosphorus Level and Growth Characters of *Pterocarpus marsupium* Seedlings for 120 days (5 replications)

Treatment	Plant height (cm)	Dry. wt. of shoot (g)	Root length (cm)	Dry weight of root (g)	Nodule number/ plant	Nodule dry wt. (mg/plant)	Root/shoot ratio (g/plant)
NM/NR	6.38	1.92	4.51	0.34	0.09	0.31	0.83
M	11.2	2.93	7.31	0.44	5.24	0.86	0.58
R	9.4	2.07	7.11	0.37	6.52	0.92	0.42
M+R+P$_0$	14.3	3.88	10.25	1.76	9.91	35.11	0.44
M+R+P$_{15}$	19.54	5.36	11.63	1.77	14.38	8.02	0.39
M+R+P$_{30}$	27.63	8.42	15.32	1.93	14.72	41.72	0.34
L.S.P. (P=0.05)	0.14	0.02	1.17	0.01	0.41	0.03	0.07

Table 31.3 : Effect of the *Rhizobium* and Phosphorus Levels as Nutrient Content in Shoots *Pterocarpus marsupium* seedlings for 120 days (5 replications)

Treatments	N mg/plant	Percent VMF colonization	Spore number/50 g. soil	P mg/plant
NN/MR	2.14	–	–	22.1
M	7.89	47.3	87	31.4
R	3.17	34.5	29	28.2
M+R+P$_5$	5.78	46.1	59	33.5
M+R+P$_{15}$	5.91	44.2	63	36.4
M+R+P$_{30}$	9.13	49.7	63	44.2
L.S.D. (P=0.05)	1.24	14.2	11	0.13

for the symbiotic association (Munns and Mosse, 1980; Allen, 1992). These results revealed that *G. fasciulatum* inoculation alone influence in the increase of mycorrhizal colonization in the roots and spore number in the rhizosphere soils. The lower level of rock phosphate, *Rhizobium* with mycorrhizal inoculation does not influence on *Pterocarpus* biomass production. Varying levels of phosphate application with *G. fasciculatum* and *Rhizobium* plants showed different growth and yield parameters. Lower levels of phosphate were observed to beat par or sometimes better than the higher levels in producing vigorous growth and yield of inoculated plants (Cruz *et al.*, 1988; Michel Sen and Resendahl, 1990).

Nitrogen uptake and phosphorus up take in shoots increased with increase in rock phosphate plus *Glomas fasiculatum* and *Rhizobium*. This may be because of the slowly dissolved available phosphorus in the soil and phosphorus being the essential element for establishment of healthy symbiosis between *Rhizobium* and *G. fasiculatum* (Lakshman, 1999).

The nutrient content increased significantly in rock phosphate amended with mycorrhizal and *Rhizobium* inoculated plants as compared to control set (Table 31.3) similarly, nitrogen and P up take increased in VAM +*Rhizobium* + rock phosphate treatments. Therefore, these studies indicated that mycorrhizal inoculation helps in the effective utilization of rock phosphate by changing in to available form, which later is taken up by the plants for better growth and development.

Acknowledgements

Authors are thankful of SAP II and COSIST New Delhi for the financial support.

References

Allen, M.F. 1992. Mycorrhizal functioning; An Integrative plant fungal process. Champion and Hall, New York. London.

Chhabra, M.L. and Jalai, B.L. 1991. Effects of VA mycorrhizal fungi *G. mossea* inoculation on utilization of rock phosphate in wheat. Indian. J. Mycol and plant pathology. 21(1) : 104-105.

Cruz, R.E.D., Manalo, M of, Aggngan, N.S. and Tambalo, T.D. 1988. Growth of three legme trees inoculated with mycorrhizal fungi and *Rhizobium* plant and soil 108 : 111-115.

Gerdmann, J. W. and Nicolson, T.H. Spores of mycorrhizal endogone species extracted from soil by wet sieving and decanting method. Trans Brit. Mycol. Soc. 46 : 235-245.

Jackson, M.L. 1973. Soil chemical analysis. Prentice Hall, New Delhi.

Jasper, D.A. 1994. Bioremediation of agriculture and forestry soils with symbiotic micro organisms. Aust. J. of Soil Research. 32 : 301-1309.

Koide, R.T. and Schreiner, R.P. 1992. Regulation of the VAM symbiosis. Annual review of plant physiology and plant-Molecular Biology. 43 : 557-581.

Lakshman, H.C. 1999. Beneficial micro organisms and their interaction on *Pterocarpus marsupium* a timber plant. Asain. J. Biotech and Environmental protection. 78 : 105 - 112.

Michelson, A. and Rosendahl, S. 1990. The effect of VA-mycorrhizal fungi, phosphorus and drought stress on the growth of Acacia Milotica and lencaena Lenocephala seedlings. Plant and soil. 124 : 713.

Mosse, B., Powell, C.L. and Hayman, D.S. 1976. Plant growth response to vesicular arbuscular mycorrhiza. IX. Interactions between VA-mycorrhiza, rock phosphate and symbiotic nitrogen fixation. New Phytol. 76 : 331-342.

Munns, D.S., and Mosse, B. 1990. Mineral nutrition of legume crops. In : Advances in legume science., PP. IIs edited by R.J. Summer field and A.H. Bunting. University of Reading England.

Munns, D.N. and Mosse, B. 1980. Mineral nutrition of Legume crops. In : Advances in Legume Science summer field. R.J. and Bunting, A.H (eds). Proceeding of the International legme conference, kew pp. 115-125.

Olsen, S.R., (eds., C.V.,. Watanabe, F.S and Dean, L.A.) 1954. Estimation of available phosphorus in soils by extraction with sodium bicarbonate. Circular 939. United States Department of Agriculture Supplement of documents. Washington, DC. US Government printing office, p. 19.

Phillips, J.H and Hayman, D.S. 1970. Improved method for clearing roots and staining parastic and vesicular arbuscular mycorrhizal fungi for rapid assessment of infection. Traus, Brit, Mycc. Soc 55 : 158-161.

Singh, H.P. 1994. Response to inoculation with *Branyrhizobium*, vesicular arbuscular mycorrhiza and phosphate solubilizing bacteria soybean in a Mol. Sc. Indian. J. Micbic./34 : 27-31.

Tilak, K.V.B.R. 1993. Associative effects of vesicular arbuscular mycorrhizae with nitrogen fixtures, proceedings of Indian National Science Academy B59 (3 and 4) : 325-332.

Trappe, J.M. 1977. Selection of fungi for endomycorrhizal in nurseries. Ann Review of phytopathol. 1.15 : 203-222.

Trappe, J.M. 1982. Mycorrhizae and productivity of arid and semi-arid rangelands. In advances in food producing system for arid and semiarid lands. New York; Academic Press. 581-599.

Walkely, A. and Black, T.A. 1934. An estimation of digital method for determining soil organic matter and a proposed modification of the chromic acid titration method. Soil Science. 37 : 29-38.

ENVIRONMENTAL BIOTECHNOLOGY

Edited By :   **Professor (Dr.) Arvind Kumar**

Published By :   **DAYA PUBLISHING HOUSE**

32

MASS PROPAGATION OF BAMBOO (*DENDROCALAMUS HAMILTONII* NEES AND EX MUNRO) IN RESPONSE TO PLANT GROWTH REGULATORS AND FERTILIZATION

• *S.K. Kaushal and Usha Rana*

College of Basic Sciences, CSKHPKV, Palampur-176 062

Abstract

Dendrocalamus hamiltonii Nees and Ex Munro like other bamboo is a multipurpose, fast growing and high yielding variety with large sized and thick walled culms having excellent fodder quality. It has been propagated exclusively by vegetative means but the produced plants are of the same physiological age. Hence, they are bound to flower terminally leading to large scale death. It is imperative to raise new planting material from seeds well in time. Therefore, studies were conducted to improve seedling vigour by applying hormones and fertilizers for the mass multiplication of elite seedlings. Different concentrations of growth regulator GA_3 and single dose of kinetin were sprayed on one year old seedling growing in plots. Foliar application of growth regulators (50 ppm GA_3 and 2×10^{-6} g/l kinetin in seedlings significantly increased the number of tillers, internode length and tiller heights to almost double. A significant reduction in sprouting, rooting rhizogenesis days and an increased percentage of rooting and sprouting was observed with farm yard manure (FYM 1.5 kg/cutting) in combination with different doses of nitrogen (10 g/cutting) and phosphorus 10 g/cutting. Plants raised from seedlings usually require a period of ten years to atttain normal size. Therefore, to reduce this long period, cutting segments from juvenile seedlings were taken with single, bi and trinodal segments and grown horizontally. The culms raised from these segments were very healthy and single node segments showed best results in culm emergence.

Key Words : Vegetative propagation, Juvenile seedlings, Dendrocalamus hamiltonii.

Introduction

Dendrocalamus hamiltonii (Nees and Ex Munro) a native of tropical Eastern Himalaya and Nepal, is cultivated in Himachal Pradesh and has been put to so many uses. It has entered the culture of hilly peoples who call it "Maggar" or "Phargalu" (Sharma and Kaushal, 1985). It is planted from 350 to 1500 m altitude and gives its best performance in Kangra valley of H.P. Marked by 2000-2800 mm annual rainfall mostly concentrated during rainy season. Once established, it is tolerant to moisture stress and has adaptability to wide range of edaphic conditions, except water logging. It is a good source to improve the economy of the people, conserve environment, prevent soil erosion, proper utilization of waste and marginal lands to reduce pressure over valuable timber and generate local employment as a cottage industry.

Bamboo is monocarpic and bears sporadically flowering culms after 40-50 years. The existing clones in Himachal Pradesh are physiologically very old. Plant growth regulators are used in improving relative growth rate (Guttridge and Thompson, 1964; Dijkstra *et al.*, 1990). Vegetative propagation of bamboo is performed by seedling division, seedling multiplication, culm cuttings and nodal segment cuttings (Bennett and Gaur, 1990; Singh, 1999). Bamboo improvement has been initiated in Arunachal Pradesh (Banwal and Singh, 1988) but no work has been taken up on the selection and clonal multiplication of elite individual of seminal origin. Naturally growing seedlings never survive due to overcrowding with grassy weeds, post monsoon dry spell and severe winter. However, such seedlings were transplanted in the nursery and for their healthy growth present investigation was designed to find out the efficacy of different treatments (plant growth regulators and fertilizers) to improve relative growth rate of seedling to attain normal culm size in short duration and to use appropriate plantation technology to multiply elite seedlings on a large scale.

Material and Methods

Viable seeds produced during April-June in sporadically flowering culms germinate during June-July were taken out from the field and were transplanted in nursery where they perform well under care. Different doses of growth regulators of GA_3 (10, 25, 100, 300, 400 and 500×10^{-6} g/l) and single dose of kinetin (2×10^{-6} g/l) were sprayed on one year old seedlings growing in plots. Fifteen seedlings of uniform height and tillers were selected for each treatment and replicated thrice. For growth hormones were sprayed on seedlings in the month of March and observations on growth were taken in April and July.

To study the effect of different fertilizers on the growth, one year old culm cuttings were given following treatments.

T_1 (Control), T_2 (FYM alone), T_3 (FYM+N_1P_1), T_4 (FYM+N_1P_2), T_5 (FYM+N_1P_3), T_6 (FYM+N_2P_1), T_7 (FYM+N_2P_2), T_8 (FYM+N_2P_3), T_9 (FYM+N_3P_1), T_{10} (FYM+N_3P_2), T_{11} (FYM+N_3P_3), T_{12} (FYM+N_4P_1) T_{13} (FYM+N_4P_2) and T_{14} (FYM+N_4P_3). Levels of nitrogen (urea g/cutting) applied were $N_1 = 2.5$, $N_2 = 5$, $N_3 = 10$, $N_4 = 15$ and P (SSP g/cutting), $P_1 = 5$, $P_2 = 10$, $P_3 = 15$. The fertilizer trials were conducted in RBD with plot size 1 m × 1 m with 10 cuttings per treatment and three replications. The nitrogen (urea) and phosphorus (SSP) were applied in two equal split doses, first half at the time of rooting and another half at rhizogenesis stage. The cuttings were planted in March and observations were taken till July.

For the vegetative propagation, seedlings raised from seeds were bisected in such a way that each part comprises root, rhizomes and tillers. Seedlings were placed in poly bags and kept in shade/polyhouse to avoid direct sunlight on young seedlings. The seedlings were kept hydrated till their division performed another seedlings division and so on and thus one seedling produced 3-4 plants in a year. Seedlings were transplanted after two years and allowed to proliferate.

For the mass multiplication and to obtain normal size of bamboo culm from seedling in short duration, single, bi and trinodal segment cuttings of juvenile elite seedlings were performed. The nodal segments were grown horizontally in FYM @ 1.5 kg alongwith different doses of nitrogen (3, 6, 9, 12 g/cutting) and phosphorus (5 g, 10 g/cutting).

Nitrogen was applied as half dose during planting and another half at rhizogenesis stage. Phosphorus was applied as single dose at the time of planting.

To evaluate the performance of bamboo culms from juvenile seedlings, various growth parameters like, number of tillers, internode length, tiller height, number of new culms, wall thickness, culm girth and number of nodes of new sprouts of bamboo raised from horizontally planted seedlings (15 cm depth) were recorded.

Results and Discussion

Response of seedlings to foliar application of growth regulators

Number of tillers

In April, one month after spray, non-significant changes in the number of tillers were observed (Table 32.1). However, in July a significant increase in tillers was found with all treatments as compared to control. Maximum tillers were produced by 50×10^{-6} g/l GA_3 that were significantly higher than control and other treatments. Lower concentration were more effective than higher treatments as compared to control. On the other hand, earlier workers (Nanda *et al.*, 1973) reported non-significant differences in the number of tiller after foliar applications of regulators. These differences in the results may be attributed to different timings of spraying.

Table 32.1 : Response of Different Foliar Applications of GA₃ at Constant Concentration of Kinetin (2×10⁻⁶ g/l) on the Growth of Seedlings

GA_3 (10^{-6} g/l)	Number of tillers			Internode length (cm)			Average tiller height (cm)		
	March	April	July	March	April	July	March	April	July
10	3.2	3.2	4.9	3.6	4.1	8.7	21.5	22.6	41.7
25	3.2	3.3	5.1	3.6	4.8	8.5	20.2	23.0	47.5
50	3.5	3.5	6.0	3.5	4.8	10.6	20.1	28.4	71.9
100	3.5	3.5	6.1	3.7	4.2	7.2	21.2	27.5	46.4
200	3.5	3.7	6.0	3.8	4.7	7.7	21.4	22.8	46.3
300	3.3	3.3	6.0	3.7	4.8	6.4	20.8	21.6	37.2
400	3.4	3.4	5.8	3.6	4.2	6.2	21.6	24.3	37.2
500	3.2	3.2	5.7	3.5	4.6	6.2	20.4	24.2	38.1
Control	3.5	3.5	4.5	3.6	4.4	5.1	21.3	22.1	35.4
CD at 5%									
Months		0.29			0.32			2.16	
Concentrations		0.62			0.68			4.56	
Interactions		0.88			0.97			6.46	

Internode length

Average length of internode increased significantly in April over March by the application of all concentration of GA_3. However, there were non-significant differences among different doses of hormone. On the other hand, in July the internode length increased significantly over April. All the concentrations exhibited significant increase in the internode length over control, whereas GA_3 (50×10^{-6} g/l) showed the hightest increase.

Average height of tillers

All the concentration showed significant increase in the average tiller heights in the month of April over March and maximum height of tiller was observed with the application of GA_3 (50×10^{-6} g/l). In the month of July, tiller height increased significantly over April in all the treatments. Again, GA_3 (50×10^{-6} g/l) was most promotory as compared to other treatments and control. The higher concentration (more than 200×10^{6} g/l) were not effective in stimulating the tiller height. Nandi *et al.* (1996) have also observed increase in average height with foliar application of GA_3.

Response of culm cuttings to fertilizer application

Days to sprouting

While the untreated took 16.2 days to sprout, a significant reduction in sprouting days was observed with all the treatments (Table 32.2). T_{10} caused maximum decrease over the control and other treatments. The effect of T_2 to T_8 was significantly lower than T_{10} to T_{14}.

Table 32.2 : Effect of Fertilizer Application on the Growth of One-year Old Culm Cuttings

Treatment	Days to sprouting	Days to rooting	Days to rhizogenesis	Rooting (%)	Sprouting (%)
T_1	16.2	47.4	74.8	51.2	82.6
T_2	13.4	44.2	75.1	54.2	84.4
T_3	13.6	45.2	74.1	53.1	85.1
T_4	12.9	45.3	68.3	60.2	84.9
T_5	12.2	43.1	68.1	60.6	85.7
T_6	12.8	38.7	65.8	63.4	85.1
T_7	11.3	38.4	65.9	68.6	88.2
T_8	10.2	35.1	63.2	68.6	87.6
T_9	10.5	35.4	60.2	72.1	85.4
T_{10}	7.2	30.1	58.2	86.4	90.2
T_{11}	7.8	30.8	58.6	85.2	91.2
T_{12}	9.4	32.2	61.2	85.8	86.2
T_{13}	7.6	34.2	60.2	82.1	90.6
T_{14}	7.8	33.5	60.4	86.6	88.4
CD at 5%	2.1	4.3	4.7	4.1	4.3

Days to rooting

Treatments, T_{10} and T_{11} took significantly lesser days for rooting as compared to control and T_2 to T_9 treatments. T_{12}, T_{13} and T_{14} also caused a significant reduction over the control and their effect were at par with T_{10} and T_{11}. T_2 to T_5 treatments resulted in non-significant decrease in rooting days over the control.

Days to rhizogenesis

A significant reduction in days to rhizome initiation was observed over control with T_4 to T_{14} treatments. T_{10} and T_{12} were most effective and had the same impact for inducing early rhizome formation. The effect of T_9 and T_{12} to T_{14} were at par with T_{10} and did not differ significantly among themselves.

Rooting

Higher doses of nitrogen and phosphorus (above T_3) significantly increased the rooting in the cutting. Maximum rooting (86.6%) was observed with T_{14}, which was non-significantly different from T_{10} and T_{12} treatments but was significantly higher than control (51.2) and T_2 to T_9 treatments.

Mass multiplicaton of elite seedlings

The elite seedlings selected over the period of last ten years has been multiplicated using vegetative propagation technology. Bisected division of the seedling were repeated to increase the number of elite seedlings for mass multiplication. Juvenile seedlings were also used for mass multiplication because they are thinner and easier to cut. Single, bi and trinodal segmental cuttings were used to evaluate root proliferation by growing them horizontally. It has been observed that single node cutting gave best result as flow of nutrient was more concentrated on one potential node than the segment having many buds and node. Plants raised from such selection were genetically improved, live for many years and provide propagating material of young physiological age.

References

Banwal, B.S. and N.B. Singh, 1998. Role of promoting substances in Bamboo cultivation. *Indian Forester.* 114(9) : 549-559.

Bennett, B.S. and R.C. Gaur, 1990. Twenty six bamboo spp. Cultivated in India. Forest Research Institute, Dehra Dun, India.

Dijkstra, P.H., Ter Reegen and P.J.C. Kuiper, 1990. Relation between relative growth rate, endogenous gibberellins, and the response to applied giberellic acid for *Plantago major. Physiol. Plant.* 79 : 629-634.

Guttridge, C.G. and P.A. Thompson, 1964. The effect of gibberelins on growth and flowering of *Fragaria* and *Duchenea. J. Exp. Bot.* 15 : 631-46.

Nanda, K.K., P. Kumar and V. Kocchar, 1973. Role of auxins, antioxins and phenols in the production and differentiation of callus on stem cuttings of *Populus robusta. NZ. J. For. Sci.* 4 : 338-346.

Nandi, S.K., L.M.S. Palani and H.C. Rikhari, 1996. Chemical induction of advantitious root formation in *Taxus baccata. Plant Growth Regulation* 19 : 117-122.

Sharma, O.P. and S.K. Kaushal, 1985. Exploratory propagation of *Dendrocalamus hamiltonii* Ex Munro by one node culm cuttings. *Indian Forester* 111(3) : 135-139.

Singh, Balbir, 1999. Studies on macro propagation of bamboo with relation to plant growth regulators MSC. Thesis. H.P. Krishi Vishvavidyalaya Palampur (H.P.), India.

ENVIRONMENTAL BIOTECHNOLOGY

Edited By :   **Professor (Dr.) Arvind Kumar**

Published By :   **DAYA PUBLISHING HOUSE**

33

DEVELOPMENT AND REPRODUCTION OF APTERAE AND ALATES OF *PENTALONIA NIGRONERVOSA COQ.* (HOMOPTERA : APHIDIDAE) IN FIELD CONDITIONS

• *C. Padmalatha*, A.J.A. Ranjit Singh and C. Jeyapaul*

Dept. of Biology, Sri Paramakalyani College, Alwarkurichi-627 412, India
* Dept. of Zoology, Rani Anna Govt. College for Women, Tirunelveli-627 008, India

Abstract

Longevity, fecundity and reproductive potential of apterous and alate forms of banana aphid *Pentalonia nigronervosa coq.* were studied in field and laboratory conditions. Third and fourth instar of alatiform nymphs needed longer time to become the next stage than the apterous nymphs. The 3rd instar of apteriform nymph took 18 ± 0.3 days to become the next instar whereas alatiform nymph took 2.3 ± 0.6 days to complete the 3rd instar stage. The duration of fourth instar of apteriform nymph was 2.8 ± 0.1 days, whereas the alatiform nymph took 3.1 ± 0.3 days to complete the 4th instar stage. The total life period of alates was shorter than apterous form. The mean progeny left by an apterous adult was higher (43.7 ± 0.7) than that by an alate adult (mean 33.8 ± 0.6). A maximum of four or five young ones were produced daily and the population did not reach the very high densities unlike many other aphids.

Key Words : Alates, Apterae, Pentalonia nigronervosa aphids, Banana, Viral vector.

Introduction

Pentalonia nigronervosa coq. (Aphididae) transmits "Bunchy Top" viral diseases in banana plantations of Kerala (Wardlaw, 1961). The reproductive rate of the aphid in the field is a key factor for population growth. Reproduction and life cycle of aphids were extensively reviewed (Kennedy and Stroyan, 1959 and Hille, 1966). Differences exist in the reproductive potential of both apterous and alate forms. Lal (1950) noted that the fecundity of apterous *Myzus persicae* was always higher than that of alate. Basant and Roy (1987) reported the influence of seasonal changes on the reproductive ability of a monoecious aphid *Greenideoida ceyloniae.* The kind of host plant and conditions under which it was grown affect the aphid fecundity and reproductive rate. Wearing (1967) noticed increasing fecundity of aphids when their food plants were water stressed. Alate forms played a major role in the distribution of population of *P. nigronervosa* (Hardy, 1941). Hence, a field study can help to contain *P. nigronervosa* and incidence of 'Bunchy Top' disease in banana cultivars. Hence a field study on reproduction and development of apterae and alates of *P. nigronervosa* was made.

Materials and Methods

For the present study on reproduction and development of aphid in the field, healthy banana plants were planted in experimental fields in Kariavattam of Kerala State. They were watered regularly. The axils of young leaves in the crown region of those plants were selected for the site to introduce first instar nymph. The nymphs were introduced gently by means of camel hairbrush. Some space was given for the easy and free movements of the aphid. A barrier was made to tightly packing the two sides with cotton. Then the upper side of the cotton was sealed with cellotape, so that the aphid would be in and around the location. Care was taken to prevent the entry of ants and predators. Observation was made daily. The number of days taken for each instar to become the next, the number of young ones produced per day were noted and young ones were removed subsequently. Ten individuals of each morph were studied from the time of birth. Fecundity in field conditions were compared with laboratory reared aphids.

Results and Discussion

Females of *P. nigronervosa* reproduce only by parthonegenitic viviparity. At the time of giving birth to young ones, the mother firmly held the substrate and slowly eject out the nymph from her genital openning. The time required for the complete ejection of a nymph varied from 5-15 minutes. The new born nymph remained protruded from the genital openning until it was capable of movement.

Four nymphal stages were found in the post-embryonic development of *P. nigronervosa* (Table 33.1). The first instar period varied from 2 to 4 days with a mean of 2.5 days. The duration of second instar nymph varied from 3 to 4 days with a mean 3.2 days. The third instar period of aptera extended from 1-3 days with a mean of 1.8 days. Fourth instar duration of aptera needed 2 to 4 days with a mean of 2.8 days. Of the four instars the third instar period was the shortest as it took 1.8 days to become the fourth and the second instar period was the longest as it took 3.2 days to become the third instar. A latiform nymphs took 2.3 days to complete the third and 3.1 days to complete the 4th instar.

Table 33.1 : The Development of the Instar Stages of Apterae and Alates Morphs of *Pentalonia nigronervosa* coq.

Life stages	No. of specimens examined	Mean ± SD duration of development for different instar	Mean ± SD duration of development for different instar
I instar	10	2.5 ± 0.4	2.4 ± 0.3
II instar	10	3.2 ± 0.5	3.2 ± 0.1
III instar	10	1.8 ± 0.3	2.3 ± 0.6
IV instar	10	2.8 ± 0.1	3.1 ± 0.3

The longevity of an apterous adult varied from 10-26 days with a mean of 19.6 days (Table 33.2). The longevity of an alate varied from 6-18 days with a mean period of 12 days. The total life period varied from 25-36 days with a mean of 30.7 days in the case of apterous adults. In the case of alates the total life period varied from 19-28 days with a mean of 23.8 days.

Table 33.2 : Frequency and Longevity Statistics of *Pentalonia nigronervosa* coq. in Field Conditions

Aphid form	Adults examined	Mean ± SD Progency (No. of offsprings)	Mean ± SD Longevity (days)	Mean ± SD Total life period (days)
Apterous	15	43.7 ± 0.7	19.6 ± 0.3	30.7 ± 0.4
Alate	15	33.8 ± 0.6	12.0 ± 0.2	23.8 ± 0.6

Table 33.3 : Efficiency of Reproduction Corresponding to Number of Days in *Pentalonia nigronervosa coq.* in Field Conditions

No. of Days	Young ones produced			
	Apterae		Alates	
	Total	Percentage	Total	Percentage
First 15 days	558	85.19	103	86.6
Next 15 days	67	10.23	11	9.2
Remaining days	30	4.58	6	4.2

The number of offsprings laid by a single apterous female varied from 32-55 with a mean of 43.7 offsprings. The fecundity range of alate was found to be 22-47 with a mean of 33.8 offsprings. The present investigation revealed that alates produced fewer offsprings than apterae as it was reported in other aphids (Weed, 1927; Toba, 1964). The apterous adults began to reproduce 24 hours after reaching adulthood. The alate began to reproduce 48 hours after reaching adulthood. The maximum number of offsprings produced by a single female during a period of 24 hours was four. An apterous adult produced about 85.19% of young ones during the initial 15 days, 10.23% young ones were produced during the next 5 days and 4.58% young ones were produced during the remaining six days. The fecundity rate in the field conditions was compared with laboratory-reared aphid. Under the laboratory conditions 87.89% young ones were produced in the first 15 days, 11.06% young ones were produced in next 5 days and 1.05% young ones in last three days. In the laboratory reared aphids the longivity of progeny is limited. The total progeny left by an adult apterous female under laboratory condition was 21.9% whereas the total progeny left by an adult apterous female under field condition was 43.7%. The decrease in mortality in natural conditions confirms their higher degree of adaptability to field situations.

Acknowledgement

I am thankful to the Principal, Rani Anna Government College for Women for providing encouragement.

References

Basant, K., Agarwala and Roy, 1987. Seasonality and reproduction in *Greenideoida ceylonia* (Homoptera : Aphididae) *Entamon* 12(2) : 109-111.

Hardy, G.H. 1941. Apididae in Australia. *Proc. Roy. Soc. Qd.* 1 : 36-40.

Hille Ris Lambers, 1966. Polymorphism in Aphididae. *Ann. Rev. Entomol.* 11 : 47-70.

Kennedy, J.S. and Stroyan, H.J.G. 1959. Biology of aphids. *Ann. Rev. Entomol.* 4 : 139-160.

Lal, R. 1950. Biology and control of *Myzus persicae* (Sulzer) as a pest of potato at Delhi Indian *J. Agric. Sci.* 20 : 87-100.

Toba, H. 1964. Life history studies of *Myzus persicae* in Hawaii. *J. Econ. Entomol.* 57 : 290-291.

Wardlaw, C.W. 1961. Banana diseases, Longman, Green and Co. Ltd. pp. 84-110.

Wearing, C.H. 1967. Studies on the relations of insects and host plant. *Nature* (London) 213 : 1052-1053.

Weed, A. 1927. Metamorphosis and reproduction of apterous of *Myzus persicae* as influenced by temperature and humidity. *J. Econ. Entomol.* 20 : 150-157.

34

EFFECT OF FREE VOLUME AND INTERNAL PRESSURE ON ION-SOLVENT INTERACTION OF SOME AQUEOUS ELECTROLYTIC SOLUTIONS

• *A.N. Kannappan and V. Arumugam*

Department of Physics, Annamalai University, Annamalai Nagar-608 002

Abstract

Sound velocity, viscosity and density were measured on sodium chloride, sodium formate and sodium acetate in the mixtures of 10, 20 and 30% DMF with water for different salt concentrations at the temperatures 303, 308, 313, 318 and 323 K. The free volume and internal pressure were computed using the sound velocity, viscosity and density. The results were discussed in terms of the structural changes around the ion and the effect of solvent. For any given concentration the variation of internal pressure and free volume obeys the relationship $\pi_i V_f^x = K$ in accordance with the previous results.

Key Words : Free volume, Internal pressure, Ion-solvent interaction.

Introduction

Electrolytic solutions in aqueous medium is useful for obtaining information on the arrangement of matter. Such studies on binary liquids have been made by several workers (K. Krishnamoorthy, S.O. Pillai and J. Kuppusami, 1977; A. Dhanalakshmi and V. Lalitha, 1985; D.K. Jha and B.L. Jha, 1989; C.V. Suryanarayana, 1986). Since it is impossible to understand the structural properties and the type of interaction involved in the binary system, the study of ternary electrolytes is gaining much importance nowadays. These considerations led us to undertake the present work of evaluating free volume (V_f and internal pressure (π_i) for measurements at different concentrations and temperatures. The results are discussed in terms of structural changes of solvent. For any given concentration the variation of internal pressure and free volume obeys the relationship $\pi_t V_f^x = K$ in accordance with previous results.

Materials and Methods

Fresh double distilled conductivity water was used for making the water-DMF mixtures. The homogeneous system is allowed to attain the room temperature. The required quantity of the sodium chloride for a given molality is then dissolved. This is repeated for other concentration of sodium chloride and also for different compositions. The ultrasonic velocities of the above solutions were measured at 303, 308, 313, 318 and 323 K using an ultrasonic interferometer (Mittal Enterprises, New Delhi) operating at a frequency of 2 MHz with an accuracy of ± 0.1 %. The required temperature has been maintained using an electronically controlled thermostat having an accuracy ± 0.1° C. They were measured densities of the solutions at these temperatures using a specific gravity bottle of 5 ml/capacity. The viscosities of the solutions at this temperature were measured using an Ostwald's viscometer and are accurate to ± 0.001 Nsm^{-2}. The same procedure were adopted for determining sound velocity, viscosity and density of sodium formate and sodium acetate in DMF-water mixtures.

Theory

The free volume and internal pressure were calculated using the relations (C.V. Suryanarayana and J. Kuppusami, 1976; D.P. Shoemaker and C.W. Garland, 1967)

$$V_f = \left[\frac{MU}{K\eta}\right]^{3/2} \tag{1}$$

$$\pi_i = bRT \left[\frac{K\eta}{U}\right]^{1/2} \frac{\rho^{2/3}}{M^{7/6}} \tag{2}$$

where M is the effective molecular weight, U is the sound velocity, η is the viscosity, K is a constant having a value of 4.28×10^9 independent of temperature, b is the space packing factor C.V. Suryanarayana (1979) equal to 2, R is the gas constant, T is the absolute temperature and D is the density.

Table 34.1 : Values of Free Volume (V_f), Internal Pressure (π_i) and the Arbitrary Constants a, b, c and d of Sodium Chloride in DMF Water Mixtures

Molality of salt (m)	$V_f \times 10^8$ m^3 mol^{-1} Temperature (K)					$\pi_i \times 10^{-6}$ Pa Temperature (K)					ax 10^{-10} Pa	bx10^3 K^{-1}	cx10^3 m^{3}mol^{-1}	dx10^3 K^{-1}
	303	308	313	318	323	303	308	313	318	323				
0.0000	1.93	2.33	2.75	3.22	3.74	2582	2465	2366	2278	2196	–	–	–	–
0.2058	1.96	2.36	2.77	3.23	3.78	2562	2451	2351	2267	2180	3.46	8.55	5.42	34.65
0.4029	1.93	2.30	2.72	3.17	3.67	2579	2471	2371	2283	2206	3.13	8.23	7.19	33.65
0.5986	1.89	2.28	2.65	3.13	3.62	2604	2481	2397	2301	2222	3.06	8.15	5.67	34.36
0.7991	1.88	2.23	2.59	3.08	3.57	2608	2501	2416	2312	2231	2.09	7.97	9.27	32.79
1.0001	1.89	2.23	2.55	3.05	3.54	2603	2501	2426	2318	2237	2.60	7.63	13.77	31.52
DMF – Water (10 : 10%)														
0.0000	1.69	2.04	2.43	2.90	3.37	2550	2432	2328	2227	2148	–	–	–	–
0.2031	1.69	2.03	2.42	2.83	3.24	2552	2440	2332	2230	2146	3.42	8.53	3.42	35.71
0.4037	1.67	2.00	2.38	2.83	3.33	2563	2441	2338	2245	2155	4.07	9.11	4.07	36.01
0.6017	1.64	1.98	2.34	2.79	3.23	2581	2459	2359	2256	–	4.45	9.39	4.45	35.63
0.8071	1.62	1.98	2.33	2.72	3.25	2587	2466	2362	2274	2173	4.19	9.18	4.19	36.46
1.0014	1.92	1.92	2.29	2.72	3.17	2592	2482	2377	2277	2193	3.45	8.59	3.45	32.90
(20 : 80%)														
0.0000	1.54	1.85	2.21	2.69	3.18	2473	2361	2258	2142	2052	–	–	–	–
0.2003	1.55	1.84	2.20	2.26	3.14	2467	2367	2261	2157	2068	4.64	9.59	4.64	37.67
0.4073	1.46	1.77	2.13	2.54	3.00	2519	2399	2285	2185	2097	6.16	10.60	6.16	37.84
0.5983	1.44	1.76	2.11	2.53	2.97	2529	2404	2293	2189	2102	4.78	9.71	4.78	37.98
0.8067	1.41	1.73	2.06	2.47	2.92	2550	2419	2314	2206	2114	5.00	9.82	5.00	37.23
1.0008	1.38	1.67	2.01	2.39	2.83	2572	2449	2334	2234	2139	4.87	9.75	4.87	37.07

Table 34.2 : Values of Free Volume (V_f), Internal Pressure (π_i) and the Arbitrary Constants a, b, c and d of Sodium Chloride in DMF Water Mixtures

| Molality of salt (m) | $V_f \times 10^8$ m³ mol⁻¹ | | | | | $\pi_i \times 10^{-6}$ Pa | | | | | a × 10⁻¹⁰ | b×10³ | c×10³ | d×10³ |
| | Temperature (K) | | | | | Temperature (K) | | | | | | | | |
	303	308	313	318	323	303	308	313	318	323	Pa	K⁻¹	m³mol⁻¹	K⁻¹
						DMF – Water (10 : 90%)								
0.1981	1.92	2.28	2.71	3.16	3.68	2584	2479	2375	2286	2205	5.41	8.20	3.73	30.81
0.3976	1.86	2.22	2.62	3.06	3.58	2609	2500	2401	2317	2223	5.46	8.25	6.38	33.89
0.6006	1.81	2.16	2.54	2.95	3.47	2631	2519	2423	2338	2423	5.51	7.49	6.91	33.59
0.7945	1.75	2.08	2.48	2.90	3.39	2656	2546	2440	2347	2258	5.54	8.33	5.92	34.04
1.0079	1.71	2.05	2.38	2.82	3.27	2674	2557	2467	2365	2282	5.68	9.91	6.37	33.77
						(20 : 80%)								
0.1958	1.68	2.00	2.39	2.83	3.29	2553	2445	2340	2242	2161	3.51	8.59	4.40	34.62
0.4003	1.66	1.91	2.32	2.73	3.21	2560	2482	2358	2267	2177	3.06	8.14	6.80	33.20
0.6007	1.58	1.89	2.25	2.65	3.13	2603	2487	2381	2285	2192	4.11	9.09	3.96	34.96
0.7996	1.54	1.82	2.18	2.60	3.03	2620	2516	2406	2299	2214	3.31	8.44	4.98	34.23
0.9815	1.44	1.76	2.06	2.48	2.90	2679	2543	2249	2334	2243	9.37	3.07	3.31	35.68
						(30 : 70%)								
0.2014	1.49	1.79	2.15	2.57	3.02	2496	2383	2279	2173	2088	4.09	9.13	2.31	36.25
0.4014	1.42	1.72	2.07	2.48	2.92	2547	2423	2309	2207	2117	7.38	11.07	1.69	37.31
0.5924	1.35	1.68	1.98	2.39	2.83	2575	2434	2335	2225	2130	4.81	9.67	1.46	37.79
0.7975	1.31	1.60	1.92	2.30	2.73	2578	2466	2354	2248	2155	4.35	9.34	1.45	37.77
0.9964	1.22	1.54	1.85	2.20	2.65	2660	2500	2381	2284	2176	7.26	10.92	0.98	38.91

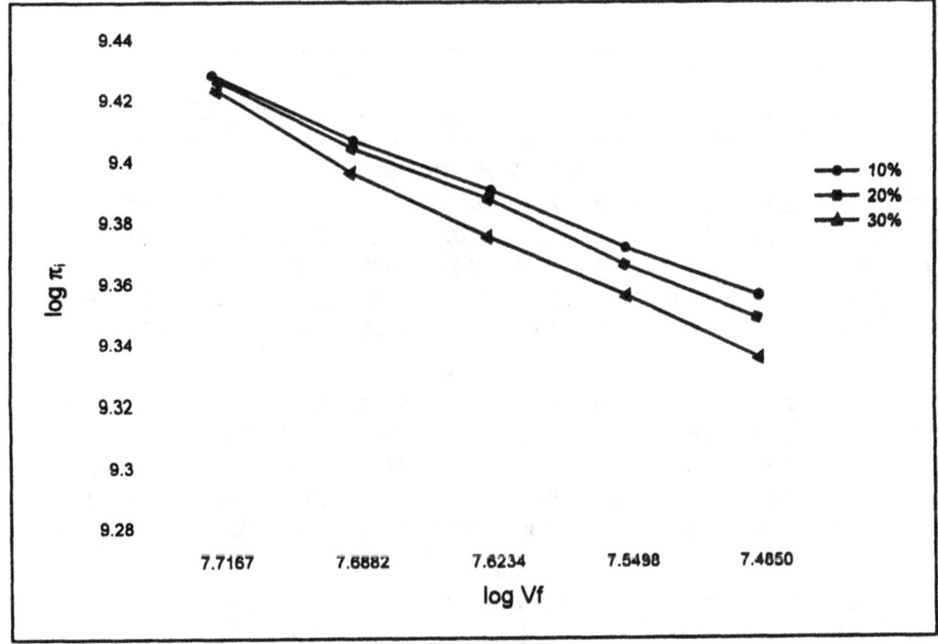

Fig. 34.1

Results and Discussion

The values of free volume and internal pressure for sodium chloride, sodium formate and sodium acetate in 10, 20 and 30% DMF-water mixtures at different molalities and temperatures have been calculated using the equations (1) and (2) and are presented in Tables 34.1-34.3. It is inferred from the tables that the free volume decreases with increasing molality of the salt as well as DMF content in the mixtures. The decrease in free volume with salt concentration is due to the presence of tightly packed solvent molecules around the ions at higher molalities. Thus there is a significant interaction between the ions and solvent molecules, which causes a decrease in free volume. As the DMF content is increased more and more DMF-water molecules are formed Das, P .B. (1977), which results in further reduction of free volume in the mixtures. Also there is a progressive decrease in internal pressure with increase in molality for all the three salts in 10, 20 and 30% DMF content. As the molality of the electrolyte increases, ion-solvent interaction increases resulting in overall increase in π_i. As π_i is a measure of interaction Anwar Ali and Anil Kumar Nain, (1994), its value is found to decrease with temperature at all the molalities in all the electrolytic solutions. It may be due to the thermal disassociation of some of the solvent molecules from the solvated ions. Further as DMF content increases, the internal pressure is found to decrease, as the complex formation is enhanced due to strong ion-solvent interaction. The variation of internal pressure and free volume with temperature, C.V Suryanarayana and J. Kuppusami (1981) at any given concentration can be represented by

Table 34.3 : Values of Free Volume (V_f), Internal Pressure (π_i) and the Arbitrary Constants a, b, c and d of Sodium Chloride in DMF Water Mixtures

Molality of salt (m)	$V_f \times 10^8$ m³ mol⁻¹ Temperature (K)					$\pi_i \times 10^{-6}$ Pa Temperature (K)					a× 10⁻¹⁰ Pa	b×10³ K⁻¹	c×10³ m³mol⁻¹	d×10³ K⁻¹
	303	308	313	318	323	303	308	313	318	323				
DMF – Water (10 : 90%)														
0.1981	1.89	2.24	2.65	3.10	3.62	2596	2490	2389	2297	2210	3.18	8.27	6.79	33.75
0.3976	1.78	2.13	2.51	2.97	3.47	2638	2523	2421	2323	2236	3.32	8.36	8.24	35.30
0.6006	1.70	2.03	2.39	2.84	3.31	2671	2556	2456	2352	2266	3.40	8.40	5.24	34.27
0.7945	1.59	1.92	2.28	2.69	3.15	2721	2592	2488	2385	2295	3.74	8.66	3.45	34.51
1.0079	1.53	1.85	2.20	2.60	3.05	2747	2617	2505	2405	2309	4.37	9.13	1.58	37.85
(20 : 80%)														
0.1989	1.60	1.94	2.30	2.76	3.20	2591	2468	2366	2255	2176	3.18	8.30	1.85	37.52
0.4100	1.55	1.86	2.22	2.63	3.10	2613	2384	2381	2284	2192	4.02	9.02	4.58	34.45
0.5999	1.45	1.77	2.11	2.51	2.90	2664	2527	2418	2314	2235	4.99	9.67	2.50	36.22
0.8062	1.38	1.70	2.01	2.41	2.87	2679	2555	2448	2338	2235	3.58	8.57	1.07	38.80
0.9862	1.31	1.59	1.89	2.27	2.68	2735	2603	2493	2380	2281	4.68	9.37	1.59	37.24
(30 : 70%)														
0.1989	1.45	1.80	2.18	2.62	3.07	2514	2378	2260	2153	2067	5.34	10.10	0.74	40.24
0.3985	1.35	1.67	2.02	2.41	2.85	2571	2433	2312	2210	2117	7.21	11.00	0.97	39.10
0.5936	1.28	1.56	1.87	2.25	2.69	2612	2482	2369	2255	2154	5.17	9.85	1.26	38.05
0.7967	1.21	1.46	1.77	2.15	2.55	2650	2529	2400	2283	2184	5.25	9.85	5.6	33.14
1.0011	1.14	1.41	1.70	2.05	2.45	2694	2544	2429	2310	2209	7.09	10.80	7.61	31.97

$$\pi_i = a \exp(-bT) \tag{3}$$

$$\pi_i = c \exp(dT) \tag{4}$$

where a, b, c and d are arbitrary constants and their values are listed in Table 34.1-34.3. This investigation also reveals that the plots of log π_i vs log V_f (Fig. 34.1) for a given concentration at different temperatures are linear and parallel (a sample plot of sodium chloride is represented in the figure satisfying the relationship C.V. Suryanarayana and J. Kuppusami (1981).

In eqn. (5) K is a constant and X_{rms} values for sodium chloride, sodium formate and sodium acetate are computed to be 0.22, 0.250 and 0.256 respectively. The corresponding K_{rms} values are $(28.3 \pm 1.5)\ 10^6$, $(27.9 \pm 3)\ 10^6$ and $(26.3 \pm 3)\ 10^6$. The computed X_{rms} values agree well with those reported by earlier workers (A.N. Kannappan and V. Rajendran, 1992; A. Dhanalakshmi, 1980; S.O. Pillai and S. Natarajan, 1988).

ENVIRONMENTAL BIOTECHNOLOGY

Edited By :   **Professor (Dr.) Arvind Kumar**

Published By : **DAYA PUBLISHING HOUSE**

35

Aerobiology and Epidemiology of Certain Diseases of Groundnut

• *S.K. Aher, S.V. Thite* and B.N. Pande***

New Arts, Commerce and Science College Parner, Dist. Ahmednagar-414 302 (M.S.)
* Shri Chhatrapati Shivaji College, Shrigonda, Dist. Ahmednagar (M.S.)
** Dept. of Environmental Science, Dr. B.A. Marathwada University,
Aurangabad-431 004 (M.S.)

Abstract

Airspora over groundnut (*Arachys hypogea* L. Var. SB11) fields at Ghargaon (Tal. Shrigonda) Dist. Ahmednagar was studied for a period of two kharif seasons from 24th July, 1996 to 16th November, 1996 and 20th July, 1997 to 8th November, 1997 using a continuous Tilak Air Sampler installed with its orifice at 0.5m above the ground level. Total 81 types of airborne components were trapped out of which 2 belonged to Myxomycotina, 7 to Zygomycotina, 25 to Ascomycotina, 3 to Basidiomycotina, 43 to Deuteromycotina and 4 to other types. The pathogenic spores of *Alternaria, Cercosporidium, Helminthosporium, Aspergillus, Myrothicium* etc. showed their prevalence during the period of present investigation. Some spore types showed marked seasonal variations in their diurnal periodicity pattern. Most of the spore types occurred in higher numbers when the temperature was in range of 28-33° C and relative humidity 60-75%. Rain had a profound effect on the composition air spora.

Introduction

Aerobiological survey conducted in various parts of India revealed the richness of air spora (Sreeramulu, 1967), but most of these studies employed 'culture plate method' of spore trapping which was shown by Gregory and Stedman (1953) to have very low efficiency. Standing vegetation has a great influence on air spora of any place and it changes with changes in weather (Gregory, 1973). Sreeramulu (1967) stressed the need for extensive studies on airspora of out-door atmosphere which provides useful information regarding the dispersal of plant pathogens, allergens, etc. Groundnut, an important oil seed crop, is subjected to various airborne fungal diseases. In view of the prevalence of tikka disease and leaf spot disease, the spore content of air over groundnut fields was studied for a period of two kharif seasons.

Materials and Methods

Tilak's continuous volumetric spore trap was employed for the present studies (Tilak and Kulkarni, 1970). The spore trap was operated from 24th July to 16th November, 1996 in first kharif and from 20th July to 8th November, 1997 in second kharif sason respectively over the groundnut fields for two different seasons at Ghargaon (Tal. Shrigonda, Dist.

Ahmednagar). The present investigations were carried out in order to study the correlation between airborne microbial content, weather parameters, growth stages of the crop plants and subsequently their effect on disease incidence on the crop.

The identification of the spore's caught was based on visual characters of spores such as shape, size, colour, wall structure, ornamentation, etc. The observed features were tallied with figures given in the publications of Barnett (1955), Subramanian (1971), Tilak (1989) and also by comparing with the culture characters.

Results and Discussion

The spore content of air over the groundnut fields in both kharif seasons, was rich both qualitatively and quantitatively. A list of fungal spore types and other components of air, their total concentration and percentage contribution to the total airspora is given in Table 35.1. All the trapped airborne mycofloral types have been categorized as 'spore types'. In addition hyphal fragments, insect parts and pollen grains were also trapped and considered as artificially formed group 'other type'. It is evident from the observed total spore types that 43 spores belonged to Deuteromycotina, 25 to Ascomycotina, to Zygomycotina, 3 to Basidiomycotina, 2 to Myxomycotina and 4 to other types. During both the kharif seasons of 1996 and 1997 Dueromycotina contributed the highest percentage (66.19%) to the total airspora followed by Ascomycotina (20.87%), Zygomycotina (7.02%), Basidiomycotina (2.61%), other types (2.57%) and Myxomycotina (0.70%) (Table 35.2).

The most regularly occuring spore types, which contributed in a considerable number, were *Alternaria, Cercospordium, Cladosporium, Hypoxylon, Bispora, Leptosphaeria, Aspergillus, Nigrospora, Curvularia, Ceratophorum, Mucor, Lophiostoma, Helminthosporium* etc. From among all, *Alternaria* (17.64 and 17.10%), *Cercosporidium* (6.52 and 17.26%), *Cladosporium* (5.73 and 12.90%), *Bispora* (3.68 and 1.55%) etc. were the most dominant spore types in occurance in the air over groundnut field in both the kharif season (Table 35.1). *Alternaria* spores stood first in contribution to the total airspora in both the kharif seasons. Highest spore incidence (1106 and 1260/m³ of air) was observed on September 16, 1996 and September 29, 1997 when there was a record of 24.6° C temperature, 72% relative humidity, 7 km/hr. wind velocity and 0.3 mm rainfall in the first season where as 24° C temperature, 77% relative humidity, 5 km/hr. wind speed, no record of rains. But two days before there was an incidence of 10.4 mm rainfall. All these weather parameters might have favoured in increasing the spore concentration in the air.

The circadian periodicity studies indicated that the spore concentration was maximum and peak was obtained at 10.00 hrs. The concentration was found decreased immediately from 11.00 to 14.00 hrs. Similar observations were also recorded by Ramchander Rao (1987), Aher (2003), Meredith (1962), Di Menna (1955), etc.

The spores of *Cercosporidium, Aspergillus, Helminthosporium, Myrothocium* pathogenic to groundnut contributed 6.52 and 17.26%, 3.53 and 4.03%, 2.47 and 1.85% and 0.33 and 0.42% in the first and second kharif seasons respectively (Table 35.1). Highest concentration of *Cercosporidium* spore (644 and 1568/m³ of air) during the first and second kharif crop seasons was noticed on October 26, 1996 and September 23, 1997 when weather parameters were recorded like 23° C temperature, 72% relative humidity and no rainfall during first kharif season and 25° C temperature, 81% relative humidity with 43.3 mm rainfall. The occurrence of *Aspergillus* spores in the air was regular. Their highest incidence (2730/m³ and 4956/m³ of air) was recorded in the months of September 1996 and August 1997 in the first and second kharif seasons respectively. High humid conditions and low temperature favoured for their occurrence in the air.

Table 35.1 : Total Concentration and Percentage Contribution of Each Airborne Component Over Groundnut Fields During Two Kharif Seasons (1996 and 1997)

Sl. No.	Spore Type	Concentration/m³		% Contribution		Average%
		I	II	I	II	
1.	Alternaria	36932	36344	17.61	17.14	17.37
2.	Cercospora	13664	36680	6.52	17.30	11.91
3.	Cladosporium	11998	27412	5.72	12.93	9.32
4.	Leptosphaeira	11760	10884	5.61	5.13	5.37
5.	Hypoxylon	8484	7602	4.05	3.58	3.82
6.	Bispora	7714	3304	3.68	1.56	2.62
7.	Aspergillus	7406	8568	3.53	4.04	3.79
8.	Curvularia	5726	2996	2.73	1.41	2.07
9.	Nigrospora	5348	3724	2.55	1.76	2.15
10.	Helminthosporium	5180	3934	2.47	1.86	2.16
11.	Sclerospora	4844	0	2.31	0.00	1.15
12.	Lophiostoma	4452	910	2.12	0.43	1.28
13.	Circinella	4214	1232	2.01	0.58	1.30
14.	Ceratophorum	3892	630	1.86	0.30	1.08
15.	Smuts	3136	4788	1.50	2.26	1.88
16.	Heterosporium	2856	1148	1.36	0.54	0.95
17.	Cheatomium	2422	1092	1.15	0.51	0.83
18.	Massaria	2366	1470	1.13	0.69	0.91
19.	Hysterium	2310	2926	1.10	1.38	1.24
20.	Pollengrain	2156	3850	1.03	1.82	1.42
21.	Xylaria	2058	812	0.98	0.38	0.68
22.	Oidium	2058	686	0.98	0.32	0.65
23.	Plasmopara	1918	0	0.91	0.00	0.46
24.	Candida	1890	2828	0.90	1.33	1.12
25.	Oithia	1856	266	0.88	0.13	0.51
26.	Pleospora	1778	1386	0.85	0.65	0.75
27.	Sporidesmium	1736	1232	0.83	0.58	0.70
28.	Nosulosphaeria	1722	2366	0.82	1.12	0.97
29.	Melanospora	1694	518	0.81	0.24	0.53
30.	Periconia	1680	1736	0.80	0.82	0.81
31.	Cephaliophora	1666	1162	0.79	0.55	0.67
32.	Pleomassaria	1638	0	0.78	0.00	0.39
33.	Rhizopus	1610	3444	0.77	1.62	1.20
34.	Beltrania	1596	0	0.76	0.00	0.38
35.	Cunninghamella	1540	0	0.73	0.00	0.37
36.	Albugo	1526	3570	0.73	1.68	1.21
37.	Massarina	1428	1470	0.68	0.69	0.69
38.	Mucor	1414	3780	0.67	1.78	1.23
39.	Valsaria	1386	602	0.66	0.28	0.47
40.	Disymosphaeria	1352	1582	0.64	0.75	0.70

Contd...

Table 35.1 – Contd...

Sl. No.	Spore Type	Concentration/m³		% Contribution		Average%
		I	II	I	II	
41.	*Exosporium*	1330	896	0.63	0.42	0.53
42.	*Memnoniella*	1316	0	0.63	0.00	0.31
43.	*Diplodia*	1232	1554	0.59	0.73	0.66
44.	*Urediniospores*	1204	1260	0.57	0.59	0.58
45.	*Botrytis*	1106	140	0.53	0.07	0.30
46.	*Epicoccum*	1078	854	0.51	0.40	0.46
47.	*Drechslera*	1064	308	0.51	0.15	0.33
48.	*Macrophoma*	1050	406	0.50	0.19	0.35
49.	*Stemonites*	1008	658	0.48	0.31	0.40
50.	*Popularia*	980	896	0.47	0.42	0.44
51.	*Cloridium*	966	476	0.46	0.22	0.34
52.	*Fusarium*	966	0	0.46	0.00	0.23
53.	*Hyphal fragments*	938	1386	0.45	0.65	0.55
54.	*Bitrimonospora*	938	798	0.45	0.38	0.41
55.	*Pyricularia*	896	1596	0.43	0.75	0.59
56.	*Physarium*	784	532	0.37	0.25	0.31
57.	*Monascus*	728	644	0.35	0.30	0.33
58.	*Pithomyces*	728	0	0.35	0.00	0.17
59.	*Haplosporella*	714	1232	0.34	0.58	0.46
60.	*Myrothecium*	700	896	0.33	0.42	0.38
61.	*Pringsheimia*	658	0	0.31	0.00	0.16
62.	*Chaetomella*	644	2114	0.31	1.00	0.65
63.	*Pestalotia*	630	882	0.30	0.42	0.36
64.	*Insect parts*	630	728	0.30	0.34	0.32
65.	*Dendryphiopsis*	588	462	0.28	0.22	0.25
66.	*Corynespora*	574	826	0.27	0.39	0.33
67.	*Chlamydomyces*	574	0	0.27	0.00	0.14
68.	*Batryodiplodia*	518	350	0.25	0.17	0.21
69.	*Claviceps*	518	0	0.25	0.00	0.12
70.	*Plant parts*	448	728	0.21	0.34	0.28
71.	*Ganoderma*	448	210	0.21	0.10	0.16
72.	*Cordana*	406	448	0.19	0.21	0.20
73.	*Trematorphaeria*	364	882	0.17	0.42	0.29
74.	*Trichothecium*	350	756	0.17	0.36	0.26
75.	*Sordaria*	350	714	0.17	0.34	0.25
76.	*Beltraniella*	322	0	0.15	0.00	0.08
77.	*Lacellina*	322	0	0.15	0.00	0.08
78.	*Bombardia*	280	0	0.13	0.00	0.07
79.	*Gloeosporium*	266	56	0.13	0.03	0.08
80.	*Dictyosporium*	210	784	0.10	0.37	0.23
81.	*Amphisphaerella*	140	182	0.07	0.09	0.08
82.	*Begnisiella*	140	0	0.07	0.00	0.03
83.	*Apiorhyncostoma*	112	0	0.05	0.00	0.03
84.	*Chloridium*	98	476	0.05	0.22	0.14

Table 35.2 : Total Concentration and Percentage Contribution of Each Spore Group Over Groundnut Fields During Two Kharif Seasons (1996 and 1997)

Sl. No.	Spore Group	Concentration/m^3		% Contribution		Average%
		I	II	I	II	
1.	Myxomycotina	1792	1190	0.85	0.56	0.70
2.	Zygomycotina	17066	12516	8.15	5.89	7.02
3.	Ascomycotina	50974	36960	24.35	17.39	20.87
4.	Basidiomycotina	4788	6258	2.28	2.94	2.61
5.	Deuteromycotina	130494	148778	62.35	70.03	66.19
6.	Other types	4172	6734	1.99	3.16	2.57

Although weather parameter were congenial for the initiation of leaf spot disease, the crop variety sown in the field "SB 11" was found to be highly resistant to airborne inoculate of *Cercosporidium*, *Aspergillus*, *Helminthosporium* and *Myothecium* to bring about the disease development during the periods when the crop was growing in the fields in both the kharif seasons.

References

Aher, S.K., B.A. Patil and B.N. Pande (2003). Circadian periodicity curves of fungal spores over groundnut field. Int. J. Mendel, Vol. 20(1-2), 17-18.

Barnett, H.L. (1955). Illustrated genera of imperfect fungi. (Minneapolis : Burgess Publishing Co.) p. 218.

Di Menna, M.E. (1955). A quantitative study of air borne fungal spores in Dunedin, New Zealand; Trans. Br. Mycol. Soc. 38 : 119-129.

Gregory, P.H. (1973). Microbiology of the atmosphere (London : Leonard Hill Publications).

Gregory, P.H. and O.J. Sledman (1953). Deposition of air-borne Lycopodium spores on plane surfaces; Ann. Appl. Biol. 40 : 651-674.

Meredith, D.S. (1962). Some components of air spora in Jamaican banana plantations, Ann. Appl. Biol. 50 : 557-594.

Ramchandra Rao, K.S. (1987). Aerobiological studies at Aurangabad. Ph.D. Thesis, Marathwada Uni., Aurangabad.

Subramanian, C.V. (1971). Hyphomycetes : An account of Indian species except Cercosporae (New Delhi : ICAR) p. 930.

Tilak, S.T. (1989). Airborne pollen and fungal spores, Vaijayanti Prakashan, Aurangabad. pp. 424-440.

Tilak, S.T. and R.L. Kulkarni (1970). A New Sampler, Experientia. 26 : 443-444.

ENVIRONMENTAL BIOTECHNOLOGY

Edited By :   **Professor (Dr.) Arvind Kumar**

Published By : **DAYA PUBLISHING HOUSE**

36

BIOFERTILIZER EFFECT OF *DICTYOTA DICHOTOMA* ON GROWTH AND YIELD OF *ABELMOSCHUS ESCULANTUS* L. (MOENCH)

• *K. Sasikumar and R. Panneerselvam*

Department of Botany, Annamalai University,
Annamalainagar - 608 002, Tamil Nadu, India

Abstract

Seaweed extract of *Dictyota dichotoma* was tested at different concentration (12.5, 25, 50, 75, 100%) on growth and yield attributer of *Abelmoschus esculantus*. In general, the seaweed extract enhanced biomass, growth of roots and shoots, leaf area index, number of roots, leaves, flower and fruits, fruit length, fresh and dry weight of fruit, and maturity time, total chlorophyll content, and yield. Seaweed extract at 25% enhanced growth and yield at the maximum.

Key Words : *Abelmoschus esculantus, Foliar spray, Seaweed liquid fertilizer, Dictyota dichotoma, Growth, Yield.*

Introduction

The role of seaweed extracts, as biofertilizer on growth and yield of crops is known (Smith and Staden, 1984; Sheekh and EL. Saied, 1999; Anantharaj and Venkatesalu, 2002; Bhosale *et. al.*, 2002). These extracts enhance seed germination, growth, disease resistance capacity to pathogenic microbes and insects, yield and uptake of nutrients by the plants. Seaweed extracts are now available commercially under the names, such as maxi corp (sea born), Algifert (marinue), Germar GA 14, Seaspray, Seasol, S.M.3, Cylex and sea crop 16 (Jeanin *et al.*, 1991). Booth (1969) observed that the value of seaweeds as fertilizers was not only due to nitrogen, phosphorus and potash content, but also because of the presence of trace elements and metabolites. Aqueous extract of *Sargassum wightii* when applied as a foliar spray on *Zizipus mauritiana* showed an increased yield and quality of fruits (Rama Rao, 1991). In the present study, the brown seaweed, *Dictyota dichotoma*, abundantly occurring along the coastal area of Tamil Nadu, has been tested for its fertilizer effect on growth and yield of *Abelmoschus esculantus*.

Materials and Methods

The marine alga *Dictyota dichotoma* was collected freshly from the Rameshwaram coastal region, and washed thoroughly with seawater and then with freshwater in order to remove the sand particles and epiphytes. They were shade-dried for 4 days, followed by oven-drying for 24 h at 60° C. The dried material was taken for the preparation of seaweed liquid fertilizer (SLF) as per the method of Rama Rao (1990). The SLF was dried in an oven at 60° C for 36 h. The residue so obtained was considered as 100% extract. This extract was diluted to 12.5, 25, 50, 75 and 100% (v/v) using distilled water.

Seeds of the F_1 hybrid variety of *Abelmoschus esculantus* L. (moench) were collected from the SPIC Bio-technology Division, Chennai, Tamil Nadu. Surface sterilized with 0.1% $HgCl_2$ for 1 min and thoroughly washed with distilled water. Seeds presoaked for 12 hrs in distilled water were sown in sterilized vermicultate, moistened with distilled water.

After 20 days germination, the seedlings of uniform length 15 ± 2 cm were transferred to pot containing garden soil. After 40 days of establishment in the soil, SLF was sprayed on the leaves at different concentration at rate of 2 ml/seedling each concentration of SLF was sprayed on 20 seedlings. Control seedlings were also maintained control seedling sprayed with distilled water were also maintained under the same field condition. After 20 days of treatment, the experiment was terminated. Measurements were recorded on fresh and dry weights of whole plant, root and shoot length, number of lateral roots, leaf area, number of leaves, flower and fruits, fruit length, and fresh and dry weight of fruit. The time taken for maturity fruit maturity was also noted. Total chlorophyll content was recorded for the central leaflet (Hiscox and Israelstam, 1979).

Results and Discussion

The analysis of seaweed extract of *D. dichotama* revealed that amongst macro nutrients the values of nitrogen was maximum followed by calcium, sodium, and potassium. Nitrate formed the major part of microelements. The content of various nutrients in the SLF from *D. dichotoma* was higher than that SLF of *Sargassum* except phosphorus which was 15.1 mg g^{-1} and 20.0 mg g^{-1} respectively. Total nitrogen in *D. dichotoma* was 196.40 mg g^{-1} as compared to 8.48 mg g^{-1} in *Sargassum* (Rama Rao, 1992). The concentrations of phytohormones were also high i.e. 197 µg g^{-1} for auxins and 199 µg g^{-1} cytokinin (Table 36.1).

In the present investigation, the leaf of *A. esculantus* treated with the SLF from *D. dichotoma* showed good growth at 25% concentration

Table 36.1 : Mineral Composition of Seaweed Extract, *D. dichotoma*

Macro nutrients*	
Nitrogen	196.40 ± 0.72
Calcium	192.01 ± 0.63
Sodium	176.00 ± 0.79
Potassium	120.20 ± 0.12
Micro nutrients*	
Zinc	1.22 ± 0.98
Copper	0.98 ± 0.43
Nitrate	11.18 ± 0.27
Manganese	2.13 ± 0.19
Plant growth hormones*	
Cytokinin	199.00 ± 0.43
Auxin	197.00 ± 0.77

* mg g^{-1} dry wt,
** µg g^{-1} dry wt,
$\overline{X}$ ± SE, n = 20

(Table 36.2). The seaweed extract of *Ecklmia maxima* showed a significant effect on the growth of tomato seedlings, when used as a soil drench (Crouch and Vaustaden, 1993). At this concentration, dry weight of the plant material was 3-fold higher than control. The effect of SLF was more pronounced on the length of root, shoot and fruit. The leaf as well the content of total chlorophyll area was also doubled at the treatment with 25% SLF. Whapham *et al.* (1993) observed that the application of SLF obtained from brown seaweed *Ascophyllum nodosum* extract increased the chlorophyll levels of cucumber and tomato. This indicated that the extract of *D. dichotoma* enhanced not only the leaf area but also the quantities of photosynthetic pigments which finally lead to increase in the plant biomass.

Tay *et al.* (1985) showed the presence of cytokinins in the extract of *Durvillea potatorum* and Wightman (1964) the presence of auxin in the extract.

Table 36.2 : Effect of Seaweed Extract, *D. dichotoma* on the Growth and Yield of *A. esculatus* (n=20) (P<0.05)

Parameters	Control	Seaweed extract concentration (%)					Critical difference at 0.05% level
		12.5%	25%	50%	75%	100%	
Total chlorophyll (mg/g fresh wt)	0.431	0.546	0.889	0.623	0.382	0.242	0.026
Whole plant fresh wt (g)	38.363	56.81	70.64	40.2	35.41	20.1	0.640
Whole plant dry wt (g)	11.42	18.12	29.51	10.35	9.36	8.42	0.415
Shoot length (cm)	45.3	58.9	86.30	42.1	39.24	36.36	1.270
Root length (cm)	16.5	20.1	29.4	15.4	13.1	11.4	2.515
Leaf area index (cm^2)	28.6	34.6	56.8	26.2	24.8	20.2	2.805
Lateral root (n)	65.0	74.0	89.0	62.0	60.0	56.0	20.980
Number of leaves (n)	10.0	14.0	19.0	9.0	7.0	6.0	13.260
Number of flowers (n)	4.0	8.0	11.0	4.0	3.0	2.0	11.090
Fruit length (cm)	12.5	17.5	21.3	14.4	11.4	19.0	2.950
Fruit dry wt (g)	12.06	14.91	21.20	12.0	10.2	8.4	8.385
Maturity time	7.0	6.0	5.0	7.0	7.0	8.0	2.150
Yield (kg h^{-1})	9.0	11.0	15.0	8.0	6.0	15.0	55.83

In the present study, we could show the cytokinin content of 199 µg and auxin content of 197 µg in the SLF from *D. dichotoma*. The plant *A. esculantus* treated with 25% SLF also showed 90% increase in the concentration of cytokinin and auxin over control, which may either be due to the uptake of cytokinin and auxin present in the seaweed extract or to the SLF - induced synthesis of cytokinin and auxin in the plant. Backett and Van Staden (1990) concluded that mineral elements present in the commercial seaweed extract of 'kelpak' were partly responsible for the enhancement of the growth of wheat plants. In the present study, addition of SLF enhanced the accumulation of N and K. The plants treated with 25% SLF had shown an increase of about 15% and 35% of N and K respectively when compared to control (Table 36.2).

In conclusion, *A. esculantus* preferred 25% SLF of *D. dichotoma* for its good growth and yield. The possibility of using *D. dichotoma* for the commercial production of seaweed liquid fertilizers is suggested.

Acknowledgement

We thank the authorities of Annamalai University for providing facilities.

References

Anantharaj, A. and Venkatesalu, V. 2002. Studies on the effect of seaweed extract on dolichos biflorus. Seaweed Res. Utiln., 24(1) : 129-137.

Backett, R.P. and Staden, J. van. 1990. The effect of seaweed concentrate on the yield of nutrient stressed wheat. Bot. Mar. 33 : 147-152.

Bhosale, A.B., Wagh, A.N., Kharbade, S.B. and Pacharni, D.P. 2002. Effect of plant growth regulators on physiological disorders, quality and yield of summer tomato (*Lycopersicon esculentum* Mill). J. Curr. Sci. 2(2) : 173-176.

Booth, E. 1969, The manufacture and properties of liquid seaweed extracts. Proc. Int. Seaweed Symp. 6 : 622-655.

Crouch, I.J. and Staden, J. van. 1993. Effect of seaweed concentrated from *Ecklonia maxima* (Osbeck) Papenfuss on *Melodogyne incognita* Infestation on tomato. J. Appl. Phycol. 5 : 37-43.

Featonby-Smith, S.C. and Vanstaden, J. 1984. The effect seaweed concentrate and fertilizer on growth and the endogenous cytokinin content of phaseolus vulgaris. S-Afr. Tydskr. Plantk. 1984, 3 : 375-379.

Jeanin, I., Lescure, J.C. and Morot-Gaudry, J.F. 1991. The effects of aqueous seaweed sprays on the growth of Maize. Bot. Mar. 34 : 469-473.

Hiscox, J.D and Israelstam, 1979. A method for the extraction of chlorophyll from leaf tissue maceration. Can. J. Bot. 57 : 1332-1334.

Humphries, E.C. 1956. Mineral components and ash analysis, In : *Modern Methods of Plant Analysis* I Ed. Peach, K. and M.V. Tracey, Springer-verlag, Berlin. pp. 468-502.

Mostafa, M. El Sheekh and Alaa Eldin F. EL Saied, 1999. Effect of seaweed extracts on seed germination, seedling growth and some metabolic processes of fabe beans (vicia faba L.) Phykos. 38. 1 and 2 : 55-64.

Rama Rao, K. 1990. Preparation, properties and use of Liquid Seaweed Fertilizer from *Sargassum*. In : Seaweed Research and Utilization Association workshop on Algal products and seminar on Phaeophyceae in India. 4th-7th June, Madras, pp. 7-8.

Rama Rao, K. 1991. Effect of aqueous seaweed extract on *Zizyphus mauritiana* Lam. J. Indian Bot. Soc. 71 : 19-21.

Rama Rao, K. 1992. Seweeds as biofertilizers in Indian Horticulture, Seaweed Res. Utilin. 14 : 99-101.

Syono, K. and Torrey, J.G. 1976. Identification of cytokinins of root nodules of the Garden Pae, *Pisum sativum* L. Plant Physiol. 57 : 602-606.

Tay, A.A.B., Macleod, J.K., Palni, L.M.S. and Letham, D.S. 1985. Detection of cytokinins in a seaweed extract, Phytochemistry 24 : 2611-2614.

Umbreit, W.W., Burries, R.H. and Staurrer, J.F. 1972. *Manometric and Biochemical Techniques.* Burgess publishing co., Minnesota. pp. 259-260.

Whapham, C.A., Blunden, G., Jenkins, T. and Hankins, S.D. 1993. Significance of betines in the increased chlorophyll content of plants treated with seaweed extract. J. Appl. Phycol. 5 : 231-234.

Wightmanf, 1964. Pathways of tryptophan metabolism in tomato plants. In : Regulateurs Naturals dela croissance vegetable canbre National de la Recherche Scientifique, Paris, pp. 191-211.

Yoshida, S.D., Farno, A., Cock, J.H. and Gonez, K.A. 1976. *Laboratory manual for physiological studies of rice.* Third edition. International Rice Research Institute, Philippines.

ENVIRONMENTAL BIOTECHNOLOGY

Edited By :   **Professor (Dr.) Arvind Kumar**

Published By :   **DAYA PUBLISHING HOUSE**

37

EFFECT OF TEMPERATURE ON THE VIRULENCY OF *FLAVOBACTERIUM* SPP. ISOLATED FROM INDIAN CATFISH (*CLARIAS BATRACHUS*) OF HIMALAYAN AND SUB-HIMALAYAN REGIONS

• *Prasad*, Yogendra and Verma, Vinay*

* Aquatic Biotechnology and Fish Pathology Lab.,
Department of Animal Science, M.J.P. Rohilkhand University, Barelly-243006

Abstract

During one year period (Oct. 2000 to Sept. 2001) about 2100 live catfishes were examined at three sampling stations namely: Nanak sagar, Sharda reservoir and Nakatia river (Sub-Himalayan region) and Nainital (Himalayan region). First two sampling stations Nanak sagar and Sharda reservoir are fed with Sharda river, originated from Himalayan region. Fish were sampled with the help of fishermen by using nets and hooks. Only alive fish were examined for being symptomatic. They were placed in plastic bags filled with hand pipe water and transported to the laboratory. Out of 2100 fish examined only 110 fish were showing symptoms like the presence of shallow ulcerations on the general surface of the body and rottening at the base of the fins. Inocula were taken from the ulcered, rotten regions and the diferent organs like liver, kidney etc. of the symptomatic fish and inoculated on to Cytophaga agar. Twenty isolates (F_1-F_{20}) of *Flavobacterium* were isolated from symptomatic fish and pure cultures of the strais were obtained onto the same medium. Virulence nature of the isolates was determined by direct challenge. Experimentally infected fishes showed the symptoms of haemorrhages at the site of injection which developed in to a shallow ulcer after 21 to 30 days of incubation. Reisolated strains from artificially infected fish were similar in their morphological and biochemical characteristics. Impact of temperature on the virulency of *Flavobacterium* is presented in this paper.

Introduction

Bacterial fish pathogens are the major constraints of aquaculture. They adversely affect the fish production and cause great economic loss. Among bacterial fish pathogens *Flavobacterium* is important pathogen because it causes various diseases in freshwater (Farkas and Olah, 1986 and Holt *et al.*, 1993) and marine (Wakabayashi *et al.*, 1984 and Bernardet *et al.* 1990) fishes. The eutrophic water bodies provide a natural habitat for the growth and proliferation of bacteria due to the availability of organic and inorganic substances in that water bodies. Different strains of *Flavobacterium* viz. *F. columnare*, *F. branchiophilum* and *F. psychrophilum* are ubiquious in aquatic ecosystem and cause columnaris disease in *Clarias batrachus* (Prasad and Qureshi, 1995); gill disease in *Oncorhynchus mykiss* (Decostere *et al.*, 1999) and cold water disease in *O. mykiss* (Madsen and Dalsgaard, 1999) respectively. The peduncle disease in fish is also reported to be caused by *F. psychrophilum* Van Duijn (1973); Kimara *et al.* (1978); Wakabayashi (1989); Ekman *et al.* (1999) and Wiklund *et al.* (2000). In fact *Flavobacterium* is known by various synonemes i.e., *Cytophaga, Flexibacter* but the present name is given by Bernardet *et al.* (1996) which has been finally accepted. Unfortunately they have not been placed in the approved list of bacterial names till today.

In columnaris disease fish infected regions get highly necrotised and small flax of the tissue are observed dangling in the water. Occurrence of columnaris disease has been reported from both tropical and temperate regions and stressors such as water temperature, pH, alkalinity and stocking density play a significant role in the incidences of this disease (Wakabayashi, 1991). However, this disease is mostly studied in aquarium fish "cotton wool" disease particularly in black and red mollies (*Poecilia sphenops*) and guppies (*Poecilia reticulata*) (Decostere and Haesebrouck, 1999). The main significance of this paper is to evaluate the impact of water temperature on the virulency of *Flavobacterium* spp.

Materials and Methods

2100 fishes have been sampled from different sampling stations, namely : Nanak sagar, Sarda dam, Nakatia river (Sub-Himalayan region) and Nainital (Himalayan region). Out of which only 110 fishes were found to be symptomatic. Live and moribund fish were used in this study to obtain the bacterial inocula. Moribund fishes were sacrificed at the sampling sites and their organs like liver, kidney, heart etc. were collected in a saline solution (0.3% of NaCl) where as alive symptomatic fishes were placed in plastic bags/bucket filled with hand pipe/bore well water and were carried to the laboratory where they were kept in 500 L plastic pools ("3x2.5") filled with unchlorinated static water. Few symptomatic fishes were sacrificed and their organs were collected in saline solution. Sampled organs were homogenized in 0.3% NaCl solution and homogenate was centrifuged at 6000 rpm for 15 minutes and supernatant was used for the isolation of bacteria on Cytophaga agar/broth (Anacker and Ordal, 1959) containing 0.05% tryptone, 0.05% yeast extract, 0.02% sodium acetate, 0.02% beef extract and 9% of Agar/liter. The pH was adjusted to 7.2-7.4. All the inoculated plates were incubated at $37\pm1°$ C for 24 hrs.

All the isolated strains of bacteria have been identified by following the different biochemical tests such as Gram staining, shape, presence of flexirubin type pigmentation, aerobic growth, oxidase reaction, indole production, catalase production, gelatin degradation and H_2S production, etc.

LD$_{50}$

Isolated bacteria were enumerated by using spread plate count (SPC) and Pour plate count methods. For this purpose 24 hrs. old colonies were used and they have been quantified by plating them on Cytophaga agar. LD$_{50}$ was determined by inoculating 0.2 ml suspension containing inocula of 2×10^3, 2×10^4 and 2×10^5 cfu/ml and the temperature were maintained at 15 to 35° C (15, 20, 25, 30, 35° C) by using thermostat. Mortalities were recorded after 24 hrs of incubation for about 30 days. They were confirmed by reisolating the bacterium from the sample of liver/kidney collected from freshly dead experimetnal fish.

Experimental Infection Trial

10 *C. batrachus* per concentration were injected intramuscularly with *Flavobacterium* spp. @ 2×10^3, 2×10^4 and 2×10^5 cfu/ml and were kept in 500 L plastic pools filled with clean bore well static water. The water temperatures were maintained between 15 to 35° C for 30 days.

Results

During one year of survey about 2100 fishes have been screened from different water bodies/sampling stations. The symptomatic fishes were mostly recorded during summer months when the water temperature was ranging between 15-25° C in Himalayan region and 25-35° C in Sub-Himalayan region (Table 37.1a and 37.1b). The symptomatic fishes showed the clinical signs of ulceration, rottening, erosin and necrotization of the skin on the darso-latral side of the body.

Table 37.1a : Showing Number of *Flavobacterium* Isolates in Catfish Sampled from Different Water Bodies of Himalayan Region

Case No.	Year	Water bodies	Fish	Water temp. °C	No. of strains isolated
1.	2000-2001	Nainital (Uttranchal) Sampling station (A)	Catfish (*Clarias batrachus*)	15-20	2
2.	2000-2001	Nainital (Uttranchal) Sampling station (B)	-do-	15-20	3

Table 37.1b : Showing Number of *Flavobacterium* Isolates in Catfish Sampled from Different Water Bodies of Sub-Himalayan Region

Case No.	Year	Water bodies	Fish	Water temp. °C	No. of strains isolated
1.	2000-2001	Nakatia river Bareilly, U.P.	Catfish (*Calrias batrachus*)	20-30	2
2.	2000-2001	Sarda dam Udham Singh Nagar (Uttranchal)	-do-	20-30	5
3.	2000-2001	Nanak sagar Udham Singh Nagar (Uttranchal)	-do-	20-30	8

Bacteriological investigations revealed the presence of long, slender, Gram negative rods, Oxidase positive and showed the gliding movement on Cytophaga medium. More or less similar colonies have also been isolated from differnt organs like liver, kidney, heart, spleen and other body organs of the symptomatic fishes. Some biochemical characteristics of the strains isolated have been presented in Table 37.2. All isolates were ineubated at temperture of 37° C and colonies appeared were found to be flat and yellow-green (Flexirubin type pigmentation) within 24 hrs. of incubation. LD_{50} tests showed that 0.2 ml suspension containing 2×10^3, 2×10^4 and 2×10^5 cfu/ml were significant because that dilutions gave 60-80% mortalities at 25, 30 and 35° C respectively.

Experimental infection trials revealed that challenged fishes exhibited the symptoms of haemorrhages and necrotization at the site of injection. In case of fishes inoculated with 0.2 ml suspension of *Flavobacterium* spp. containing 2×10^3, 2×10^4 and 2×10^5 cfu/ml and were kept in 500 L plastic pools at the water temperature 15, 20 and 25° C exhibited the mild symptoms of the disease on the onset of 30, 26, 23 and 28, 25, 21 and 27, 24 and 20 days, respectively (Table-37.3a). Similarly in case fishes inoculated with same doses of *Flavobacterium* spp. and kept at water temperature 25, 30 and 35° C exhibited the symptoms of the disease on the onset of 26, 23, 20 and 24, 22, 19 and 23, 20 and 18 days respectively (Table-37.3b).

Table 37.2 : Some Bio-chemical Characteristics of *Flavobacterium* spp. Isolated from *Clarias batrachus*

S.No.	Characters	Bacteria (*Flavobacterium* = F)			
		F_{1-5}	F_{6-10}	F_{11-15}	F_{16-20}
1.	Gram reaction/shape	–/R	–/R	–/R	–/R
2.	Motility	+	+	+	+
3.	Colour of colony	Y-O	O	Y-O	Y-G
4.	Growth at different temperature	37°C	25°C	35°C	37°C
5.	Growth in different salinity				
	0.5%	+	+	+	+
	1.0%	–	–	–	–
6.	Oxidative (O)/Fermentative (F)	O	O	O	F
7.	Oxidase reaction	+	+	+	+
8.	Indol production	–	–	–	–
9.	H_2S production	+	+	+	+
10.	Catalase production	+	+	+	+
11.	Orinithin decarboxylase	–	–	–	–
12.	Phasphatase	+	+	+	+
13.	Gelatin degradation	+	+	+	+

Table 37.3a : Showing Experimental Infection Trials at the Different Temperatures and Inocula of *Flavobacterium* in *C. batrachus*

Viable Cell/ml.	2×10^3			2×10^4			2×10^5		
Temp (0° C)	15	20	25	15	20	25	15	20	25
No. of fish used	10	10	10	10	10	10	10	10	10
Mean wt. (gm)	90	90	90	90	90	90	90	90	90
Dose (ml.)	0.2	0.2	0.2	0.2	0.2	0.2	0.2	0.2	0.2
No. of days	30	28	27	26	25	24	23	21	20
Lesions	Haemorrhage			Haemorrhage			Haemorrhage		
Mortality	10/0			10/1			10/2		
Isolation of Bacteria	10/0			10/0			10/0		

Table 37.3b : Showing Experimental Infection Trials at the Differnt Temperatures and Inocula of *Flavobacterium* in *C. batrachus*

Viable cell/ml.	2×10^3			2×10^4			2×10^5		
Temp (0° C)	25	30	35	25	30	35	25	30	35
No. of fish used	10	10	10	10	10	10	10	10	10
Mean wt. (gm.)	90	90	90	90	90	90	90	90	90
Dose (ml.)	0.2	0.2	0.2	0.2	0.2	0.2	0.2	0.2	0.2
No. of days	26	24	23	23	22	20	20	19	18
Lesions	Haemorrhage			Haemorrhage			Haemorrhage		
Mortality	10/2			10/6			10/8		
Isolation of Bacteria	10/2+			10/6+			10/8+		

Discussion

Occurrence of columnaris disease in *Clarias batrachus* is recorded in 5.65% of fishes examined. It suggests that incidences of this disease is significant in the Nanak sagar and Sarda dam. It is interesting that though the 5.65% of fishes were showing the symptoms of columnaris disease but *Flavobacterium* spp. were found persistantly only in 0.95% of the fish sampled. It is evident from *in vitro* culture that the isolates of this bacterium prefer $35°C^+$ for proper growth. Simultaneously, occurance of this disease was recorded mostly during the summer months when the water temperatures were above 25° C. It forces us to predict that catfishes in the Nanak sagar and the Sarda reservoir are prone to *Flavobacterium* infection above 25° C. Which coincide with the view of Holt (1975), Decostere and Haesebrouchk (1999), and Prasad and Qureshi (1995).

Persistant occurence of the strains were recorded from the different organs as well as necrotized region of infected fish. It suggests that this bacterium is not only infecting the external body surface or gills but it also gets into the internal organs which is contradicting with the findings of Snieszko (1981) who reported that this bacterium infects the skin and gills while other organs remain uninfected. The symptoms such as necrotization, haemorrhages and erosion of skin, exhibited by the symptomatic and artificially infected fish are similar to those described by Bullock *et al.* (1971), Ekman *et al.* (1999). Moreover, the columnaris disease is mostly studied in salmo trout (Holt *et al.*, 1989), Eel (Chen *et al.*, 1982) and Carps (Bootsma and Clerx, 1976) and minimum attention has been paid on the Indian catfish.

The bacterium is long, slender, Gr -ve rods, oxidase +ve and shows gliding movement on Cytophaga agar. Appearance yelow green colonies indicate the presence of Flexirubin type pigmentation that is the characteristic feature of *Flavobacterium*. Similar view has been expressed by Bernardet *et al.* (1996). Experimental infection trials suggest that this organism is highly pathogenic in nature. Temperatures play a significant role on the pathogenesis of this bacterium because when the same doses (0.2 ml) of $2x10^3$, $2x10^4$ and $2x10^5$ concentration were enoculated in the same length and weight groups of fishes and were kept at different temperature they showed clinical sign in minimum dose on the onset of 18 days at 35° and they exhibited the clinical sign on the onset of 30 days when kept at 15° C. This finding is similar to Morrison *et al.* (1981) who reported that outbreaks of columnaris disease in atlantic salmon is being more severe at higher end of the range. Holt *et al.* (1975) found that mortality in experimental *F. columnaris* infection increases with increasing temperature. On contrary *F. psychrophilum* causes severe outbreaks of bacterial gill disease (rottening and fusion of gill filaments) when water temperature is below the 10° C (Hilger *et al.*, 1991 and Holt *et al.*, 1993).

Result also indicates that 25-30° C more appropriate for the growth of *Flavaobacterium*. Same results have been given by Soltani and Burke (1994) and Prasad and Qureshi (2000) who reported the optimum growth of this micro organism is ranging between 20-30° C. It is evident from the challenge tests that appearance of symptoms of columnaris disease was more or less similar to that of natural infection. Thus it can be concluded that *F. columare* grows more rapidly at temperature ranging between 25-35° C.

Acknowledgement .

The authors are thankful to Indian Council of Agricultural Research, New Delhi for granting research project. Thanks are also due to Dr. T.A. Qureshi for his constructive criticism and healthy suggestion rendered in the preparation of this manuscript.

References

Anacker, R.L. and Ordal, E.J. (1959). Studies on the myxobacterium *Chondrococcus columnaris*. 1. Serological typing. J. Bacteriol 78 : 25-32.

Bernadet, J.F., Campbell, A.C. and Buswell, J.A. (1990). *Flexibacter maritimus* is the agent of black patch necrosis' in Dover sole in Scotland. Dis. Aquat. Org. 8 : 233-237.

Bernardet, J.F., Segers, P., Vancanneyet, M., Berthe, M., Kersters, K. and Vandamme, P. (1996). Cutting a Gordian Knot. Emended classification and description of the genus *Flavobacterium*. Emended description of the family *Flavobacteriaceae* and proposal of *Flavobacterium hydatis* nom. nov. (basonym *Cytophaga aquatilis* strohl and Taist 1978). Int. J. Syst. Bacteriol. 46 : 128-148.

Bootsma, R., Clerx, J.P.M. (1976). Columnaris disease of cultural carp. *Cyprinus carpio* L. characteristics of the causative agent. Aqua. 7, 371-384.

Bullock, G.L., Conroy, D.A. and Snieszko, S.F. (1971). Bacteria diseases of fishes. In : Disease of Fishes, Book 2A(Ed), Snieszko, S.F. and Axelrod H.R., Neptune, T.F.H. Publishers, N.J. p. 151.

Chen, C.R.L., Chung, Y.Y., Kuo, G.H. (1982). Studies on the pathogenicity of *Flexibacter columnaris*. I. Effect of dissolved oxygen and ammonia of the pathogenicity of *Flexibacter columnaris* to eel *Anguilla japanica*. CAPD Fisheries series No. 8, Reports on fish Dis. Res. 4 : 57-67.

Decostere, A. and Haesebrouck, F. (1999). Outbreak of columnaris disease in tropical aquarium. Vet record 144 : 23-24.

Decostere, A., Haesebrouck, F., Charlier, G. and Ducatelle, R. (1999). The association of *Flavobacterium columnare* strains of high and low virulence with gill tissue of black mollies (*Poecilia sphenops*). Vet. Microbiol 67 : 287-298.

Ekman, E., Borjeson, H. and Johansson, N. (1999). *Flavobacterium psychrophilum* in Baltic salmo of salmo brood fish and their offspring. Dis. Aquat. Org. 37 (3) : 159-163.

Farks, J. and Olh, J. (1986). Gill necrosis - acomplex disease of carp. Aquaculture 58 : 17-26.

Hilger, I., Ullrich, S. and Anders, K. (1991). A few ulcerative flexibacteriosis like disease (Yellow pest) affecting young Atlantic cod *Gadus morhua* from the German Wadden Sea. Dis. Aquat. Org. 11 : 19-29.

Holt, R.A., Amandi, A., Rohovec, J.S. and Fryer, J.L. (1989). Relation of water temperature to bacterial cold water disease in Coho salmon, Chinook salmon and rainbow trout. J. Aquat. Anim. Health. 1 : 94-101.

Holt, R.A., Rohovec, J.S. and Fryer, J.L. (1993). Bacterial cold water disease. In. V. Inglis R.J. Roberts and N.R. Bromage (Editors). Bacterial disease of Fish. Blackwell Scientific Publications, Oxford pp. 3-22.

Holt, R.A., Sander, J.E., Zinn, J.L., Freyer, J.L. and Pilcher, K.S. (1975). Relation of water temperature to *Flexibactor columnaris* infected in steel head trout (*Salmo gairneri*), Coho (*Oncorhynchus kisutch*) and Chinook (*O. teshawytscha*) Salmon. J. Fish. Res. Board Can. 32 : 1553-1559.

Kimura, T., Wakabayashi, H. and Kudo, S. (1978a). Studies on bacterial gill disease in salmonids. 1. Selection on bacterium transmitting gill disease. Fish pathol. 12 : 233-242.

Madsen, L. and Dalsgard, I. (1999). Reproducible method for experimental infection with *Flavobacterium psychrophilum* in rainbow trout (*Oncorhynchus mykiss*) Dis. Aquat. Org. 36(3) : 169-176.

Morrison, C., Cornick, J., Shum G. and Zwicker, B. (1981). Microbiology and histopathology of Saddleback disease of underyearling Atlantic salmon *Salmo solar* L.J. Fish Dis. 4 : 243-258.

Prasad, Y. and Qureshi, T.A. (1995). Incidences of columnaris disease in *Clarias batrachus*. Aquaculture for 2000 A.D. : 275-282.

Snieszko, S.F. (1981). Bacterial gill disease of freshwater fishes. Fish Disease leaflet No. 62, Fish and Wildlife service, Washington, DC, p. 11.

Soltani, M. and Burke, C.M. (1994). Responses of fish-pathogenic *Cytophaga/Flexibacter* like bacteria (CFLB) to environmental conditions. Bull. Eur. Assco. Fish Pathol. 4 : 185-187.

Van Duijn, C. and C. Jr. (1973). Disease of fish (e.d.), Thomas, C.C. springfield gllinois, p. 372.

Wakabayashi, H. (1991). Effect of environmental conditions on the infectivity of *Flexibacter columnaris* to fish J. Fish Diseases. 14 : 279-290.

Wakabayashi, H., Egusa, S. and Fryer, J.L. (1980). Characteristics of filamentous bacteria isolated from a gill disease of salmonids. Can. J. Fish. Aquat. Sci. 37 : 1499-1505.

Wakabayashi, H., Huh, G.J. and Kimura, N. (1989). *Flavobacterium branchiophilia* sp. nov. a causative agent of bacterial gill disease of fresh water fishes. International J. Sys. Bacteriol 39 : 213-216.

Wiklund, T., Madsen, L., Bruun, M.S. and Dalsgaard, T. (2000). Detection of *Flavobacterium branchiophilia* from fish tissue and water samples by PCR amplification. J. App. Microbiol. 88 (2) : 299-307.

ENVIRONMENTAL BIOTECHNOLOGY

Edited By :   **Professor (Dr.) Arvind Kumar**

Published By : **DAYA PUBLISHING HOUSE**

38

SOLUBILITY OF SERICIN AS INFLUENCED BY PROPERTIES OF DIFFERENT SOURCES OF WATER

• *C. Doreswamy and Ramakrishna Naika*

Sericulture College, Chintamani-563 125

Abstract

The results of the studies on the solubility of Sericin as influenced by properties of different sources of water revealed that significantly highest sericin solubility (25.88 per cent) was observed in tap water (Bangalore) and it was lowest in distilled water (23.00 per cent) in NB_{18} race whereas in case of $PMxNB_{18}$ the highest sericin solubility of 25.88 and 25.38 per cent was recorded in tap water and bore well water (Mamballi) respectively which are found significantly high compared to 22.00 per cent in open well water (Siddlaghatta).

Introduction

The silk fibre composed of two proteins namely fibroin and sericin. The fibroin is the actual core of the silk, which is enclosed by a gummy material sericin. For easy unwinding of silk fibre, the gummy substance sericin has to be softened. The extent of the loss of the sericin during cooking and reeling depends the quality of water and for reeling and duration of cocoon cooking, temperature of cocoon cooking and reeling. If the water is alkaline, sericin solubility will be more whereas in the case of soft water the sericin solubility is optimum (Shamachary and Gowramma, 1988).

As the silk is pertinacious fibre, high pH hydrolysis the peptide linkage of fibroin to some extent, cocoon becomes soft and gets damaged during brushing and filaments attached to each other which does not unwind easily at the time of reeling and suffers from a defect known as plastering. The solubility of sericin is least when the pH of water is low and slowly increases at a pH 5-8 and it is rapid from 8.00 onwards. Acidity of water has a negative correlation with sericin solubility, which decreases with increase in water acidity. Sericin being a protein does not dissolve easily in the water with low concentration of neutral salts and solubility is greater in the presence of sodium bicarbonates than in the calcium bicarbonate (Kim, 1989). Permanent hardness inhibits swelling and solubility of sericin. Unsuitable water may not loosen the cocoon structure thereby leading to poor extraction of silk (Natheshwar Sah, 1983, Sundaramurthy and Narasingh, 1992 and Halliyal *et al.*, 1987).

Materials and Methods

Reeling water samples of different sources like open well water (Siddlaghatta), Bore well water (Mamballi), Bore well water (Santhemaralli), tap water (Bangalore) from traditional silk reeling areas and distilled water from laboratory were collected in plastic cans and few drops of touludine was added to avoid the major changes of water properties and analyzed their chemical constituents like pH, EC, Hardness, Ca, Mg and bicarbonates as per procedure (Kim, 1989).

Two races of cocoons namely PMxNB$_{18}$ and NB$_{18}$ were collected, cut open the cocoon, removed the pupae and moulted skin and shell weight was recorded. Cocoon shell was treated with 2 per cent potassium hydroxide solution at 70-80° C for few minutes with constant stirring till the cocoon shell becomes cotton like fluffy material. The treatment was stopped and washed with tap water and treated with glacial acetic acid 1 ml in 1 lt. of water to neutralize the alkali. Material was thoroughly washed in tap water and dried in hot air oven at about 100° C till constant weight was obtained and their weight was deducted from shell weight. The solubility of sericin was calculated using the formula.

Sericin content (x)=shell weight of cocoon--weight after treatment

Sericin solubility (%)=X x 100

Results and Discussion

The sericin solubility was 25.88, 27.87, 26.50, 28.50 and 23.00 per cent in open well water (Siddlaghatta), Bore well water (Mamballi), Bore well water (Santhemaralli), tap water (Bangalore) and Distilled water respectively in NB$_{18}$ race and 22.00, 25.38, 23.88, 25.88 and 22.62 per cent in open well water (Siddlaghatta), Bore well water (Mamballi), Bore well water (Santhemaralli), tap water (Bangalore) and distilled water, respectively in PM x NB$_{18}$ race. Highest sericin solubility 25.88 per cent was observed in tap water and was found significantly high over all other water samples and lowest sericin solubility of 23.00 per cent in distilled water in NB$_{18}$ race. In the case of PM x NB$_{18}$, highest sericin solubility of 25.88 and 25.38 per cent in tap and Bore well water (Mamballi) respectively found significant compared to 22.00 per cent in open well water (Siddlaghatta).

The sericin solubility is significantly high in tap water due to the akaline pH (8.00), hardness (420 ppm), Ca (27.50 ppm), Mg (16.00 ppm) and Bicarbonates (23.80 ppm) whereas sericin solubility is optimum in PM x NB$_{18}$ and NB$_{18}$ due to optimum pH (7.20), hardness (150 ppm), Ca (20.75 ppm) Mg (11.75 ppm) and Bicarbonates (192 ppm). But the sericin solubility is significantly low in distilled water which was recorded acidic P^H (6.70) and free from all other salts. These results are supported by Shamachary and Gowramma (1988). The extent of loss of sericin depends on the quality of water. If the water is alkaline sericin solubility will be more whereas in the case of soft water the sericin solubility is optimum. Natheshwar Sah (1983) stated that slightly alkaline water has a good solvent effect on dried and hard sericin of the cocoons. Kim (1989) reported that sericin solubility slowly increases at a level of pH 5-8 while it is rapid from 8 onwards. Sericin solubility is also greater in the presence of bicarbonates as we could observe in tap water whereas acidity of water has a negative correlation with sericin solubility. It decreases with increases in water acidity and sericin being a protein does not dissolve easily in the water with low concentration of salts as we could observe in distilled water (6.70).

Table 38.1 : Properties of Different Sources of Water Collected from Traditional Silk Reeling Areas

Sources of water	pH	EC (M.mhos)	Hardness (ppm)	Calcium (ppm)	Magnesium (ppm)	Bicarbonates (ppm)
Open well water	7.20	505.00	150.00	20.75	11.75	192.00
(Siddlaghatta)		(22.48)	(12.28)	(4.66)	(3.63)	(13.89)
Bore well water	7.90	1307.50	327.00	23.00	12.00	288.00
(Marnballi)		(36.16)	(18.10)	(4.89)	(3.59)	(16.98)
Bore well water	7.50	1050.00	303.25	24.75	14.25	288.00
(Santhemaralli)		(32.41)	(17.45)	(5.07)	(3.90)	(16.98)
Tap water	8.00	704.00	420.00	27.50	16.00	238.00
(Bangalore)		(26.54)	(20.50)	(5.33)	(4.12)	(15.47)
Distilled water	6.70	00.00	00.00	00.00	00.00	00.00
		(1.00)	(1.00)	(1.00)	(1.00)	(1.00)
F-test	*	*	*	*	*	*
SE m±	0.02	3.95	0.824	1.31	1.04	0.28
C.D. at 5%	0.06	11.92	2.483	1.97	1.56	0.84

Note : Values in parenthesis are square root transformed values

Among the two races of cocoons, the sericin solubility is comparatively more in NB_{18} cocoons canpared to $PM \times NB_{18}$ race irrespective of water samples tried. This shows that NB_{18} cocoons are very sensitive to variations of water properties like pH, hardness, EC etc., whereas $PM \times NB_{18}$ is quite resistant to variations of water properties.

The fibroin protein was varied from 74.12 to 79.56 per cent and 71.50 to 77.00 per cent in $PM \times NB_{18}$ and NB_{18} respectively. The fibroin protein was 76.12, 78.00, 74.62, 74.12 and 79.56 per cent in open well water (Siddlaghatta), Bore well water (Mamballi), Bore well water (Santhemaralli), tap water (Bangalore) and distilled water, respectively in $PM \times NB_{18}$ and 73.50, 74.12, 72.13, 71.50 and 77.00 per cent in open well water (Siddlaghatta), Bore well water (Mamballi), Bore well water (Santhemaralli), tap water (Bangalore), and Distilled water respectively in NB_{18}.

Table 38.2 : Solubility of Sericin as Influenced by Properties of Different Sources of Water

Sources of water	Sericin solubility (%)		Fibroin (%)	
	$PM \times NB_{18}$	NB_{18}	$PM \times NB_{18}$	NB_{18}
Open well water (Siddlaghatta)	23.88	26.50	76.12	73.50
Bore well water (Mamballi)	22.00	25.88	78.00	74.12
Bore well water (Santhemaralli)	25.38	27.87	74.62	72.13
Tap water (Bangalore)	25.88	28.50	74.12	71.50
Distilled water	20.44	23.00	79.56	77.00
F-test	*	*	*	*
SE m±	0.46	0.18	0.17	0.21
C.D. at 5%	1.40	1.71	0.53	0.63

Highest fibroin protein (79.56) was observed in distilled water which was found significantly high over all other treatments and the lowest fibroin content was 74.12 per cent in tap water in PM x NB_{18} race whereas in case of NB_{18} highest fibroin per cent was observed in distilled water (77.00) which was found significant over all other water samples and significantly lowest fibroin per cent (71.50) was observed in tap water. These results could be supported by the statements made by Natheshwar Sah (1983). Alkaline pH hydrolysis the peptide linkage of the fibroin protein as it is protenacious in nature as we could observe in tap water. Whereas sericin being a protein does not dissolve easily in the water with low concentration of salts in distilled water leads to incomplete sericin solubility which accounts to more fibroin percentage. Alkaline pH apart from dissolving sericin which also hydrolysis the fibroin protein whereas acidic pH leads to incomplete sericin solubility and protects the fibroin. Since water quality is dynamic in nature, periodical analysis of water for suggesting suitable correcting measures is very much essential for mobile water testing laboratory should be ensured to reeler units.

References

Halliyal, V.G., Shankar, A.G., Mangannavar, D.R., Vijayeendra, M.K. and Nadiger, G.S. 1987. Studies on the effect of water quality on reeling performance and raw silk. *Proceedings of the seminar on prospects and problems of sericulture in India*, pp. 321-337.

Kim, B.H. 1989. *Filature water engineering*, Associated business centre Ltd., Colombo, Sri Lanka, p. 23.

Natheshswar, Sah 1983. The water and silk industry part-1 and II, *Indian silk*, 22 (4&5) : 3-13.

Shamachary and Gowramma, 1988. Loss of sericin during the process of cooking and reeling, *Indian silk*, 27 (5) : 52.

Sundaramurthy, T.S. and Narasingh, K.A.L. 1992. Cocoon boiling technique, *Indian silk*, 30 (3) : 31.

ENVIRONMENTAL BIOTECHNOLOGY

Edited By :   Professor (Dr.) Arvind Kumar

Published By : DAYA PUBLISHING HOUSE

39

OPTIMIZATION STUDY FOR THE PRODUCTION OF GAMMA LINOLENIC ACID BY *MUCOR SP.*

• *A. Arun & G.K. Thilaka*

P.G. Unit of Microbiology, Thiagarajar College,
Madurai (Tamil Nadu), India

Abstract

Gamma Linolenic acid is a high value 18:3 ω6 poly unsaturated fatty acid which acts as hormone balancer and health promoter is produced intracellularly by all members of Zygomycota especially Mucorales. Mucor species was isolated and identified from soil sample. Optimization study for GLA production in different temperature and pH conditions was carried out both in still and shaking conditions. The presence of GLA was primarily identified by Sudan Black staining method. The GLA production was indirectly studied by analyzing the dry weight and oil content of the cell grown in different conditions. The fatty acid analysis by using Gas chromatography method was done for the best dry weight and oil content results obtained in different pH and temperature conditions, the results were compared with 10% GLA pure form. The study revealed that the Microbial Production of the fatty acid (GLA) has been considered as an economical way for the higher production of Gamma Linolenic Acid.

Key Words : Gamma Linolenic Acid.

Introduction

Fatty acids are long chain (hydrocarbon) carboxylic acids and are the compounds of many lipids including glycerides. Because they are one of the building blocks used in making complex lipids, they have considerable medicinal and nutritional value (Weete and Gandhi, 1992). For example, GLA is an essential fatty acid in the omega-6 family. Essential fatty acids are needed for normal brain function, growth and development, bone health, stimulation of skin, regulation of metabolism and maintenance of reproductive process (Bruinsma, K.A. Taran D.L., 2000).

Owing to the medicinal and nutritional potential of fatty acids, there has been considerable interest in their production. Traditionally GLA has been harvested from plant such as Evening Primrose oil. Recently there has been considerable interest in the microbial production of GLA (Ratledge, 1993). Microbial production of fatty acids has been considered as an economical way to produce large quantity of GLA.

GLA is a high value 18: 3 ω6 polyunsaturated fatty acid (PUFA) that is produced intracellularly by all the members of Zygomycota (Vander Westheizen *et al.*, 1994) especially the Mucorales. GLA is the precursor for the physiologically active hormones, the prostaglandins.

GLA being the health promoter and hormone balancer (Weete and Gandhi, 1992) it is used in the treatment of various illness such as rheumatoid arthritis (Jantti *et al.*, 1989) multiple sclerosis (Barber, 1993), Schizopharania, premenstrual symptoms (Horrobin, 1983). It can offer relief from those irritating symptoms such as cramps, breast tenderness and irritability, it decreases viral infection, it is believed that tumour growth and metastasis can quell with GLA especially breast cancer (Kenny, F.S., Pinder, 2000).

It also stops the progression of nerve disease and helps for good nerve functions. It also acts as cancer fighters, terminally ill patients suffering from pancreatic cancer tripled their life expectancy after taking excessive dose of GLA (Stoll, B.A., 1998). GLA actually increase the cell resilience and moistens the fatty layer beneath the skin. Thus delivering a multitude of benefits as health promoters (Graham, 1984; Ratledge, 1992).

In the present work *Mucor sps* was isolated and used in the optimization study for the production of GLA.

Materials and Methods

The *Mucor sps* was isolated from the soil sample. The soil sample was serially diluted, plated on PDA and isolated. The GLA production by *Mucor sps* was studied in different temperatures and different pH conditions both in still and shaking conditions. The isolated *Mucor sps* was inoculated in two separate production media one was incubated in shaking conditions. After incubation the presence of lipid (GLA) was primarily confirmed by Sudan Black B staining method (Kock and Ratledge, 1993) and the dry weight and oil content of the cell was analysed (Max J Kennedy *et al*) optimization study for GLA production by *Mucor sps* was done in different temperature conditions (32° C, 34° C, 37° C, 39° C, 41° C) both in still and shaking conditions. For pH studies *Mucor* species was inoculated in two separate batches of production media with different pH (2.5, 3.5, 4.5, 5.5, 6.5, 7.5, 8.5). One was incubated in still condition and another in shaking conditions at the temperature in which growth is maximum.

After incubation the dry weight and oil content of the cell was analysed.

The fatty acid analysis (Max J Kennedy *et al*) was done for the best dry weight and oil content results obtained in pH (5.5) and temperature (37° C) conditions. The result was

Table 39.1 : Dry Weight and Oil Content of the cell Mass in Still and Shaking Conditions at Different Temperature

	Dry weight (gm/l)		Oil content (ml/gm)	
Temperature	Still condition	Shaking condition	Still condition	Shaking condition
32°C	0.11	0.24	0.008	0.01
34°C	0.19	0.31	0.009	0.02
37°C	0.28	0.44	0.01	0.04
39°C	0.16	0.26	0.007	0.03
41°C	0.009	0.14	0.006	0.01

Table 39.2 : Dry Weight and Oil Content of the Cell Mass in Still and Shaking Conditions at Different pH.

	Dry weight (gm/l)		Oil content (ml/gm)	
pH	Still condition	Shaking condition	Still condition	Shaking condition
2.5	0.15	0.31	0.009	0.01
3.5	0.12	0.28	0.008	0.03
4.5	0.28	0.48	0.01	0.03
5.5	0.6	1.1	0.02	0.05
6.5	0.48	0.92	0.01	0.02
7.5	0.02	0.19	0.008	0.01
8.5	0.01	0.12	0.008	0.02

compared with 10% GLA (pure form) obtained from A to Z Pharma Pvt. Ltd., Chennai. The peaks obtained both in 10% GLA (standard) and the test sample (still and shaking conditions) was compared with compound matching software "LEO" and confirmed.

Results and Discussion

Influence of Temperature for Dry Weight and Oil Content of the Cell

Table-39.1 shows the dry weight and oil content of the cell mass in still and shaking conditions in different temperature. The maximum dry weight (0.44 gm/l) was obtained in 37° C in shaking condition and the maximum oil content (0.04 ml/gm) was obtained in 37° C in shaking condition. The dry weight and oil content was high in shaking conditions when compared with dry weight and oil content in still conditions.

Influence of pH for Dry Weight and Oil Content of the Cell

Table-39.2 shows the dry weight and oil content of the cell mass in still and shaking conditions at different pH. Based on the optimization study at different pH the maximum dry weight (1.1 gm/l) was obtained in pH 5.5 in shaking condition and the maximum oil content (0.05 ml/gm) was obtained in pH 5.5 in shaking condition. The dry weight and oil content was high in shaking conditions when compared to dry weight and oil content in still conditions.

Fatty acid analysis was done for the high dry weight and oil content obtained in different pH and temperature conditions. The results were compared with 10% GLA (pure form). The peaks obtained both in 10% GLA (standard) and the test samples (at pH 5.5 and temperature 37° C in still & shaking conditions) were compared with compound matching and software "LEO" and GLA compound was confirmed according to the Gas chromatography results. The maximum GLA Peak was observed in the sample (at pH 5.5 and temperature 37° C) obtained in shaking conditions.

Based on the dry weight, oil content and fatty acid analysis the maximum GLA production was seen in pH 5.5 at the temperature 37° C in shaking condition.

References

Bruinsma, K.A., Taren, D.L. Dieting, essential fatty acid intake and depression. Nutrition Rev. 2000.

Barber, A.J. 1988. Pharmaceut. J. 240, 723-725.

Graham, J. (ed) 1984. Evening primrose oil. Wellingborough, UK: Thorsons.

Horrobin, D.F. 1979. The Lancet 1, 529-531.

Jantti, J., Seppala, E., Vapaatalo, H., Isomaka, H. 1989. Evening Primrose oil and oil in treatment of Rheumatoid arthritis. Clin. Rheumatoid. 8 : 238-244.

Kenny, F.S., Pinder, S.E., Ellis, I.O. *et al.* Gamma Linolenic acid with tamoxifen as primary therapy in breast cancer. Int J. cancer 2000; 85 : 643-648.

Kock, J.L.F., and Botha, A., 1993. Acetic acid a novel source of cocoa butter equivalent and Gamma Linolenic acid. South African Journal of Science 89, 465.

Max, J., Kennedy, Sarah, L., Reader and Julian Davies, R. 1993. Fatty acid production characteristics of fungi with particular emphasis on GLA production. Biotechnol and Bioengg 42, 625-634.

Ratledge, C. 1992. Microbial lipids: Commercial relatives or academic curiosities, pp. 1-25.

Stoll, B.A. Breast cancer and the western diet: role of fatty acids and antioxidant vitamins. Eur J cancer 1998; 34 (12) : 1852-1856.

Vander Westheizen, J.P.J., Kock, J.L.F., Botha, A. and Botes, P.J. 1994. The distribution if the ω-3 and ω-6 series of cellular long chain fatty acids in fungi. Systemic and Appl. Microbio 17, 327-345.

Weete, J.D. and Gandhi, S. 1992. Potential for fungal lipids in biotechnology. In : Handbook of applied Mycology, Fungal biotechnology, D.K. Arora, R.P. Elander and K.G. Mukerji, eds. 4, pp. 377-400, New York: Marcel Dekker Inc.

ENVIRONMENTAL BIOTECHNOLOGY

Edited By :   **Professor (Dr.) Arvind Kumar**

Published By :   **DAYA PUBLISHING HOUSE**

40

A STUDY ON THE ROLE OF ASCORBIC ACID IN INDUCED LEAD TOXICITY IN ALBINO MICE : A SPECTROSCOPIC STUDY

• *J. Arjun*, M. Das**, S. Dey*

* Department of Zoology, Lumding College Lumding-782 447
** Department of Zoology, Gauwahati University, Gauwahati-781 014
R.S.I.C. N.E.H.U. Shillong

Abstract

The present communication deals with effect of Vitamin C on the induced lead toxicity in blood of albino mice. Four groups of mice were subjected to 30, 60, 90 and 120 days of experiment. One group serves as normal, the second group administered orally with lead acetate, the third group was treated simultaneously with ascorbic acid followed by administering and dose of lead acetate for 30, 60, 90 and 120 days. Atomic absorption spectrophotometric analysis of blood revealed reduction in blood lead level in 120 days simultaneously treated group. The reduction group indicating antitoxic role of viamin-C in removing lead from blood of mice.

Key Words : Vitamin C, Lead toxicity.

Introduction

Lead represents one of the most alarming element among the inorganic environmental pollutant which in present day environmental situation is posing a great threat to the animal and human health. The absorption of environmental lead occurs through gastro intestinal tract and interfere with all the normal physiological process. Lead, as absorbed in the body and recorded by a constant rise in blood lead level. A small amount of volatile lead is absorbed through lungs and which is also manifested by a sharp elevation in blood lead level. In recent years the environment has become highly contaminated with heavy metal particularly with lead. The input of this metal in the environment has increased immensely during last decade which has created serious health hazard leading to permanent neurological and hematological disorder with deposition of metal in important target organs like liver, kidney and skeletal system resulting in metabolic disjunction leading to morbidity and lethal effect (Friberg *et al.*, 1986; Seiler *et al.*, 1986).

Healthy albino mice of the same age group and weight (100+10 gm) were selected from laboratory stock and were divided into four groups, each comprising 20 individuals One of the groups was given normal diet and this group was taken as normal/control group. The remaining three groups were administered with oral sublethal dose of lead acetate (5 mg/kg body weight). One of these three groups was subjected to ascorbic acid pretreatment (8 mg/kg body weigh) for 30, 60 and 120 days, which was followed by administering oral dose of lead for 30, 60 and 120 days. The other two groups were treated simultaneously with ascorbic acid (8 mg/kg body weight) and lead acetate (5 mg/kg body weight) for a period of 30, 60 and 120 days.

Blood was collected from the experimental and control group of albino mice and was subjected to atomic absorption spectroscopy for evaluation of lead concentration in blood.

For analyzing amount of lead in blood of different experimental group 0.3 ml of blood from control and experimental groups were dried and treated with hydrogen peroxide for over night the samples were evaporated to dryness and hydrochloric acid. The digested samples were assayed for lead in Perkin Elmer 2380 Atomic Absorption Spectroscopy.

Results

In 30 days lead exposure, the individuals with simultaneous treatment of lead and vitamin C showed a decline of blood lead level (0.4 ppm) as compared to those receiving the sublethal dose of lead acetate (0.8 ppm) alone. The vitamin C pre treatment group on the other hand showed further showing a gradual increase up to the end of experiment period lead exposed animals.

In vitamin C pre treatment group the blood lead level was found to be declining (0.1 ppm) equal to that of simultaneous treated group in 60 days of treatment whereas following 90 days of treatment the lead acetate treated group recorded blood lead level of (0.18 ppm) as compared to vitamins and lead treated group which recorded (0.1 ppm) in both experimental set in response to vitamin C supplementation. In exposed period of 120 days the lead acetate treated group recorded a blood lead level in higher range (2 ppm) whereas in simultaneous treated group it showed decline in blood lead level to (0.10 ppm) and vitamin C pretreated group further in blood lead level to (0.5 ppm). The result of the analysis is presented in the table showing the recovery response with vitamin C following chisquare response analysis.

Table 40.1 : Recovery Response of Vitamin C Treatment on Lead Toxicity in Albino Mice

Time of Exposure in days (N)	Lead Acetate Treatment (X)	Lead Acetate & Vitamin-C (simultaneously)	Vitamin C pretreated followed by Lead Acetate
7	0.5 ppm	0.5 ppm	0.2 ppm
14	0.6 ppm	0.5 ppm	0.2 ppm
30	0.8 ppm	0.4 ppm	0.15 ppm
45	0.12 ppm	0 3 ppm	0.12 ppm
60	0.15 ppm	0.10 ppm	0.10 ppm
90	0.18 ppm	0.10 ppm	0.10 ppm
120	2.00 ppm	0.10 ppm	0.05 ppm
N 07	4.35	2.00	0.92
$\dfrac{(O-E)^2}{E}$ =	$\dfrac{(4.93)^2}{2.42}$	$\dfrac{-1.58}{2.42}$	$\dfrac{(1.50)^2}{2.42}$
x^2	10	1.0316	0.9297

$$P > 9.21 \ (0.05)$$

When (response) chisquare test applied to assess late antitoxic role of Vitamin C in induced lead toxicity, the result was found significant (P>0.05). This indicated good responses of the treatment. The individual treatment as observed from the present response analysis was found to be highest in recovery of lead toxicity, when Vitamin C applied followed by lead treatment, as

$$P>9.21 \text{ at } 6 \text{ d.f}$$

Simultaneous application of lead and Vitamin C also found to be significant but at lower level compared to pretreatment of Vitamin C. However the time lag did not show comparative toxicity of lead (Pb) in this experiment.

Discussion

At the cell level (erythrocyte) the effect of lead are multiple and complex (Goyer *et. al.*, 1988; Deyctal, 1997) and it is known that metal has particular affinity for carboxyl groups of glutamic acid and aspartic acid the sulphydryl group of cysteine and phenoxy group of tyrosine prevent the protein of cell membrane or enzyme system (Goyer, 1988). In this context it is to be noted that ascorbic acid also has carboxyl group for lead binding. This probably explain the increase of blood lead level at 30 and 120 days of experiment with treatment of lead acetate alone.

In simultaneous lead and Vitamin C treated group a gradual decline in record lead was recovered from 30 days of supplementation with Vitamin C up to the end of experimental period which indicated the antimetal detoxicating role of Vitamin C to reduce lead toxicity where as in experimental group pre-treated with dietary Vitamin C followed by lead exposure showed no remarkable rise in blood lead level during the entire experimental period. Further lead level found to be reduced to 0.05 ppm at the end of the experimental period which definitely suggested that the pre-treatment of Vitamin C exert better protective role in induced lead toxicity due to high ascorbic acid level already existing in blood promotes the binding of lead with carboxyl group of vitamin which may act as a chelating agent and play a role on detoxification of lead. The use of ascorbic acid as antidote for acute intoxification of certain metal such as Vanadium (Gorski and Zaprowska, 1982) etc are well documented. The antitoxic role of vitamin C has been further established from statistical analysis of data.

The finding of the present investigation definitely revealed that vitamin C treatment exhibit a revival and neutralising action in response to lead induced toxic impact. The role of ascorbic acid in detoxification of various chemical pollutant was reported by many authors (Kutsky, 1973; Baker *et. al.*, 1971) some authors also describe its role as potent antitoxic factor and anitoxident (Young and New Berne, 1981; David *et. al.*, 1988) which have the property to neutralize the chemical stress in the physiological level and act as a potent antioxidant in cellular level to prevent cell injury.

References

Baker, Hudges, Hood Sutierkich March and Conham. 1971. David J, Wise and J.R. Tomasso 1988. Ascorbic acid inhibition of nitrate induced methenoglobinemia in channel cat fish. The progression fish culturist 5077-80.

Day, S., Arjun, J., Das, M. 1997. Erythrocyte membrane dynamics in albino meice offspring born to female with lead toxicity during pregnancy (Biomedical letters, U.K, 1997).

Friberg, L., Nordberg, G.F., and Vank, V.B. 1986. Hand book of toxicity on metals Vol.II Elecvier Science Publisher Amstardam.

Goyer, R., 1988. Lead. In Handbook on toxicity of inorganic compound (H.G. Seilor *et. al.*, eds) 1988. Marcel Dekker. Inc. New york and Basel.

Gorski, M. and Zaprowska, H. Med. Sadowey, Kryminol, 32, 91. (1982). In Handbook on toxicity of inorganic compounds (H.G. seilor *et. al.*, eds) 1988.

Marcel Dekker Inc New York and Basel pp. 360-382.

Jones, M.M., Besigner, M.A. 1988. Journal Toxical. Env. Health 12 749 (1983) In Handbook on toxicity of inorganic compounds (H.G. Seiler *et. al.*, eds) Marcel Dekker Inc New York and Basel.

Kutsky, R.J. 1973. Ascorbic acid. In Handbook of Vitamins and Hormones, pp. 24-29 Nostrund, co New York.

Seiler, H.G., Seigel, H., Sieges, A. 1988. A Handbook on toxicity of inorganic compounds. Inc New York and Basel.

Young, V.R. and P.N., Newberne. 1981. Vitamins and Cancer prevention in issue and dilemma cancer, 47, 1226-1240.

41

BIODEGRADATON OF CYANIDE IN GOLD PROCESSING WASTES USING THE MICROBIAL ISOLATES

• *A. Arun & K. Thillai*

P.G. Unit of Microbiology, Thiagarajar College,
Madurai, Tamil Nadu, India

Abstract

Treatment of cyanide wastes with biological methods are recognized to be cheaper and faster than other conventional methods. Cyanide containing gold processing waste samples were collected from different gold processing industries in Madurai city. The cyanide degrading microorganism was isolated by enrichment method and identified to be *Pseudomonas* sp. It was then subjected to biodegradation study on cyanide sample in enrichment media without carbon source, with different carbon source and then with different percentage of inocula on the raw gold processing samples. Cyanide content was determined before and after treatment. The isolate utilizes cyanide as its nitrogen source and forms ammonia as a byproduct of cyanide degradation. Highest degradation efficiency of the isolate was analysed and it was dependent on the nutrient supplemented and concentration of CN in the sample. The study revealed the potential use of *Pseudomonas* sp. in detoxifying cyanide compounds.

Key Words : Cyanide degradation, detoxification, toxic waste.

Introduction

In modern era, many toxic compounds have entered into the environment as a result of human activities but these compounds itself have turned to be lethal and hazardous to the mankind. One such example of the toxic pollutant is cyanide compounds. Cyanide compounds are used extensively in many industries (Dhillon and Shivaraman, 1999) and it is estimated that 3 million tons of cyanide per year are used world wide. Cyanide almost acts instantly as a poison, nearly 50-60 milligrams turns to be lethal dose.

Gold processing and mining industries are highly responsible for releasing large quantities of cyanide in wastewater (Kishore, 1999), as cyanide is used in the form of KCN for purification of gold. Owing to the importance of gold and toxicity of cyanide, waste water must be treated before release (Marianne, 1996). Biological treatment methods are found to be cheaper and effective than chemical and physical detoxification methods (Paul Meyers, 1991).

A number of microorganisms have been reported to posses the ability to produce assimilate or detoxify cyanide by various enzymatic pathways (Desai, 1998). In the present work the microbes are isolated from the gold processing waste by enrichment method and was then used for biodegradation study. The cyanide content in gold processing waste before and after treatment was determined and the ammonia released as a byproduct of cyanide degradation was also estimated after 72 hrs and thereby the degradation efficiency of the isolate (*Pseudomonas* sp.) was determined at regular intervals.

Materials and Methods

Four different gold processing waste samples A, B, C and D were collected from gold processing industries at various sites in Madurai. The unknown cyanide content in the sample was determined by plotting the O.D. values measured at 590 nm on the standard graph of cyanide (KCN) drawn by phenylenepthalien copper method. Then 1 ml of gold processing waste sample was enriched in the presence of different mM KCN in 50 ml of mineral salt's media (MSM) and cyanide microdegrader was isolated (Michael, D. Adjei and Yoshiyuki Ohta, 1999) and identified to be *Pseudomonas* sp. according to the Bergey's manual of Bacteriology (Holt, 1997). The isolate was inoculated into 50 ml enrichment media (MSM) with 1 ml of gold processing waste sample A, B, C and D and to another set of mineral salt's medium with 1 ml of gold processing waste was supplemented with 1 % glucose 1 % fructose 1% lactose and 1% sucrose. Then 10 ml of gold processing waste samples were supplemented with 2%, 4%, 6%, 8% of inocula and these were incubated in shaker at 37° C then the degradation efficiency of the isolate was determined byanalyzing the cyanide content at regular intervals (24, 48, 72, 96 hrs) and the Ammonia released as a by product after 72 hours was determined by plotting the O.D. values measured at 420 nm on the standard graph of ammonium chloride drawn by Nessler's method.

Results and Discussion

Table 41.1 shows the amount of cyanide in the gold processing waste sample A, B, C and D in ppm/ml. sample D is found to be highly toxic with increased cyanide content.

Table 41.2 shows that time consumed by sample D is found to be 96 hours whereas other samples were completely degraded with short time period indicating the time consumed by the isolate to degrade cyanide solely depends on the cyanide concentration of the sample. Ammonia liberated merely depends on the amount of cyanide degraded by the isolate with respect to time intervals.

Table 41.3 and 41.3(a) indicate carbon source also imparts (or) increases the degradation efficiency of the *Pseudomonas* sp. Here the ammonia liberated depends upon the percentage of cyanide degradation.

Table 41.1 : Estimation of Unknown Concentration of Cyanide in Gold Processing Wastes

Sample	OD	Cyanide concentration ppm/ml
A	0.63	570
B	0.85	780
C	0.79	720
D	1.23	1140

Table 41.2 : Biodegradation of Cyallide Containing Wastes in Enrichment Media

Sample	Percentage of Degradation*				Conc. Ammonia in μg/ml liberated after 72 hours
	24 hrs	48 hrs	72 hrs	96 hrs	
A	91.2	100	–	–	65
B	89	92	100	–	43
C	90	93	100	–	59
D	91.2	92.1	–	100	60

* cyanide degradation at regular time intervals in hours.
Conc-concentration

Table 41.3 : Biodegradation of Cyanide Containing Wastes in Enrichment Media With Different Carbon Source (at regular time intervals)

Sample	1 % Glucose			1 % Fructose			1 % Lactose			1 % Sucrose		
	24hr	48hr	72hr	24hr	48hr	72hr	24hr	48hr	72hr	24hr	48hr	72hr
A	92.9	100	–	91	100	–	100	–	–	100	–	–
B	89.7	94.8	–	88.4	93.5	100	94.8	100	–	96.4	100	–
C	90.2	94.4	100	88.8	94.4	100	94.4	100	–	95.8	100	–
D	91.2	93.4	97.0	91.2	94.7	98.2	92.1	97.3	100	92.9	98.2	100

hr-hours.

Table 41.3(a) : Estimation of Ammonia After 72 hours

Sample	Ammonia Conc. in µg/ml			
	1% Glucose	1% Fructose	1% Lactose	1% Sucrose
A	70	66	82	82
B	46	43	67	66
C	53	51	74	71
D	52	30	44	43

Table 41.4 : Biodegradation of Cyanide in Gold Processing Waste with Different Percentage of Inocula

Sample	% of inocula	% of cyanide degradation at regular time intervals*					
		24 hrs	48 hrs	72 hrs	96 hrs	120 hrs	144 hrs
A	2	100	–	–	–	–	–
	4	100	–	–	–	–	–
	6	100	–	–	–	–	–
	8	100	–	–	–	–	–
B	2	92	94	97.4	100	–	–
	4	93.5	96.1	98.4	100	–	–
	6	94.8	97.4	100	–	–	–
	8	94.8	98.7	100	–	–	–
C	2	93	97	100	–	–	–
	4	94.4	98.6	100	–	–	–
	6	95.8	100	–	–	–	–
	8	95.8	100	–	–	–	–
D	2	80.7	81.5	86.8	92.9	97.3	100
	4	82.4	82.4	89.7	94.7	98.2	100
	6	82.4	83.3	90.3	97.3	100	–
	8	83.3	84.2	92.1	100	–	–

* time intervals in hours. hrs-hours.

Table 41.4(a) : Estimation of Ammoina After 72 hours

% of Inocula	Concentration of ammonia*			
	A	B	C	D
2	116	95	106	75
4	118	96	106	74
6	118	98	108	77
8	119	99	109	82

* ammonia released after 72 hours in microgram/ml.

Table 41.4 and 4(a) show as the percentage of inocula got increased the percentage of cyanide degradation was also found to be increased, similarly the ammonia content was found to be increased.

Based on the study, the cyanide biodegradation was found to be dependent on the carbon source that was supplemented and the percentage of inocula that was incorporated to the sample.

Sample D was found to be possessing higher concentration of cyanide and rapid degradation of sample D was observed in the degradation procedures with supplementation of carbon source in enrichment media when compared to others. Regarding the ammonia liberation, it depends on the degradation percentage of cyanide within 72 hours. The present study investigates the complete detoxification of cyanide in the industrial waste which would certainly be helpful to overcome the hazardous effect of cyanide on the environment.

References

Desai, J.D. 1998. Microbial degradation of cyanides and its commercial applications. Jr. Scien. & Ind. Research 57: 441-453.

Dhillon, J.K. and Shivaraman, N. 1999. Biodegradation of cyanide compounds by *Pseudumonas sp.* (S1) can. Jr. Microbiol 45: 201-208.

Kishore, M.E. 1999. Biodegradation of cyanide by *Bacillus Cereus Var mycoides*. Asian. Jr. of Microbiol. Biotech and Env. Sc. Vol. No. (1-2) : 37-46.

Marianne, M. Figueria 1996. Cyanide Biodegradation by an *Escherichia coli*. Strain Can. Jr. Microbiol, 42: 519-523.

Michael, D. Adjei and Yoshiyuki, Ohta 1999. Isolation and characterisation of a cyanide utilising *Burkholderia cepacia* strain. World. Jr. Mic & Biotech 15: 699-704.

Paul, R. Meyers 1991. An efficient cyanide degrading *Bacillus pumilus* strain. Jr. Gen. Microbiol, 137: 1397- 1400.

ENVIRONMENTAL BIOTECHNOLOGY

Edited By :   **Professor (Dr.) Arvind Kumar**

Published By :   **DAYA PUBLISHING HOUSE**

42

The Expository Aspects of Biotrophic Status by VAM Fungi in Vidarbha Region

• *S.U. Meshram, G.B. Shinde*, S.S. Pande, A.S. Shanware and R.R. Kamdi*

Rajiv Gandhi Vikas Biotechnology Centre, L.I.T. Premises, Nagpur (M.S.) India
* P.G. Deptt. of Biochemistry, L.I.T. Premises, Nagpur University
Nagpur (M.S.) India

Abstract

The strategic management of soil-plant environment system must be explored through sustainable agricultural and economic practices. A study on native VAM colonization in the roots of various crops and number of spores present in the rhizosphere soil samples from Chandrapur, Wardha and Nagpur Districts was carried out. A total 24 numbers of different VAM mycorrhizal spores were isolated and identified from rhizosphers soil of different crops collected from 31 field sites of Vidarbha region including waste/barren land. Out of 24 VAM fungi identified, the genus *Glomus* was found to be dominating having frequency of occurrence 45.8% followed by 25% of *Acaulospora*, 12.5% of *Scutellospora*, 8.3% of *Sclerocystis* and 3.3% of *Gigaspora*. The samples showed a wide range of variability in root colonization due to VAM fungi which ranged from 16.3% to 97% in soyabean crop of Ballarpur field site of Chandrapur District. The factors such as occurrence, distribution, significant variation in root colonization and spore density of VAM fungi in the rhizosphere soil are closely governed by various edaphological parameters of the region as well as rhizosphere effect of the host plant.

Key Words : VAM fungi, Spore density, Crop productivity, Sustainable farming

Introduction

Over the past few decades, agricultural experience in our country owing to bulk use of chemical fertilizers i.e. about 18.5 million tons of NPK during 1999-2000 (Motsara, 2000) has awakened all concern on soil productivity and long-term hazardous effects on the biosphere contributing to environmental pollution. Hence, a revolution towards ecofarming system was inevitable. The ability of VAM fungi to produce a dramatic response in plant growth is well documented (Rajan *et. al.*, 2000). The potentialities of Mycorrhizae have rightly been assessed under present investigation in backward region of Vidarbha to illustrate various edaphological factors across the soil and crop growing environments including waste/barren land.

The mycorrhizal-plant partnership is a basic, essential and integral part of plant survival and growth. Although the inoculum response varied with crops, location, season, agronomic practices, level of soil fertility and the most importantly interaction with native soil microflora whose activities largely determine the chemical and physical properties of the soil. Crop productivity is usually limited by Nitrogen, and Phosphorous availability, particularly in the tropical and subtropical climatic conditions. On this backdrop, VAM fungi have gained importance in recent years, to provide enough nutrients and because of their potential role to enhance the growth and yield of crops in marginal environments and ultimately in sustainable agriculture. Continuous monoculture, use of chemical fertilizers (Vyas and Vyas, 1995), nature of pre-existing crops and human intereferences (Reece and Bonham, 1978, Trappe, 1981) result in lower percentage of root infectivity and spore density in VAM fungi. To assess the benefits rendered by VAM fungi, studies were undertaken in Vidarbha region regarding occurrence and distribution of spore density as well as the degree and intensity of root colonization of the crops viz. Soybean, Cotton, Beans, Sorghum, Sunflower, Cowpea, Arhar, Wheat, Rice and Gram.

Materials and Methods

The rhizosphere soil samples of different crop varieties were collected in the sterile polythene bags from 8-10 different spots in each field upto a depth of 25-30 cm. The collected rhizospheric soil samples were air dried and then passed through 2 mm sieve, to collect the fine soil for fungal spore collection. About 250 gm of air dried soil samples was added to 1000 ml of water and stirred thoroughly, until all soil aggregates dispersed to leave a uniform suspension. The suspension was then passed through a sereis of sieves in decreasing pore sizes. The process was repeated 6-8 times for each soil sample, to ensure complete isolation of VAM fungal spores (Gerdmann and Nicolson, 1963). After thorough washing and sieving of the soil samples, the sieving of each portion collected on each mesh were pooled and collected on Whatmann's filter paper number 40. Further, the counting of VAM spores was performed by drawing the intersecting gridline on Whatman filter paper and numbered all the squares serially. The sieved materials were then transferred on the filter paper and counted for spore population by observing under Leitz. Binocular Stereoscopic Microscope.

The identification was carried out to enumerate genera and species of VAM fungal spores. Identifications of a total 24 numbers of the isolated VAM fungal spores were done based on spore morphology, sporocarp colour, dimension of spore, number and thickness of spore wall layer, thickness and mode of hyphal attachment (Mukerji *et al.*, 1991). The classification of the spores at generic and species level was investigated by using standard methods (Hall and Fish, 1979, Trappe, 1982, Berch and Trappe, 1987, Raman and Mohan Kumar, 1988, Schenck and Perez, 1990).

The roots of different crop varieties collected from varied localities were washed carefully under running tap water separately to remove adhered soil particles. The washed roots were cut and stained following the standard methods (Philip and Hayman, 1970, Koske and Gemma, 1989). The stained roots were examined and/or spores for determining per cent root colonization by VAM fungi, following the Gridline intersection method (Giovannetti and Mosse, 1980, Newmam, 1966).

Simultaneously, chemical analysis of soil sampes were performed for determining the effect of soil pH, electrical conductivity, organic carbon (Walkley and Black, 1934, Walkley, 1935 and Walkley, 1947), organic matter (Russel and Engel, 1928), total percentage of nitrogen (Jackson, 1973), available phosphorous (Bray and Kurtz, 1945) and soil potassium using standard procedures.

Results and Discussion

Studies regarding quantitative estimation of rhizosphere soil samples from various crops of Wardha, Chandrapur and Nagpur districts showed lot of variation in soil chemical parameters (Tables 42.1, 42.2 and 42.3). Parameshwaran and Augustin (1988) reported positive correlation between soil density and available NPK along with Zinc, Copper and Iron. Further, present investigation suggests no correlation between spore density and per cent root infection from individual sites. This is in accordance with the work carried out by Miller *et al.*, 1985, Land and Schonbeck, 1991. This is mainly because the factors which promote maximum spore development are different from that which promote spore germination and subsequent initiation of infection. Spore germination in soil is primarily governed by physical factors like temperature, moisture, oxygen and carbon dioxide contents (Schenck *et al.*, 1975, Daniel and Trappe, 1980, Koske, 1981, Le Tacon *et al.*, 1983) and by chemical characters like pH, soil nutrients (Siquerra *et al.*, 1982, Carr, 1991), while the spread of infection depends on successful entry in the host (Amijee *et al.*, 1990) and phosphate cocentration of host tissues (Sanders, 1975). VAM fungal spores also showed indigenous dormancy, under such conditions the discrepancy might arise between spore density and root infection.

A total of 24 VAM fungi, 6 species of genus *Acaulospora* (Fig. 42.1), 2 species of genus *Gigaspora* (Fig. 42.2), 11 species of genus *Glomus* (Fig. 42.3), 2 species of genus *Sclerocytis*

Table 42.1 : Chemical Analysis of Rhizosphere Soil Sample from Various Crops of Wardha District of Vidarbha Region of Maharashtra State

Sr. No.	Sample No.	Location	Crop Sample	Spore Count 30 gm	Root Coloni- zation (%)	pH	E.C. (mm hos/ cm³)	Organic Carbon (%)	Organic Matter (%)	Total Nitrogen (%)	Avail- able Phosph- orous (kg/ha)	Avail- able Pota- ssium (kg/ha)
1.	A6	Kelzar	Soybean	37	88.3	7.5	0.3	1.12	1.93	0.1	45	275
2.	A9	Kelzar	Soybean	67	89	7.4	0.25	0.6	1.03	0.05	34	454
3.	C	Kelzar	Soybean	50	52.3	7.7	0.3	0.82	1.41	0.07	31	227
4.	A5	Kelzar	Sorghum	78	83.6	7.8	0.25	1.04	0.79	0.09	20	435
5.	A10	Kelzar	Sorghum	62	–	7.5	0.3	0.34	0.58	0.03	36	155
6.	A1	Kelzar	Arhar	59	50	7.6	0.25	1.12	1.93	0.1	74	185
7.	A3	Kelzar	Arhar	37	91.3	7.4	0.3	0.92	1.58	0.08	31	545
8.	A2	Kelzar	Cowpea	73	78.6	7.2	0.4	0.72	1.24	0.06	108	175

Table 42.2 : Chemical Analysis of Rhizosphere Soil Sample from Various Crops of Chandrapur District of Vidarbha Region of Maharashtra State

Sr. No.	Sample No.	Location	Crop Sample	Spore Count 30 gm	Root Coloni- zation (%)	pH	E.C. (mm hos/ cm³)	Organic Carbon (%)	Organic Matter (%)	Total Nitrogen (%)	Avail- able Phosph- orous (kg/ha)	Avail- able Pota- ssium (kg/ha)
1.	BP3	Ballarpur	Soybean	37	–	7.2	0.35	0.9	1.55	0.08	43	535
2.	R1	Sasti	Soybean	65	97	7.4	0.3	0.94	1.62	0.08	22	360
3.	R3	Gauri	Soybean	51	92	7.7	0.3	0.94	1.55	0.08	14	484
4.	S2	Ballarpur	Soybean	31	79.6	7.2	0.25	0.6	1.03	0.05	49	504
5.	S3	Ballarpur	Soybean	41	16.3	7.6	0.3	0.26	0.44	0.02	11	434
6.	R4	Ballarpur	Cotton	53	–	7.7	0.3	0.84	1.44	0.07	38	565
7.	S1	Ballarpur	Cowpea	70	–	6.5	0.4	1.6	2.75	0.14	67	412

Table 42.3 : Chemical Analysis of Rhizosphere Soil Sample from Various Crops of Umred Region of Nagpur District in Vidarbha Region of Maharashtra State

Sr. No.	Sample No.	Location	Crop Sample	Spore Count 30 gm	Root Coloni-zation (%)	pH	E.C. (mm hos/ (cm³)	Organic Carbon (%)	Organic Matter (%)	Total Nitrogen (%)	Available Phosph-orous (kg/ha)	Available Pota-ssium (kg/ha)
1.	U5	Kalamna	Sorghum	39	86.6	7.9	0.42	0.23	0.39	0.02	17.92	940
2.	U10	Udasa	Sorghum	18	44.4	8	0.38	0.43	0.74	0.04	22.4	1,243
3.	U11	Udasa	Sorghum	14	75.5	8.1	0.4	0.3	0.51	0.03	31.36	1,366
4.	U15	Panchgaon	Sorghum	26	55.5	8	0.4	0.68	1.13	0.07	35.84	1,444
5.	U4	Kalamna	Gram	39	43.3	7.8	0.46	0.35	0.6	0.03	35.84	1,036
6.	U13	Panchgaon	Gram	17	–	8	0.35	0.38	0.65	0.03	22.4	802
7.	U14	Panchgaon	Gram	17	40.4	8	0.39	0.53	0.91	0.05	22.4	1,008
8.	U1	Kalamna	Wheat	23	62.2	8	0.42	0.3	0.51	0.03	13.44	571
9.	U12	Udasa	Wheat	18	53.3	8	0.38	0.53	0.91	0.05	17.92	1,444
10.	U2	Kalamna	Arhar	25	24.4	8	0.33	0.95	1.63	0.08	26.88	515
11.	U3	Kalamna	Arhar	15	–	7.6	0.41	0.42	0.72	0.04	13.44	638
12.	U8	Udasa	Sunflower	20	86.6	8.1	0.39	0.53	0.91	0.05	17.92	1,444
13.	U9	Udasa	Sunflower	10	68.8	8	0.36	0.69	1.18	0.06	13.44	1,444
14.	U6	Kalamna	Soybean	30	68.8	8.1	0.35	0.15	0.25	0.01	22.4	1,444
15.	U7	Udasa	Beans	34	–	8	0.36	0.07	0.12	0.1	8.96	1,444

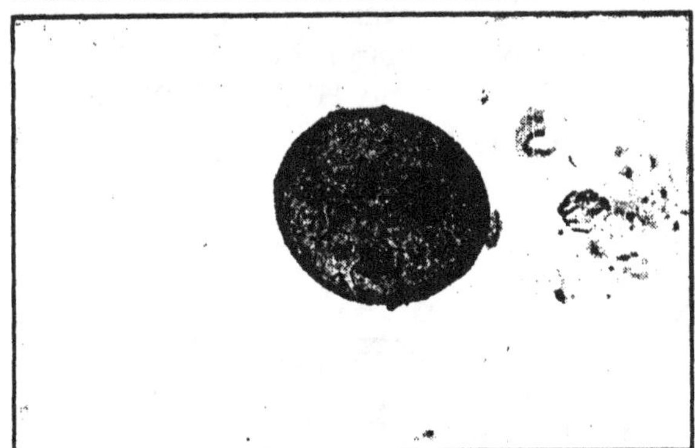

Fig. 42.1 : Spore of *Aculospora sporocarpia*.

Fig. 42.2 : Azygospore of *Gigaspora albida*.

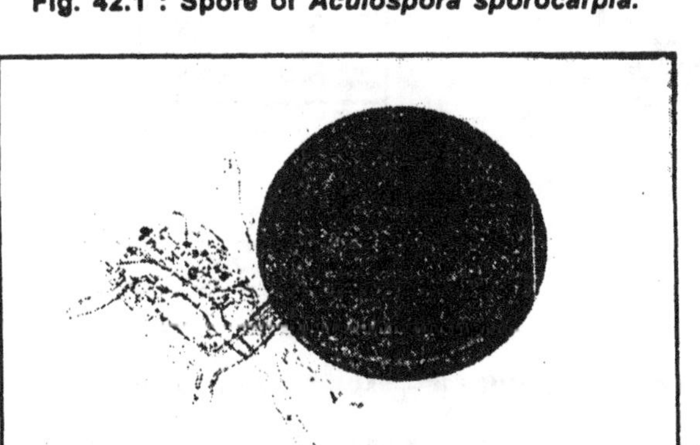

Fig. 42.3 : Chlamydospore of *Glomus etunicatum*.

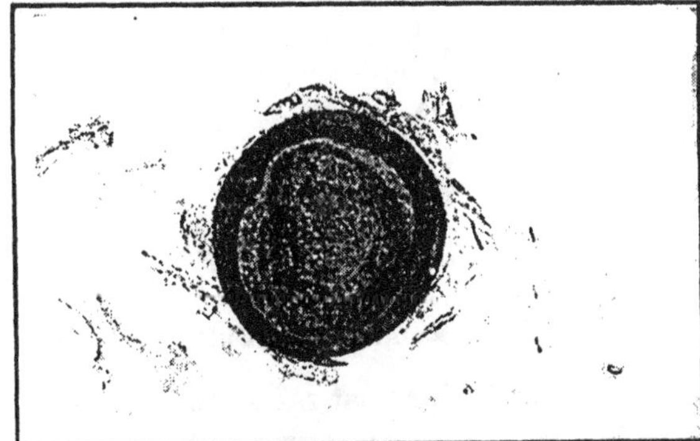

Fig. 43.4 : Spore of *Sclerocystis rubisformis*.

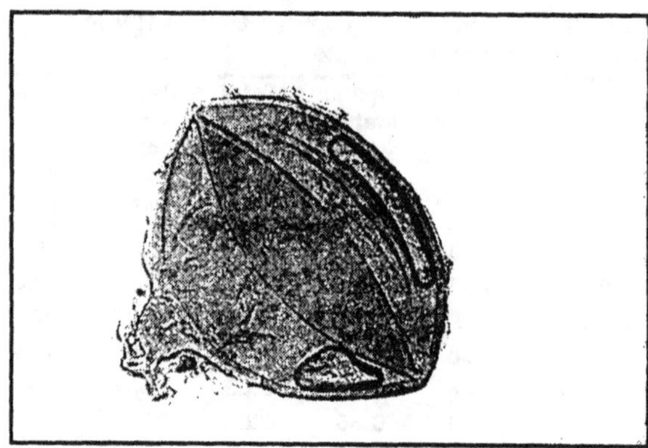

Fig. 42.5 : Spore of *Scutellospora dipaillosa*.

(Fig. 42.4) and 3 species of genus *Scutellospora* (Fig. 42.5) were isolated and identified from rhizosphere soil of different crops collected from 31 field sites of Vidarbha region including waste/barren land.

In the present study, the genus *Acaulospora* was found predominantly in the rhizospheric soils of Wardha and some of its species were found in all the three localities. The genus *Sclerocystis* was found only in Chandrapur and *Scutellospora* was observed in Wardha and Umred, while it was absent in rhizospheric soil of Chandrapur. The genus *Glomus* was present in all the three localities. Out of 24 VAM fungi identified, the genus *Glomus* was found to be dominant having frequency of occurrence 45.8% followed by 25% of *Acaulospora*, 12.5% of *Scutellospora*, 8.3% of *Sclerocystis* and 3.3% of *Gigaspora* (Fig. 42.6), the fact suggested that the genus *Glomus* was highly adaptable to various ecological niche and could make mycorrhizal associations with plants under wide range of soil factors. While the other VAM fungi viz., *Aculospora*, *Gigaspora*, *Sclerocystis* and *Scutellospora* were found to be susceptible to change in soil factors. This type of variation in the distribution were also reported earlier from the soil of Scotland (Gerdmann and Nicolson, 1963), Newzealand (Mosse and Bowen, 1968), Australia (Bevege and Richards, 1971), Florida (Schenck and Hinson, 1971), South Africa (Hattingh, 1972) and Oregon (Gerdmann and Trappe, 1974).

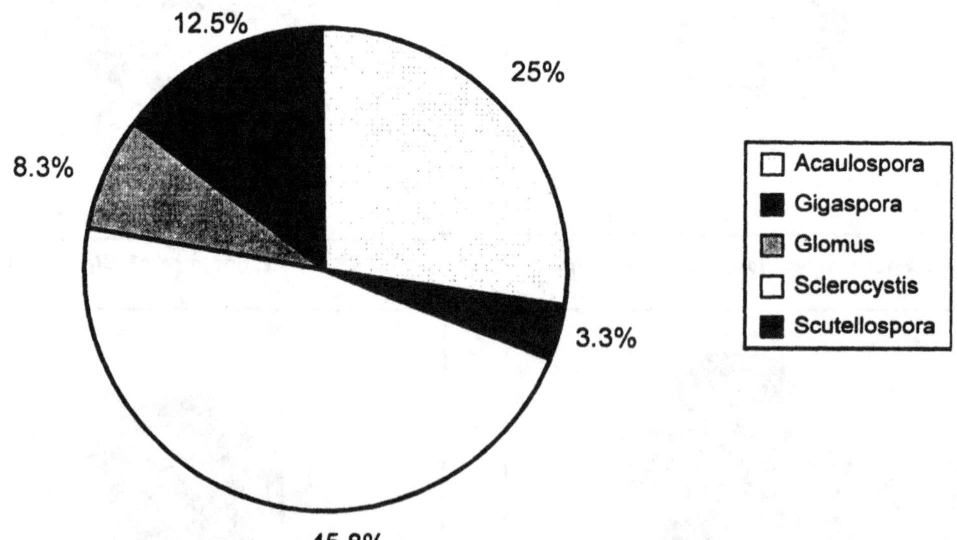

Fig. 42.6 : Occurrence of Different Genera of VAM Fungi (values are expressed in per cent).

Although the host range of the VAM spore is wide, the present studies revealed the dominance of one sporetype with particular crop variety in a specific location, while it was totally absent from the same crop variety in other localities, this observation was in conformation with the studies conducted by Bevege and Richard (1971). The samples showed a wide range of variability in root colonization and number of VAM spores which was ranging from 16.3% to 97% in Soybean crop of Ballarpur field site of Chandrapur District of Vidarbha region (Table 42.4).

Table 42.4 : Native VA Mycorrhizal Colonization in the Roots and Number of Spores Present in the Rhizosphere Soil Samples Collected from Chandrapur, Wardha and Nagpur Districts

Sr. No.	Sample No.	Host crop plant	No. of Spores/ 30 gm Rhizosphere soil	VAM root colonization (%)
*1.	A1	Arhar	59	50
*2.	A2	Cow-pea	73	78.6
*3.	A3	Arhar	37	91.3
*4.	A5	Sorghum	78	83.6
*5.	A6	Soybean	37	88.3
*6.	A8	Soybean	20	82.0
*7.	A9	Soybean	67	89.0
*8.	A10	Sorghum	62	-
*9.	C	Soybean	50	52.3
**10.	R1	Soybean	65	97.0
**11.	R2	Arhar	45	77.6
**12.	R3	Soybean	51	92.0
**13.	R4	Cotton	53	-
**14.	S1	Cow-pea	70	-
**15.	S2	Soybean	31	79.6
**16.	S3	Soybean	41	16.3
**17.	S4	Rice	50	24.3
***18.	U1	Wheat	23	62.2
***19.	U2	Arhar	25	24.4
***20.	U3	Arhar	15	-
***21.	U4	Gram	39	43.3
***22.	U5	Soybean	39	86.6
***23.	U6	Soybean	30	68.8
***24.	U7	Beans	34	-
***25.	U8	Sunflower	20	86.6
***26.	U9	Sunflower	10	68.8
***27	U10	Sorghum	18	44.4
***28.	U11	Sorghum	14	75.5
***29.	U12	Wheat	18	53.3
***30.	U13	Gram	17	-
***31.	U14	Gram	17	40.4
***32.	U15	Sorghum	26	55.5

*=Wardha   **=Chandrapur   ***=Nagpur

Overall, the exploration of varied fields including waste/barren land in regard to VAM fungi enlightened the multifarious benefits which could be conferred to enhance crop growth, grain yield and ultimately proved to be a boon for preservation of soil health for posterity.

Acknowledgement

The financial assistance rendered during the course of investigation by DBT, Ministry of Science and Technology, Government of India, New Delhi is duly acknolwedged.

References

Amijee, F., Strible, D.P. and Tinker, P.B . 1990. Soluble carbohydrates in roots of leek (*Allium porrum*) plants in relation to phosphorous supply and VA mycorrhizas. Plant and Soil. 124 : 195-198.

Berch, S.M. and Trappe, J.M. 1987. Revision of Trappe's 1982. Synaptic key of genera and species of Endogonaceae. Mycotaxan.

Bevege, D.I. and Richard, B.N. 1971. Some aspects of Endogone forming mycorrhizas with Hooppine (*Aracucaria cunninghamii* Ait). XV IVFRO Congr. Sect. pp 24.

Bray and Kurtz. 1945. Soil Sci. 59 : 39.

Carr, G.R. 1991. Use of Zwitterionic hydrogen ion buffers in media for growth tests of *Glomus caledonium.* Soil. Biol, Biochem. 23(2) : 205-206.

Daniel, B.A. and Trappe, J.M. 1980. Factors affecting spore germination of the VAM fungus *Glomus epigaeus.* Mycologia. 72 : 457-471.

Gerdmann, J.W. and Nicolson, T.H. 1963. Spores of Mycorrhizal endogone species extracted from soil by Wet Sieving and Decanting. Trans. Br. Mycol. Soc. 46 : 235-244.

Gerdmann, J.W. & Trappe, J.M. 1974. The endogonaceae in the Pacific North West. Mycologia Memoir. 5: 1-76.

Giovannetti, M. and Mosse, B. 1980. An evaluation of technology's for measuring VAM infection in roots. New phytol. 84 : 489-500.

Hall, I.R. and Fish, B.J. 1979. A Key to the endogonaceae. Trans. Br. Mycol. Soc. 73 : 261-27.

Hattingh, M.J. 1972. A note on the fungus Endogone. J.S. Afr. Bot. 38 : 29-31.

Jackson, M.L. 1973. Soil Chemical Analysis. Prentice Hall India Pvt. Ltd. New Delhi, Nitrogen deteminations for soil and plant tissue. 183.

Koske, R.E. 1981. *Gigaspora gigantea :* Observations on spore germination of a VA-mycorrhizal fungus. Mycologia. 73: 288-300.

Koske, R.E. and Gemma, J.N. 1989. A modified procedure for staining roots to detect VAM. Mycol. Res. 92(4) : 486-505.

Land, S. and Schonbeck, F. 1991. Influence of different soil types on abundance and seasonal dynamics of vesicular-arbuscular mycorrhizal fungi in arable soil of North Germany. Mycorrhiza. 1(1) : 39-44.

Le Tacon, F. Skinner, A. and Mosse, B. 1983. Spore germination and hyphal growth of a vescicular arbuscular mycorrhizal fungus *Glomus mosseae* (Gerdemann and Trappe) under decreased oxygen and increased carbon dioxide concentration. Can. G. Microbial. 29 : 1280-1285.

Miler, D.D., Domoto, P.A. and Walker, C. 1985. Mycorrhizal fungi at eighteen apple root stock plantings in the United States. New Phytol. 100: 379-391.

Mosse, B. and Bowen, G.D. 1968. A key to the recognition of some Endogone spore types. 51: 469-483.

Motsara, M.R. 2000. Potentialities of Biofertilizers in integrated management system. In : Proc. of symposium on current status of biofertilizers tecnology and its promotion in Karnataka, Tamil Nadu, Kerala and Pondicherry (W). RN Bisoyi. Regional Biofertilizer Development Centre Banglore, India. 130-142.

Mukerji, K.G., Sharma, M., Kaushik, A. and Raut, P. 1991. Taxonomy of VAM fungi. Mycorrhiza News. 3(2) : 3.

Newman, E.I. 1966. A method of estimating the total length of root in a sample. J. Apple. Ecol. 3:13

Parameshwaran, P. & Augustine, B. 1988. Distribution and ecology of VAM, in a scrub jungle. In Mycorrhizae for green Asia. (Eds. A. Mahadevan, N. Raman and K. Natrajan) Madras, India. pp. 91-94.

Phillips, J.M. and Hayman, D.S. 1970. Improved procedures for clearing roots and staining parasitic and VAM fungi for rapid assessment of infection. Transactions of the British Mycological Society. 55:158-161.

Rajan, S.K., Reddy, B.J.D. and Bagyaraj, D.J. 2000. Screening of arbuscular mycorrhizal fungi for symbiotic efficiency with *Tectona grandis* for Ecol. Manage. 126:91-95.

Raman, N. and Mohankumar, U. 1988. Techniques in Mycorrhizal Research, published by CAS in Botany, University of Madras, India. pp 79-273.

Reece, P.E. and Bonham, C.D. 1978. Frequency of endomycorrhizal infection in grazed and ungrazed blue grama plants. J. Range Manage. 31 : 149-151.

Russel and Engel. 1928. Proc. 1st Intern. Congr. Soil Sci. 4 : 343.

Sanders, F.E. 1975. The effect of foliar applied phosphate on mycorrhizal infecton of onion roots. In: Endomycorrhizas. (Eds. F.E. Sanders, B. Mosse and P.B. Tinker). Academic Press, London. pp 261-277.

Schenck, N.C. & Hinson, K. 1971. Endotrophic Vescicular Arbuscular mycorrhizae on Soybean, Florida Mycologia. 63 : 672-675.

Schenck, N.C., Graham, S.O. and Green, N.E. 1975. Temperature and light effect on contamination and spore germination of vescicular-arbuscular mycorrhizal fungi. Mycologia. 67 : 1189-1192.

Schenck, N.C. and Perez, V. 1990. Manual for the identification of VAM fungi. 3rd edition (Published by Synergistic Publication, Guinesvielle, Florida). pp 1-286.

Siquerra, J.O., Hubbell, D.H. and Schenck, N.C. 1982. Spore germination and germ tube growth of a vescicular-arbuscular mycorrhizal fungus *in vitro*. Mycologia. 74:952-959.

Trappe, J.M. 1981. Mycorrhizae and productivity of arid and semi-arid range lands. Advances in food producing system for arid and semi-arid lands. Academic Press, New York, Inc. pp 581-599.

Trappe, J.M. 1982. Synaptic Keys to the genera and Species of Zygomycetous mycorrhizal fungi. Phytopathol. 72 : 1102-1108.

Vyas, S.C. and Vyas, S.S. 1995. Influence of pesticides on mycorrhizal infection and chlamydospores formation in soybean (*Glycine* max). In Mycorrhizae : Biofertilizers for the future. Proceedings of the third national conference on mycorrhiza. (Eds. A. Adholeya and S. Singh). Tata Energy Research Institute, New Delhi, pp. 147-151.

Walkley and Black 1934. Soil Sci. 37 : 29.

Walkey. 1935. J. Agr Sci., 25 : 598.

Walkey. 1947. Soil Sci. 63 : 251.

ENVIRONMENTAL BIOTECHNOLOGY

Edited By :   Professor (Dr.) Arvind Kumar

Published By : DAYA PUBLISHING HOUSE

43

ERICULTURE – A VENTURE FOR RURAL BETTERMENT

• B. Sannappa, Ramakrishna Naika, R. Govindan and Siddagangaiah*

Department of Sericulture/Agricultural Marketing and Co-operation*
University of Agricultural Sciences, GKVK, Bangalore

Abstract

Ericulture is an employment generating industry giving gainful employment to the rural masses. It is remarkable for its low investment, quick and high returns, which makes it an ideal industry that fits into socio-economic fabric of rural India. The castor leaf available in the country from the total area under castor for rearing of eri silkworm without affecting the seed yield was estimated to be 1,38,130 tonnes creating an opportunity to produce 41,43,900 kg of eri cocoon (shell) yield with an income potential of Rs. 8,288 lakh. The return per rupee invested was also found to be highly encouraging over sole castor bean production (Rs. 1.16). Thus eri silkworm rearing provides an opportunity of two avenues of revenue to the farmers.

Key Words : *Castor, Ericulture, Employment generation, Profitability.*

Introduction

Among the commercially exploited non-mulberry silkworms, the eri silkworm, *Samia cynthia ricini* Boisduval is the only species domesticated completely and adopted to indoor rearing all through the year. Eri silk is commonly considered as poor man's silk and its production in India is considered as a small-scale backyard venture. However, it offers tremendous scope and opportunities for developing it into an industry with immense potentialities for self-employment and inturn can play a vital role in the poverty alleviation besides generating additional farm income to the cultivators (Siddique *et al.*, 1993). The other factors that can give an impetus to the progress of industry and ensure higher production are the domesticated nature of this species, minimum space requirement, adoptability to varied agro-climatic conditions and quicker returns to capital investment.

In India the production of eri silk accounts for about 6.14 per cent of total raw silk production with 970 MT and among non-mulberry raw silks, it accounted for 63.56 per cent during 1998-99. The North-Eastern parts of the country are said to be the home land for ericulture, accounting for 90 per cent of total eri silk production. Of late, it has been tried in southern states in the country to exploit its potential. Eri silkworm is a polyphagous and rearing it on leaves of castor, *Ricinus cummunis* L. alone can ensure production of good quality cocoons (Patil and Savanurmath, 1994). India is one of the world's principal producers of castor covering 8.0 lakh ha area with 7.81 lakh tonnes of castor bean with an average yield of 970 kg/ha (Anon., 1997). Keeping this in view an attempt was made to explore the potential existing in terms of the availability of castor leaf for cocoon production. The ever increasing population and decreasing land-man ratio in the country call for increasing the productivity per unit area and per unit time. Further, it also helps to supplement the additional farm income besides increasing the non-mulberry raw silk production in the country which can sustain rural cottage industries as well as the farming communities.

Materials and Methods

To explore and examine the projected issues, the published and experimental research data were gathered from different sources (Anon., 1994 and 1997). Statistical analysis was done as per Sundarraj *et al.* (1972). From the available data an attempt has been made to empirically analyse the economics of castor seed-cum-eri cocoon production.

Results and Discussion

Eri silk moth being polyvoltine 4 to 6 life cycles are possible alround the year even in semi-arid regions of the country besides the traditional North-Eastern belt. Eri silkworms are harder than mulberry silkworms and wherever castor can grow eri silkwonn rearing can be taken up. The commercial object of eri silk is equivalent to mulberry cut cocoons. Thus in addition to producing castor beans, the leaves can also be used for eri silkworm rearing providing an opportunity of two avenues of revenue to the farmer.

Table 43.1 : Distribution of Area, Production and Productivity of Castor in India (1995-96)

States	Area (000'ha)	Production (000' tonnes)	Yield (kg/ha)
Gujarat	388.1	613.4	1580
Andhra Pradesh	262.6	71.6	270
Rajasthan	38.6	43.8	1130
Karnataka	23.7	16.7	700
Tamil Nadu	50.6	16.1	320
Orissa	27.0	14.0	530
India	806.6	781.4	970

When we look at the distribution of castor area among different states in the country (Table 43.1) it reflects good scope for taking up ericulture in addition to bean production. The observations revealed that almost all the castor genotypes studied supported the different parameters of larval, cocoon and grainage and were found to be suitable for cocoon production. However, Acuna and RC-8 were found to be significantly superior to other genotypes (Sannappa and Jayaramaiah, 1999).

The prospects of ericulture succeeding as a potent agro-cottage industry in the country based on convertibility of castor leaves into silk. Keeping this in view an attempt has been made to know the income generating potential of ericulture in India (Table 43.2). A perusal of Table 43.1 indicates that, in the country a total of 8,06,600 ha of area is under castor (1995-96) with a potential of producing 55,252 tonnes of total leaf yield. Of this, the amount of leaf available for rearing i.e., 25 per cent of the total leaf yield (Devaiah *et al.*, 1984; Siddique *et al.*, 1993) without affecting the castor seed yield was estimated to be 1,38,130 tonnes. By utilizing this quantity of leaf about 1,38,13,000 DFLs can be reared from which we can expect 41,43,900 kg of cocoon (shell) yield. Thus generating an attractive additional income to the tune of Rs. 8288 lakh to the economy in addition to the regular castor bean production.

Further to supplement the income generating potential of ericulture from the point of view of farmers, the economics of eri silkworm rearing along with castor bean production using Aruna variety was worked out (Table 43.3) (Mishra, 1999). On an average farmers could realize net return of Rs. 2345 from one hectare of castor bean production with an expenditure of Rs. 14560. Also at the same time utilizing the defoliated leaves of castor if rearing of eri silkworm is taken up by investing Rs. 1440/ha, it generates an income of Rs. 9036. Thus when both the activities were taken up (castor-cum-eri cocoon production), the total per hectare expenditure was estimated to be Rs. 16000 with a net profit of Rs. 5406, giving rise to a net additional benefit of Rs. 3060. The return per rupee invested was also found to be highly encouraging with Rs. 1.34 in castor-cum-eri cocoon production compared to that of sole production of castor bean (Rs. 1.16).

Table 43.2 : Income Generating Potential of Ericulture in India

1.	Area under castor (ha)	806600
2.	Total yield of castor seed (tonnes)	781400
3.	Estimated total leaf yield (Tonnes)	552521
4.	Estimated leaf yield (tonnes/ha)	0.685
5.	Quantity of leaf available (25% of the total leaf yield) for ericulture without affecting the seed yield (tones)	138130
6.	Quantity of leaf available per hectare for ericulture (tonnes)	0.17
7.	*Total number of eri DFLs that can be reared	13813000
8.	Total expected eri cocoon yield (kg) @ 30 kg/100 DFLs	4143900
9.	Income from cocoon yield (Rs. in lakhs) @ Rs. 200/kg	8288

* Leaf consumption /DFL is about 10 kg.

Table 43.3 : Economics of Castor (Aruna) Seed-cum-eri Cocoon Production (ha)

	Particulars	Amount (Rs.)
A.	**Castor seed production**	
	1. Total expenditure	14560
	2. Revenue from castor seed @ Rs. 9.4 /kg	16905
	3. Net profit	2345
	4. Benefit : Cost ratio	1.16
B.	**Eri cocoon production**	
	1. Cocoon production cost (451 DFLs)	1440
	2. Revenue from 45.13 kg cocoon shell @ Rs. 200 /kg	9036
	3. Revenue from castor seeds @ Rs. 9.4 /kg	12380
C.	**Castor-cum-eri cocoon production**	
	1. Total revenue (cocoon shells + beans)	21405
	2. Total expenditure (both castor bean and cocoon production)	16000
	3. Net Profit	5406
	4. Benefit : Cost ratio	1.34

Conclusion

The study revealed that the vast income generating potential of eri cocoon production when we consider the area under castor in the country, the production can be increased by five times. Further, the direct and indirect benefits with employment generation potential in other related sectors in the economy particularly in rural areas are the added advantages which are also to be accounted. Like mulberry silkworm the eri silkworm is also multivoltine and farmers can take up eri cocoon production round the year. Considering the potentiality of ericulture and vast area under castor cultivation mainly for bean production in the country, there is a need for making concerted efforts in the following directions to strengthen and to sustain the growth of eri industry in the country.

- Establishment of Technical Service Centres (TSCs) in castor cultivating areas for transfer of technology among farmers for tapping the unexploited potential.
- Establishment of seed production and multiplication centres to ensure the timely supply of disease free layings.
- Development of organized sector of cocoon procurement and marketing facilities, spinning and weaving.
- Training of farmers and other human resource engaged in ericulture for better growth and management of eri industry.

References

Anonymous. 1994. *Package of practices for higher yields.* Univ. Agric. Sci. and State Dept. Agric., Bangalore, p.98.

Anonymous. 1997. *Directory of Indian Agriculture.* Centre for Monitoring Indian Economy Pvt. Ltd., Mumbai, p.402.

Devaiah, M.C., Govindan, R. and Thippeswamy, C. 1984. Economics of rearing eri silkworm, *Samia cynthia ricini* Boisduval on castor, *Ricini communis* L. Indian J. Seric. 23 : 117-120.

Mishra, S.D. 1999. Aruna variety, A unique host for the eri silkworm. Part-I. Two revenues from the same resource. Indian J. Seric. 38 : 99-101.

Patil, G.M. and Savanurmath, C.J. 1994. Eri silkworm - The poor man's friend. Indian Silk. 33(4) : 41-45. .

Sannappa, B. and Jayaramaiah, M. 1999. Interaction of castor genotypes and breeds of eri silkworm in relation to cocoon and grainage parameters. Mysore J. agric. Sci. 33: 214-223.

Siddique, A.A., Rajaram and Sengupta, A.K. 1993. Eri - A Common man's silk. Indian Silk. 32(4) : 34-36.

Sundarraj, N., Nagaraju, S., Venkataramu, M.N. and Jagannath, M.K. 1972. *Design and Analysis of Field Experiments.* Directorate of Research, UAS, Bangalore, p. 419.

ENVIRONMENTAL BIOTECHNOLOGY

Edited By :   **Professor (Dr.) Arvind Kumar**

Published By : **DAYA PUBLISHING HOUSE**

44

Photoperiod Exerted Biochemical Alterations in Ovotesticular Lysosomal Enzymes in Aquatic Snail, *Lymnaea Luteola*

• *D.L. Bharamal and S.G. Nanaware**

Department of Zoology, Shri Pancham Khemaraj Maha Vidyalaya
Sawantwadi, Maharashtra, India
* Department of Zoology, Shivaji University, Kolhapur, Maharashtra, India

Abstract

The experiment was conducted to study the short day (16 hD : 8 hL) and long day (16 hl : 8 hD) photoperiod exerted effects on histology of ovotestis and biochemical changes in ovotesticular lysosomal enzymes - Acid phosphatase and esterase in aquatic snail *Lymnaea luteola*. The study revealed that histologically at short day all the ovotesticular cell types i.e. germinal epithelium cells, nurse cells, sertoli cells, ovum and sperm showed maximum staining activity indicated active gametogenesis whereas at long day these cellular types showed minimum staining activity indicated suppression of gametogenesis. Biochemically at short day the concentration of both the lysosomal enzymes were increased and decreased after long day photoperiod. The study showed the lysosomal enzymes have intricate role in gametogenesis in this snail.

Key Words : Photoperiod, Lysosomal enzymes, Ovotestis, Snail.

Introduction

The various important metabolites and enzymes play a very significant role in the growth and reproduction in gastropods (Nanaware, 1974). In gastropods various glands and the different parts of the reproductive system have been target organs of several biochemical and histochemical investigators for long time (Grainger and Shillitoe, 1952). Light influences the maturation of gametes and thus also the rate of oviposition (Garcia, 1975). In gastropods photoperiod and its role in reproduction has been studied by few investigators (Bohlken and Joosse, 1982, Bharamal and Nanaware, 2002). But there is no account on photoperiod induced alterations on the lysosomal enzymes in the ovotestis of *Lymnaea luteola*. Based on this, our attention has been focussed to study the biochemical changes in lysosomal enzymes - Acid phosphatase and esterase activities and their possible role in ovotestis of a freshwater snail *Lymnaea luteola* under different photoperiods.

Materials and Methods

The snails collected from local Rajaram tank near Shivaji University Campus, Kolhapur, Maharashtra were acclimated to the laboratory conditions for a week. Batches of 25 snails each were transferred to three separate glass aquaria. One batch was exposed to control (12hL: 12hD), second to long day (16 hL : 8 hD) and third to the short day (16 hD : 8 hL) photoperiods for five weeks. After every week, 2-3 snails were sacrificed and their ovotestis were analysed histologically (H-E and MT) and biochemically for two lysosomal enzymes acid phosphatase (Bergmeyer, 1965) and esterase (Colowick and Kaplan, 1955).

Results and Discussion

Photoperiod exerted histological preparations of the ovotestis of *Lymnaea luteola* in, indicated at short day size and number of various ovotesticular cell types like germinal epithelium cells, nurse cells, sertoli cells, ovum and sperm showed maximum staining intensity whereas at long day this staining reactivity was low. The photoperiod exerted biochemical changes in the ovotesticular lysosomal enzymes are shown in Table 44.1, indicating concentration of lysosomal enzymes were high at short day and low at long day.

Table 44.1 : Photoperiod Exerted Changes in lysosomal Enzymes in *L. luteola*

Photoperiod	Enzyme	Duration of Photoperiod exposure in weeks				
		1st	2nd	3rd	4th	5th
Control	A.P.	1010.56 + 32	1102.78 + 30	1176.00 + 22	1206.14 + 48	1263.20 + 26
12hL : 12hD	E	93.26 + 16	97.68 + 18	106.42 + 18	118.14 + 16	126.24 + 08
Short Day 16 hD : 8 hL	A.P.	1155.14 + 30 (+ 14.30%)	1240.20 + 22 (+ 12.46%)	1346.36 + 20 (+ 14.48%)	1453.10 + 10 (+ 20.48%)	1590.12 + 14 (+ 25.80%)
	E	107.44 + 14 (+ 15.20%)	121.30 + 10 (+ 24.18%)	133.48 + 12 (+ 25.42%)	142.40 + 18 (+ 20.50%)	150.76 + 20 (+ 19.42%)
Long Day 16 hL : 8 hD	A.P.	960.22 + 20 (- 4.98%)	903.10 + 12 (- 18.10%)	864.76 + 28 (- 26.46%)	781.40 + 14 (- 35.21%)	600.14 + 28 (- 52.49%)
	E	87.40 + 10 (- 6.28%)	82.48 + 16 (- 14.53%)	72.20 + 08 (- 32.15%)	68.20 + 10 (- 42.27%)	59.14 + 10 (- 53.19%)

N.B. :- (I) Acid phosphatase activity was expressed in terms of μg-nitrophenol phosphate and Esterase activity in terms of μg-nitrophenol acetate per gram wet weight tissue of ovotestis.

Figures in paretheses represent percentage difference (+ increase; - decrease).

Bharamal and Nanaware (2002) pointed out short day photoperiod stimulated gametogenesis whereas long day photoperiod suppressed it in garden slug *Semperula maculata*. Pelluet and Henderson (1954) observed an inhibition of gametogenesis after continuous exposure to light of high intensity whereas weak illumination caused more rapid maturation of spermatozoa in snails kept under normal conditions. Complete darkness led to resorption of the garnets (Smith, 1966). Varute (1968; 1971) showed increased β - glucuronidase activity in Spermatogenesis and activity was lowered after cessation of spermatogenesis in anuran spermatogenetic cycles. Nanaware (1974) pointed out increase in β - glucuronidase, acid phosphatase and esterase activity during gametogenesis in ovotestis of garden slug *Semperula maculata*.

Our observations indicated at short day, gametogenesis was stimulated and lysosomal enzyme showed maximum concentration and at long day photoperiod gametogenesis was suppressed and the concentration of lysosomal enzymes were minimum.

Acknowledgement

One of the authors (DLB) is deeply indebted to The Principal, B. N. Katkar for encouragement.

References

Bergmeyer, H.V. 1965. In: Methods of enzymatic analysis, 2nd ed. Academic Press, New York.

Bharamal, D.L. and Nanaware, S.G. 2002 Photoperiod induced histological and histochemical alterations in ovotesticular nutrients in Semperula maculata. J. Shivaji Univ. (Science) 37: 73 - 80.

Bohlken, S. and Joosse, J. 1982. The effect of photoperiod on female reproductive activity and growth of the fresh water pulmonate snail *Lymnaea stagnalis* kept under laboratory conditions. Int. J. Invert Repord, 4: 213-222.

Colowick, S. P. and Kaplan, N.O. 1955. In: Methods in enzymology. Vol. 1. ed. Academic Press, New York.

Garcia, M.C. 1975. Composition de la lumiere et activite reproductric chez l'Escargot otala lactea muller (Moll. Gast. Rum.). Bull. Soc. Zool. Fr. 100: 147-152.

Grainger, J.N.R. and Shillitoe, A.J. 1952. Histochemical observations on galactogen. Stain Technol. 27: 81-85.

Nanaware, S.G. 1974. Biochemical and Histochemical studies on the reproductive organs of some gastropods. Ph.D. Thesis, Shivaji University, Kolhapur.

Pelluet, D. and Henderson, N.E. 1954. The effects of visible radiation on the gametogenesis of the slug *Deroceras reticulatum*. Proc. Nova Scotian Inst. Sci. 23: 420-421,

Smith, B.J. 1966. Maturation of the reproductive tract of *Arion ater* (Pulmonata: Arionidae). Malacologia 4: 325-349.

Varute, A.T. 1968. Studies on β-glucuronidase in frog breeding hibernation and matamorphosis. Ph.D. Thesis. Shivaji University, Kolhapur.

Varute, A.T. 1971. Histoenzymorphology of β-glucuronidase in the ovary of *Rana tigrina*, during seasonal breeding hibernation cycle. Acta Histoche. 41: 276-292.

ENVIRONMENTAL BIOTECHNOLOGY

Edited By : **Professor (Dr.) Arvind Kumar**

Published By : **DAYA PUBLISHING HOUSE**

45

SOWING AND CLOSING TIME EFFECT ON FLOWERING AND SEED PRODUCTION OF PERSIAN CLOVER (*TRIFOLIUM RESUPINATUM* L.) UNDER PUNJAB CONDITION

• *J.S. Kang, M.S. Tiwana and Avtar Singh*

Deparment of Agronomy and Agromet
Punjab Agricultural University Ludhiana (PB)

Abstract

Sowing time (Sept 30, Oct. 20, Nov. 10, Nov. 30, Dec. 20, Jan 10) and closing time (Feb. 28, Mar. 15, Mar. 30, Apr. 14) effect indicated that maximum seed yield of 599 and 610 kg/ha was realized from November 10 in 1997-98 and 1998-99, respectively. March 15 closed produced, maximum seed yield of 654 and 685 kg/ha during 1997-98 and 1998-99, respectively. Initiation of flowering (IF) was not influenced by sowing schedule but 50 per cent flowering (50F) was affected and prolonged from 28 to 33 days during the first and from 27 to 34 days in the second year when the sowing was delayed from September 30 to January 10. Complete flowering (l00F) followed the similar trend. Altogether different trend was reflected by closing time, where IF, 50F and 100F was prominently enhanced and took lesser number of days (from 54 to 23 days and 55 to 24 days during the first and second year, respectively for 100F) to reach these stages due to high temperature and decreased humidity. November 10 sown crop produced maximum number of seeds/head (38.3 and 41.7) and volume weight (217 and 221) during the first and second year, respectively and when the crop was closed on March 15 produced maximum number of seed/head (38.7 and 40.3) and volume weight (218 and 221) during the first and second year, respectively. Percentage of quality green seed and percentage of germinability at harvest were also better with November 10 sown and March 15 closing date.

Introduction

Persian clover (*Trifolium resupinatum* L.) is being considered better over berseem for certain reasons like resistance to stem rot (*Sclerotinia sclerotiorum*) disease and relative ability to tide over moisture stress. Being a multicut forage crop, sowing time plays an important role in increasing forage and seed yield. Mostly the crop sown for fodder is left for seed production after taking some fodder cutting. Lot of research work has been done on its competitive crop berseem (*Trifolium alexandrinum* L.) however, relatively quite less work is reported on the seed production of Persian clover. With the latest improvement in newly developed variety at Punjab Agricultural University the present studies were undertaken.

The time of leaving the crop for seed setting is another important aspect which decided about the period available for different crop phases like vegetative, flowering and seed setting because excessive/poor vegetative growth after leaving the crop for seed setting adversely affects the seed yield. The last cut for fodder should be adjusted in such a way that the blooming and seed development stages coincide with the favourable weather conditions. Thus, a judicious choice of the time of the last cut of the Persian clover for fodder and leaving the crop for seed production may help in keeping a balance between vegetative and reproductive phases for obtaining highest quality seed yield.

Material and Methods

Field experiment was conducted to know the influence of sowing time and closing time on the seed yield of Persian clover at Forage Research Farm, Punjab Agricultural University, Ludhiana during 1997-98 and 1998-99. The experimental soil was loamy sand, low in organic carbon (0.31%) and available nitrogen (220 kg/ha) medium in available phosphorous (18.2 kg/ha) and potassium (217 kg/ha). The experiment was laid out in split plot design with six sowing dates (Sept. 30, Oct. 20, Nov. 10, Nov. 30, Dec. 20 and Jan. 10) in the main plots and four time of last cut for fodder (Feb. 29, March 15, March 30, April 14) in sub-plots with four replications. Newly developed variety Shaftal 69 was sown using a seed rate of 12 kg/ha by broadcast method. A basal dose of nitrogen at 12 kg N/ha, phosphorous at 50 kg P_2O_5/ha was applied each year. The first cutting was taken 55 days after sowing and subsequent cuttings were taken at an interval of 30 - 35 days. The last harvesting was done in different plots according to the time of closing the crop for seed production. The data on fodder yield was recorded at the time of harvesting. The detailed observations on initiation of flowering, 59 per cent flowering and 100 per cent flowering were recorded under all treatments. The seed crop was harvested in the mid of June and all parameter were recorded for yield and quality determination. Weather data was also recorded.

Results and Discussion

Fodder yield

Maximum green fodder yield (780 and 763 kg/ha during 1997-98 and 1998-99, respectively) was obtained from Sept. 30 sowing (Table 45.2) and with delay in sowing it decreased significantly. December and January sown crop made less vegetative growth because of severe cold (Table 45.1) and less time for growth which resulted in less green and dry matter production.

Table 45.1 : Meteorological Data for the Crop Season

| Month | Temperature (°C) | | | | Relative Humidity (%) | |
| | Maximum | | Minimum | | | |
	1997-98	1998-99	1997-98	1998-99	1997-98	1998-99
October	30.6	28.1	16.6	16.9	65	72
November	27.1	23.7	9.5	11.4	61	73
December	21.7	15.6	4.7	8.3	62	88
January	18.4	18.6	4.7	5.6	76	74
February	21.7	21.5	7.2	9.0	69	70
March	25.9	25.3	12.4	11.1	70	68
April	32.1	34.8	16.8	18.4	56	54
May	36.2	40.2	20.9	23.6	45	38
June	35.9	37.7	24.8	26.2	61	54

Table 45.2 : Green Fodder (q/ha), Dry Matter (q/ha) and Straw Yield (q/ha) Under Different Sowing and Closing Time of Persian Clover

Treatment	Green Fodder		Dry Matter		Straw Yield	
	1997-98	1998-99	1997-98	1998-99	1997-98	1998-99
Showing Time						
Sept. 30	780	763	108	98	33.5	34.0
Oct. 20	670	647	97	84	34.2	34.7
Nov. 10	579	528	80	67	33.5	34.8
Nov. 30	347	345	51	44	31.5	33.0
Dec. 20	177	191	26	24	25.7	26.6
Jan. 10	173	184	26	23	19.0	19.7
CD at 5%	82.2	78.3	22.3	18.1	3.7	3.9
Closing Time						
Feb. 28	263	253	32	31	24.6	26.5
Mar. 15	418	412	57	53	29.8	31.0
Mar. 30	510	499	70	65	31.8	31.3
Apr. 14	626	608	100	78	32.0	33.1
CD at 5%	76.4	69.8	18.6	12.0	2.2	2.3

Flowering behaviour

Initiation of flowering (IF)

During both the years sowing time could not influence IF but closing time (Table 45.5) has greatly influenced. As the closing was delayed up to April 14 Persian clover took lesser number (10 and 9 during first and second year, respectively) of days to initiate flowering in comparison to 31 and 30 days taken when the crop was closed on 28th February.

50 per cent flowering (50F)

The period to attain 50 per cent flowering was prolonged as the sowing was delayed; whereas delay in closing enhanced it from 41 days in Feb. 28 closing to 18 days in Apr. 14 during first year and corresponding figure of 40 and 18, respectively during the second year.

100 per cent flowering (100 F)

The trend for 100 F was not different from 50F. Delay in sowing caused more days to reach 100F but reverse trend was seen in closing time, whereas delay in closing time resulted in lesser number of days from 54 to 23 days and 55 to 24 days during the first and second year, respectively for complete flowering because of immediate rise in temperature (from 259 to 32.1) and decrease in humidity.

Seed yield

Effect of sowing time

The optimum sowing time is very important to achieve the maximum potential of seed yield in multicut crop. In the present study the seed yield of Persian clover was considerably affected by sowing time (Table 45.3). November 10 sowing produced significantly higher seed yield (599 and 610 kg/ha during 1997-98 and 1998-99, respectively) as compared to all other dates. Lowest seed yield (326 and 360 kg/ha during 1997-98 and 1998-99, respectively) was obtained from January 10 sowing. The highest seed yield from November 10 was result of more number of seeds/head, volume weight (Table 45.3) and straw yield (Table 45.2). Probably the plant reserve which in case of November sowing were utilized for the production of more flower heads and number of seeds/head by more growth and higher number of cutting in September and October sowing which is evident from fodder yield data (Table 45.2). Earlier Singh *et al* (1980) had also reported similar results but in berseem.

Table 45.3 : Effect of Sowing and Closing Time on the Seed Yield (kg/ha), Volume Weight (g) and Number of Seeds/head

Treatment	Seed Yield		Number of seeds / head		Volume weight	
	1997-98	1998-99	1997-98	1998-99	1997-98	1998-99
Showing Time						
Sept. 30	491	552	34.0	36.0	212	214
Oct. 20	537	553	35.2	38.2	214	219
Nov. 10	599	610	38.3	41.7	217	221
Nov. 30	558	591	36.6	36.2	215	213
Dec. 20	467	491	33.0	34.2	211	214
Jan. 10	326	360	30.9	32.0	204	209
CD at 5%	30	40	3.0	5.3	NS	NS
Closing Time						
Feb. 28	413	451	32.8	34.3	211	214
Mar. 15	654	685	38.7	40.3	218	221
Mar. 30	629	645	38.1	39.6	214	218
Apr. 14	288	324	28.9	31.1	206	210
CD at 5%	22	30	2.7	4.0	NS	NS

December and January sowings produced the lowest seed yield in each year. Reduction in seed yield beyond November 10 might be due to low temperature (Table 45.1) prevailing in the month of December and January which adversely affected the plant growth in many ways. The low temperature condition interferes with the uptake of nutrients from the soil and their translocation within the plants which affects growth (Brady, 1980). It shows less efficient use of plant nutrients in the soil. The physiological as well as metabolic activities are retarded severely due to low temperature. The October and November sown crop during reproductive phase, faced congenial environments when the mild temperature was favourable for flowering, fertilization and seed development. This is substantiated by the data on number of seeds per head. This also indicate that there was better translocation of assimilates from the vegetative to the reproductive parts.

The flowering and fertilization phases coincided with higher temperature conditions with December and January sown crops and the flowers appeared late which were forced to premature without proper development of the seed. The adverse weather (high temperature and low humidity) at maturity reduced the period (Table 45.1) for seed development affecting the translocation of the assimilates from the vegetative parts to the reproductive parts which is indicated by low dry matter and straw yield from December and January sown crop. Similar observations were shown by Rai *et al.* (1977) who found

Table 45.4 : Effect of Sowing and Closing Time on Percentage of Yellow (Y), Green (G) and Brown (B) Seeds and Their Germinability at Harvest

Treatment	1997-98			1998-99		
	Y	G	B	Y	G	B
Showing Time						
Sept. 30	28.1 (89)	51.8 (96)	20.1(52)	27.0 (89)	52.0 (97)	21.0 (54)
Oct. 20	28.2 (90)	50.2 (97)	21.6 (51)	27.0 (89)	52.0 (96)	21.0 (52)
Nov. 10	28.0 (91)	51.0 (95)	21.0 (52)	27.2 (90)	51.7 (97)	21.1 (52)
Nov. 30	27.3 (89)	51.5 (96)	21.2 (54)	27.0 (88)	51.0 (95)	22.0 (51)
Dec. 20	25.4 (90)	51.7 (97)	22.9 (51)	27.1 (89)	49.5 (94)	23.4 (54)
Jan. 10	27.5 (90)	49.0 (96)	23.5 (53)	27.0 (90)	47.9 (96)	25.1 (52)
Closing Time						
Feb. 28	28.0 (91)	53.6 (95)	18.4 (57)	25.6 (90)	48.4 (95)	26.0 (50)
Mar. 15	29.0 (91)	52.5 (95)	18.5 (55)	29.2 (91)	52.3 (96)	18.5 (52)
Mar. 30	26.3 (92)	49.5 (97)	24.2 (57)	26.8 (90)	53.1 (95)	20.1 (53)
Apr. 14	27.0 (92)	47.8 (95)	25.2 (56)	26.4 (92)	47.5 (96)	26.1 (53)

Figures in parentheses is corresponding germinability at harvest.

Table 45.5 : Initiation of Flowering (IF), 50 Per Cent Flowering (50F) and 100 per cent flowering (100 F) Under Different sowing and Closing Time

Treatment	1997-98			1998-99		
	F	50F	100F	F	50F	100F
Showing Time						
Sept. 30	22	28	37	22	27	37
Oct. 20	22	29	38	22	28	38
Nov. 10	22	29	39	22	30	39
Nov. 30	23	29	39	23	31	40
Dec. 20	23	31	41	23	31	42
Jan. 10	23	33	43	23	34	44
Closing Time						
Feb. 28	31	41	54	30	40	55
Mar. 15	25	30	41	26	32	41
Mar. 30	24	29	40	25	31	40
Apr. 14	10	18	23	9	18	24

decreased number of flower heads with delay in flowering which also increased the number of immature seeds (Table 45.4). Decrease in number of heads/plant with delay in sowing was also reported by Singh *et al.* (1980) and Brar (1986). The results are also in conformity with earlier finding in berseem from the work of Taneja *et al.* (1990), Rana *et al.* (1992) and Taneja *et al.* (1994). Colour of seed is an important parameter for judging the quality of seed in Persian clover. More the number of brown seeds, poor is the quality of the seed produced. Delay in harvesting caused more number of brown seeds. Green seeds outnumbered (Table 45.4) both yellow and brown seeds during both the years and had more germinability at harvest. There was an increasing trend for number of brown seeds when sowing and closing was delayed.

Effect of closing time

The closing time for seed setting is very important as this will decide the extent of vegetative growth, time of flower initiation and formation and development of the seed, which further affects the yield and quality of the seed produced. The crop closed for seed on March 15 gave more seed yield (654 and 685 kg/ha during 1997-98 and 1998-99, respectively) than closing on April 14, March 30 or February 28 and was significantly (Table 45.3) superior to these treatments during both the years. The reduction in seed yield from the crop closed on April 14 may be due to the shorter span of reproductive phase of the crop, coincidence of the pollination and fertilization period with high temperature (Table 45.1) which proved deleterious to pollen growth and embryo development. Several other scientists have reported similar results on berseem seed production (Brar, 1986, Taneja *et al.*, 1994 and Tiwana *et al.*, 1997). The closing on Mar. 15 produced the lower percentage of brown, seeds which is desirable. The green seeds produced outnumbered brown and yellow seeds. Jansen (1995) was of the view that varietal aspects, environmental factors and genetic factors in Persian clover played an important role for formation of green, yellow and brown seeds.

Conclusion

The research results obtained indicate that sowing time and closing the crop for seed. has a great significance for good quality seed production. According to the present study delay in sowing prolonged the 50 and 100 per cent flowering but delay in closing the crop for seed production enhanced IF, 50F and 100F stages. November sown crop when closed for seed in mid of March is the most suitable combination for good quality high yield of seed from Persian clover under Indian conditions.

References

Brady, N.C. 1988. Nature and Properties of Soil. Eurasia Publishers, New Delhi, India.

Brar, G.S. 1986. Seed production of late sown berseem (*Trifolium alexandrinum* L.) as influenced by irrigation, cutting management and variety. Ph.D. Dissertation, Punjab Agricultural University, Ludhiana.

Jansen, P.I. 1995. Seed production quality in Trifolium balansae and T. resupinatum : The role of seed colour. Seed Sci. Technol. 23 : 353-64

Rai, S.D., Joshi, Y.P. and Malik, H.S. 1977. Effect of sowing dates and seed rates on the seed production and seed quality of berseem (*Trifolium alexandrinum* L.) var. Pusa Giant. Seed Res. 5 : 11-16.

Rana, D.S., Sheoran, R.S., Joon, R.K. and Yadav, B.D. 1992. Effect of sowing dates, seed rate and phosphorous levels on fodder and seed production of Egyptian clover (*Trifolium alexandriunum*). Forage Res. 18 : 34-36.

Singh V., Joshi, Y.P. and Verma, S.S. 1980. A note on sowing date and growth hormones on the seed yield of berseem (*Trifolium alexandrinum*). Forage Res. 6 : 137-38.

Taneja, K.D., Sharma, H.C. and Singh, D.P. 1990. Influence of last cut and fertility levels on the green forage, stover and seed yield of Egyptian clover (*Trifolium alexandriunum* L.) under late sown conditions. Forage Res. 16 : 127-37.

Taneja, K.D., Sharma, H.C. and Singh, D.P. 1994. Influence of sowing date, time of last cut and fertility level on nodulation activity in Egyptian clover (*Trifolium alexandriumum*). Indian J. Agron. 39 : 158-60.

Tiwana, M.S., Thind, I.S., Tiwana, U.S., Devinder Singh and Puri, K.P. 1997. Effect of closing date on the seed yield of berseem. Proc. International Conference on Ecological Agriculture, Vol. I, Chandigarh, India.

ENVIRONMENTAL BIOTECHNOLOGY

Edited By :   **Professor (Dr.) Arvind Kumar**

Published By :   **DAYA PUBLISHING HOUSE**

46

EFFICACY OF TANNERY EFFLUENT ON MICROBIOTA OF THE PLANT *CYMPOSIS TETRAGONALOBA*

• *S.R. Thorat and R.T. Chaudhari*

School of Environmental and Earth Sciences
North Maharashtra University, Jalgaon, (M.S.) India

Abstract

Physico-chemical and microbiological characteristics of the tannery effluent was analysed. The result were calculated on the basis of the treated and untreated tannery effluent on the plant growth of *Cymopsis tetragonaloba* and also with respect to seed germination, flowering and fruiting. Dilution effluent of 25% and 50% revealed positive impact on plant growth. There was a reduction in rhizosphere microflora of the test plant when treated with effluent at a higher concentration. The microflora was meagre and restricted only on bacteria like pseudomonas and xanthomonas.

Key Words : Physico-Chemical and Microbiological characteristics, tannery effluent (treated and untreated), Cymopsis tetragonaloba.

Introduction

Tannery industry earns a sizable foreign exchange for our coutnry. But in this industry various types of chemicals and dyes are used for the removal of fat from the skin. This process gives rise to a potentionally toxic effluent. Micro-organisms play a pivotal role in degrading complex toxic organic compounds and make them available for biological cycle. Mostly this effluent with a number of toxic metals and high BOD values are discharged directly into the riverine system. Thus it affects the native microflora and makes it a burden on the environment.

Industrial effluents are nowadays commonly used for agricultural purpose. These effluents based on the type of industry, may be beneficial or harmful to crop plants (Somasekhare *et al.*, 1984). Observation on the possibility of using industrial effluent water for raising forest plantation was made by neelay and Dhondizal (1985). Influence of bavistin and monocrotophos on seed germination and seedling growth of *Trigonella foenum - graeacun* was reported by Kamble and Sabale (1999).

The information was very scanty. Therefore, in this perspective the present investigation was undertaken to study the effect of tannery effluent on microbiota of the plant *cymopsis tetragonaloba* and also to establish whether such a study would be helpful to evaluate the effect of treated effluent on the growth of plant *cymopsis tetragonaloba* and its rhizosphere microflora.

Material and Methods

The effluent samples were collected from the tannery situated near by Aurangabad city. The standard methods (APHA, 1985) were used for the collection. The work on *Cymopsis tetragonaloba* was carried out in factory premises. The physico-chemical parameters of the tannery effluent were recorded. The bacterial density, actinomycetus, yeast and would counts of tannery effluent were enumerated by adopting microbiological techniques (APHA, 1978). The bacterial strains were identified using Birgeys manual of systematic bacteriology (1987). The effluent was treated by the process of dilution for concentration namely 25%, 50% and 75%. This diluted effluent was used to irrigated test plants namely *Cymopsis tetragonaloba*. An undiluted effluent and sterile distilled water control was also maintained. Plant growth was monitored periodically on the basis of seed germination plumule apearance, shoot and root length flowering and fruiting etc. Observed upto 45 days. At the end of the experimental period rhizosphere soil was withdrawn from test plants exposed to different effluent concentration and rhizosphere microbial load was enumeratd.

Results and Discussion

The physico-chemical parameters of the tannery effluent are presented in Table 46.1 and its bacterial density in Table 46.2. Effluent had objectionable amounts of BOD high, COD and TDS (i.e. 800 to 1500, 400 to 750, 2300 to 2750 mg/lit) against ISI standards. Thus the discharge in the riverine system requires proper pretreatment. The pH of the effluent is slightly alkaline and has an unacceptable blackish brown colour with heavy load of suspended soids. Similar results were obtained by Gourisankar *et al.* (1997).

Two bacterial species namely pseudomonas aeruginose and xanthomonas sp. are only

Table 46.1 : Physico-chemical Characteristics of the Tannery Effluent

Sr. No.	Parameters	Observation
1	Colour	Blackish brown
2	pH	6.80 to 8.98
3	Temperature (°C)	27.6 to 29.2
4	COD mg/lit	400 to 750
5	BOD mg/lit	800 to 1500
6	TDS mg/lit	2300 to 2750

Table 46.2 : Microbial Characteristic of the Tannery Effluent

Sr. No.	Parameters	Microbial Load (CUF × 10/ml.)
1	Total heterotropic bacterial load	28.0
2	Yeast	Nil
3	Mould	Nil
4	Actinomycetes	Nil

microbial survivors in this tannery effluent and it is devoid of fungi, yeast and actinomycetes. The diluted effect on plant growth is presented in Table 46.3. The raw effluent exerted very high degree of inhabitation on ceuster bean.

Cymopsis tetragonaloba germination on the other hand 25% and 75% of effluent had shown positive effect on the germination of the plant. This type of positive influence was observed in flowering and fruiting of the plant. Among various dilutions 50% was found to have better influence on germination, plumule formation and leaf dimensions. On the other hand 25% effluent concentration was observed to impart overall beneficial influence on plant growth in terms of flower and fruit numbers etc. over the control. This kind of impact of effluent on plant growth could be attributed to the presence of toxic materials excess on deficiency of micronutrients and also to its effect on soil porosity and aeration especially when the concentration of the effluent was increased (Somashekhar *et al.*, 1984).

The beneficial effects of effluents on plant growth in terms of root and shoot length in dilutions of 25% and 50% could be due to the dilution of inhibitory chemicals in the effluents which may bring a depressive action on the plant growth (Oblisamy, 1979).

Result presented in Table 46.4 indicated the inhibitory effect of effluent on rhizosphere microflora with the increase in the effluent concentration. This could also be a reason for the negative impact of effluent on plant growth in its higher concentration.

Effect of paper mill effluent on germination of tree crops was reported by Kannapira *et al.* (1977). Irrigation with pulp and paper mill effluent increased the height and collar diameter of *Eucalyptus, Canaldulensis, Pongamia pinnata, Acacia areariculiformis, Lecucaena sp.* and *dendrocalus stricuts.*

Table 46.3 : Effect of Effluent at Various Dilution on Plant Growth (*Cymopsis Tetragonaloba*)

Sr. No.	Parameters	Effluent Concentrations (%)				
		25	50	75	100	Control
1.	Seed germination (%)	70	80	40	10	80
2.	Plummule formation (%)	70	80	40	10	80
3.	Shoot length [b](cm)	22.8	23.5	12.5	11.5	13.5
4.	Root length [b](cm)	7.5	5.62	6.25	5.0	5.98
5.	Total leaf number [b](cm)	26	25	15	12	22
6.	Mean leaf length [b](cm)	3.25	5.20	3.5	3.0	3.42
7.	Mean leaf width [b](cm)	2.37	3.4	2.4	2.15	2.17
8.	Total numbers of flower[b]	34	32	16	7	36
9.	Total number of fruits[b]	27	26	11	–	29

a – Mean value of eight seeds, b – Mean value of four plants.

Table 46.4 : Effect of Treated Effluent on the Rhizosphere Microflora

Sr. No.	Micro-organisms	Effluent Concentrations (%)				
		25	50	75	100	Control
1.	Bacteria ($\times 10^6$ cfu/gm)	46	12	15	5	110
2.	Fungi ($\times 10^5$ cfu/gm)	60	46	35	7	100
3.	Actionmycetes ($\times 10^5$ cfu/gm)	25	25	–	–	70

Shanti and Sanjal (1999) observed the effect of rice mill effluent on seed germination of some cereal crops. On the basis of the preliminary work, it is suggested that the effluent from rice mill can be used for irrigation purpose after proper dilution.

A case study of galvanising plant and its effluent at Hirakud was studied by Ranjita and Sanjat (1999). They observed high concentration of TDS, TSS and Cl in the effluent and high amounts of metallic pollutants like Fe, Zn, Mn, Al, Pb, Cr, Ni, etc. treatment of the effluent is highly required to render the effluent suitable for discharge into inland surface waters or on land for irrrigation.

Thus the present investigation shows the adverse effects of tannery effluent on the riverines system and on the growth of plants as well as native microflora. Nevertheless suitable treatments such as dilution brings down its toxic effect and makes it less harmful to the environment. In fact the treated effluent was compared with that of control in supporting plant growth at appropriate dilution.

Acknowledgement

The authors wish to acknowledge their indebtedness to the technical director of the tannery industry for providing all necessary facilities to carry out this investigation.

References

Padhar, Abanti and Sahu, Sanjat K. 1999. Effect of rice mill effluent on seed germination of some cereal crops. Poll. Res. 118(2) 187-189.

APHA 1985. Standard methods for examination of water and wastewaters. 16th edition, American Public Health Association, U.S.A.

Gourisankar and Pande 1991. Microbiota of Sugar mill effluent treatment and effect of treated effluent on plant growth proc. 59th SCSS. Indian Science Cong. Part III 164 Calcutta, India.

Kamble, A.B. and Sable, A.B. 1999. Influence of bavistin and monocrotophous on seed germination and seedling growth of *trigonella foenum - graecum* L. Poll. Res. 18(1) : 61-65.

Kannapirar, S. Dhevagi, P., Ponnaih, C., Obllasani, G. 1997. Effect of paper mill effluent on germination of tree crops. Indian J. Environ Hlth. Vol. 39 No. 4, 330-332.

Neelay, V.R. and Dhondizal, I.P. (1985) : Observation on the possibility of using industrial effluent water for raising forest plantation. J. Trop. forestry, 2 : 131-139.

Pande, Ranjita and Sahu, Surjat K. 1999. A case study of Galvanizing plant and its effluent at Hirakud, Orissa Jr. of Industrial Pollution Contrl. 15(1) pp. 35-41.

Somaskekhar, R.K., Gowda, M.T.G., Shettigar, S.L.N. and Srinath, K.P. 1984. Effect of industrial effluent on crop plants. Ind. J. of Environ. Hlth 26(2) 136-146.

ENVIRONMENTAL BIOTECHNOLOGY

Edited By : **Professor (Dr.) Arvind Kumar**

Published By : **DAYA PUBLISHING HOUSE**

47

ASSESSMENT OF VESICULAR ARBUSCULAR MYCORRHIZAL (VAM) ASSOCIATION IN MINING AREA - A CASE STUDY FROM JHARIA COALFIELD, JHARKHAND

• *Dr. S.K. Mati and Chandan Shee*
Centre of Mining Environment, Indian School of Mines, Dhanbad

Abstract

Degradation of land is a part and parcel for any opencast mining activities. The overburden (OB) dumps produced by the opencast mining are very poor in quality in terms of plant growth. To reclaim these dumps biologically, a long-term nutrient cycling between spoil-plant has to be established. One of the symbiotic microbes, Vescicular arbuscular Mycorrhiza (VAM) fungi plays an important role in establishment of vegetation cover and initiation of nutrient cycling for the development of self-sustaining ecosystem in OB dumps and other mined out areas.

The aims of the present investigations were to study the VAM spore density in the rhizosphere of host plants and intensity of infection in root. The study revealed that all five species of VAM fungi were present in revegetated OB dumps. Those species are: *Glomus, Gigospora, Acaulospora, Enterophospora, and Sclerocystis. Glomus* is the most predominant genus found in the mining area. The density of VAM spore was found maximum in mining area in the range of 595 spores/5g of soil than non-mining area (400 spores/5g of soil). Nine commonly growing plants were studied, namely: *Azadirachta indica, Cassia seamea, Alstonia scholaris, Melia azadirachta, Techtona grandis, Acacia auriculoformis, Prosopis juliflora, Dalburgia sissoo* and Eucalyptus. The spore density in rhizosphere of nine host plant was found in the range of 425 to 600 spores/5g of soil. The intensity root infection was studied for all nine plants. Maximum infections was recorded in D. sissoo (99%) followed by Prosopis juliflora (95%), A. auriculiformjs and Tectona grandis (93%), Melia azadarech (83%), Alstonia scholaris (78%), Polyalthia (66%), Cassia seamea (44%) and Azadirachta indica (35%), Eucalyptus (0%). The study concluded that these host plants are suitable for the biological reclamation of OB dumps.

Key Words: OB dumps, VA Mycorrhiza, spore density, root infection.

Introduction

Mining activity is an unavoidable destructive process. Though there are problems of mine wastes in terms of erosion, environmental pollution, damage to adjoining agricultural fields, forests etc., many a time they are exaggerated. These hazards are within measurable limits and can easily be ameliorated to a significant extent by extensive research and proper planning (Soni and Vasistha, 1986). It has been estimated that about 2,800 hectares of biologically unproductive area will be produced every year. To reclaim these areas biologically, a long-term nutrient cycling between soil-plant has to be established.

Reclamation of the mined out areas and OB dumps by afforestation is the cheapest and could be easily achieved. It is obvious that one must look for useful treatments and management strategy so that vegetation can be established quickly and economically leading to a self-sustaining ecosystem.

One group of soil microorganisms is important for the development of long-term plant community structure is the "mycorrhiza fungi". Vescicular Arbuscular Mycorrhizal (VAM) fungi, by virtue of their symbiotic association with roots of most terrestrial plants, are the most significant mycorrhiza fungi. VAM are not only more efficient in utilization of available nutrients from the soil-matrix but also involved in transfer of nutrients from components of soil-minerals and organic residues to soil solution. It also enhances nutrient cycling in an ecosystem. VAM fungi also commonly know as "Endomycorrhizae", because they grow in the cortical region of the plant root.

The present investigation was carried out to study the intensity of root infection and density of VAM spores in rhizosphere of mining area.

Material and Methods

For assessment of mycorrhizal association, soil samples along with fine feeder roots were collected from different host plant species growing in the reclaimed overburden dumps. Soil samples were also collected from non-mining areas as a control. About 500g of soil sample was taken from rhizosphere of each host species. Three replications each of 5g soil was taken for isolation of VAM spores, by wet sieving and decanting techniques (Gerdemann and Nicolson, 1963). The spores of VAM fungi were identified following the key of Schenk and Perez (1988).

Root infection of VAM fungi was studied after clearing the root segments with 10% KOH and staining with trypan blue (Phillips and Hayman, 1970). For confirmation of infection, the presence of intracellular infection, vescicles, arbuscles or both characters of VAM infection were recorded and per cent root colonization was calculated following the grid line method (Giovannetti and Mosse, 1980).

Result and Discussion

A variety of VAM spores were recorded from rhizosphere of different host species in mining area. They are mainly belonging to *Glomus, Gigospora, Acaulospora, Enterophospora,* and *Sclerocystis. Glomus* is the mostly predominant genus found in mining area. The characteristic morphological features of different spores found in mine area are given in Fig. 47.1.

Table 47.1 : VAM Spore Density in Mining and Non-mining area

VAM spores found in different condition	Density of spores/5 g soil
1. Rhizosphere of non-mining area (control site)	400
2. Rhizosphere of mining area (OB dumps)	595

The number of spores found in rhizosphere of mining area was found 595 spores/5g of soil, which is more than the rhizosphere of non-mining areas (400 spores/5g of soil) (Table 47.1). This is because, the plants grown in dumps are always in nutrient and moisture stress condition. Thus the absorption of nutrients will be effective, if the plants are colonized with mycorrhizal fungus (Rhodes and Gerdemann, 1978). Due to host specificity of VAM fungi, (Habte and Munjunath 1991), few plants grown efficiently in mining area with efficient

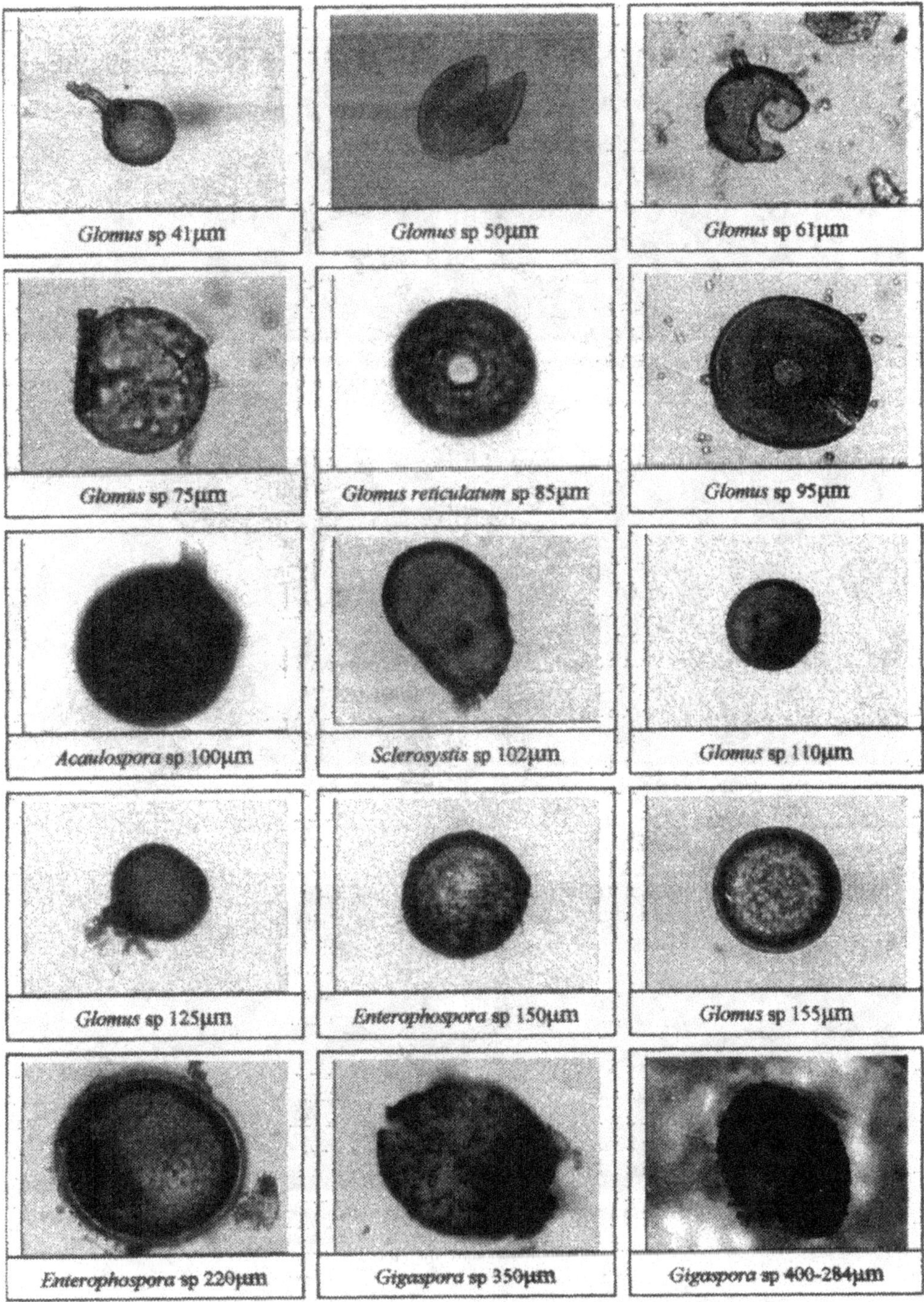

Fig. 47.1 : Morphological features of different types of VAM spore.

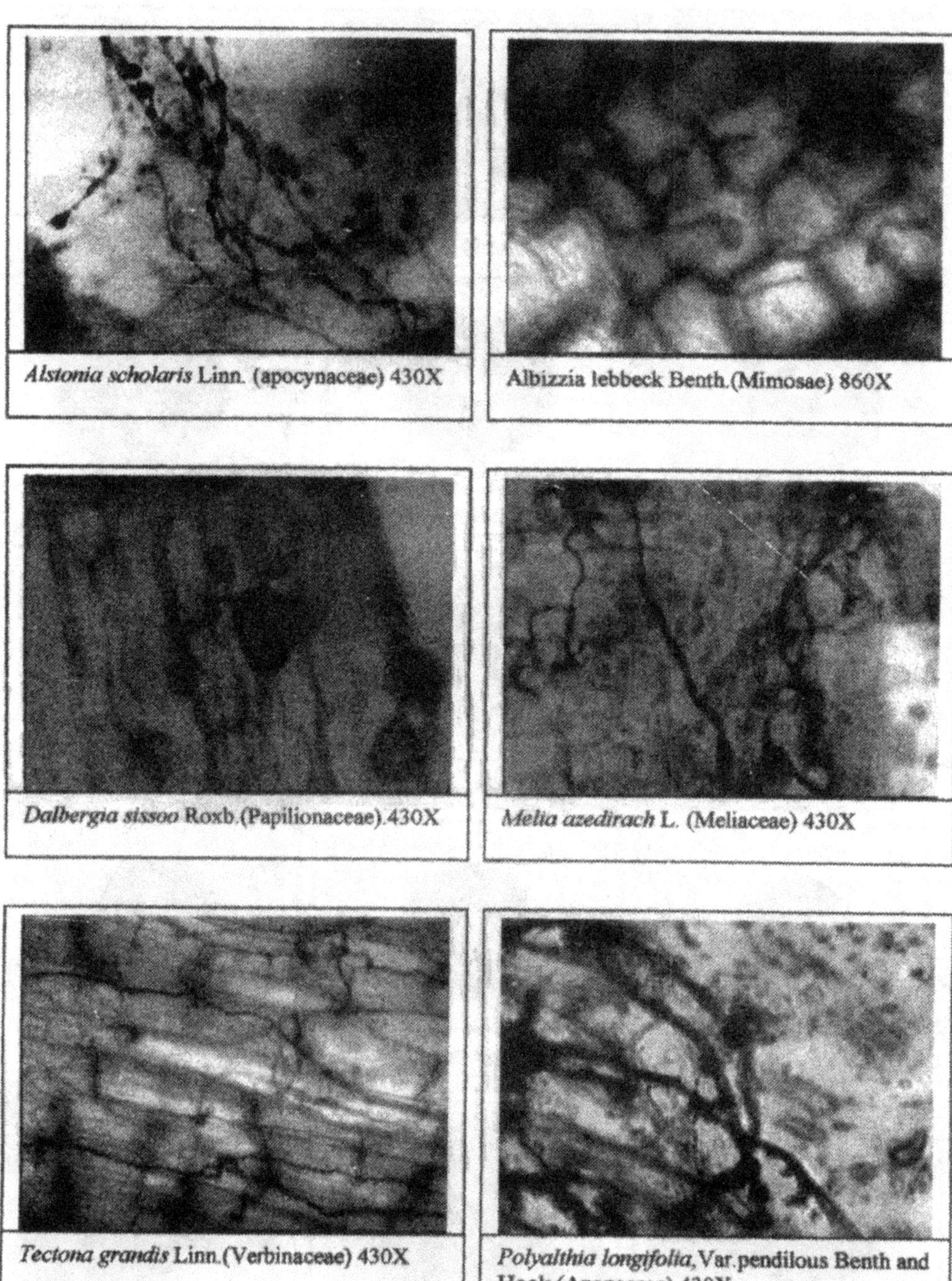

Fig. 47.2 : Mycorrhizal root infection in some plant roots of mining area.

Table 47.2 : VAM Spore Density in the Rhizosphere of Different Host Plants

Name of host plant	Spore density /5g soil
1. *Dalburgia sissoo* Roxb. (Papilionaceae)	600
2. *Prosopis juliflora* L. (Mimoseae)	595
3. *Acacia auriculoformis* A. Cunn (Mimoseae)	585
4. *Techtona grandis* L. (Verbenaceae)	500
5. *Melia azadirach* L. (Meliaceae)	467
6. *Alstonia scholaris* (Linn) R. Br. (Apocynaceae)	456
7. *Cassia seamea* L. (Papilionaceae)	445
8. *Azadiracta indica* A. Juss (Meliaceae)	425

colonization of VAM. The colonisation of VAM fungi could be easily quantified by studying the density of spores in the rhizosphere of host plant. As the VAM specificity is different, the density of the spores was also differ (Table 47.2). The quantity of spores also varies among different ages of overburden dumps in open cast mine. This is because the minimum age of dumps contain very less number of plants so the mycorrhizal association also vary less. Thus the plant density and mycorrhizal density is directly proportional.

The percentage of root colonization among different plants also varies species to species (Wilson and Martneh, 1998). Out of ten studied *Dalburgia sissoo* contain maximum infection (99%), followed by *Prosopis juliflora* (95%), *A. auriculiformis* and *Tectona grandis* (93%), *Melia azadarech* (83%), *Alstonia scholaris* (78%), *Polyalthia longifolia* (66%), *Cassia seamea* (44%) and *Azadirachta indica* (35%). However, in *Eucalyptus* no VAM infection was recorded. The VAM infection in the host plants are shown in Fig. 47.2. The percentages of infection in different plants are shown in Table 47.3.

This study will be helpful to identify the efficient mycorrhizal colonizing plants and those plants should be planted for the biological reclamation of the overburden dumps, where available nutrients and moisture contents are always very low (Helm and Carling, 1990, Maiti, 1997).

Table 47.3 : Intensity of Root Infection

Name of host plant	Type of infection			% Root colonization
	Vesicle	Arbuscle	Hyphae	
1. Azadirachta indica	–	–	+	35%
2. Cassia seamea	–	–	+	44%
3. Polyalthia longifolia	+	–	+	66.6%
4. Alstonia scholaris	+	+	+	78%
5. Melia azadirach	–	–	+	83%
6. Techtona grandis	+	–	+	93%
7. Acacia auriculoformis	+	–	+	93.5%
8. Prosopis juliflora	–	–	+	95%
9. Dalburgia sissoo	+	–	+	99%

References

Gerdemann, J.W. and Nicolson, T.H. 1963. Spores of mycorrhizal endogone species extracted from soil by wet sieving and decanting method. Transactions of the British Mycological Society 46 : 235-244.

Giovannetti, M. and Mosse, B. 1980. An evaluation of technique for measuring VAM infection in roots. New Phytologist 84 : 489-500.

Habte, M. and Munjunath, A. 1991. Categories of vescicular-arbuscular dependency of host species. Mycorrhiza 1(1) : 3-12.

Helm, D. and Carling, D. 1990. Use of On-site mycorrhizal inoculum for plant establishment on abandoned mined lands. Bureau of Mines contract report. Palmer, Alaska.

Jehne, W. and Thompson, C.H. 1981. Endogone mycorrhizae in plant colonization on coastal sand dunes at coololo, Queensland. Australian journal of ecology. 6 (3) : 221-230.

Maiti, S.K. 1997. Importance of VAM fungi in coal mine overburden reclamation and factors affecting their establishment on overburden dumps. Environment and Ecology 15 (3): 602-608.

Phillips, J.M. and Hayman, D.S. 1970. Improved procedures for clearing and staining parasitic and vescicular arbuscular mycorrhizal fungi for rapid assessment of infection. Trans. Br. Myco. Soc., 55 : 158- 161.

Rhodes, L.H. and Gerdemann, J.W. 1978. Nutrient translocation in vescicular arbuscular mycorrhizae in cellular interactions. Symbiosis parasitism (eds. C.B.Cook, P. W. Pappas and E.D. Rudolph). The Ohio State University Press, Columbus, Ohio. pp. 178-195.

Schenk, N.C. and Perez, Y. 1988. Manual for the identification of VA mycorrhizal fungi. Florida: International Culture Collection of VA Mycorrhiza Fungi, 241 pp.

Soni, P. and Vasistha, H.B. 1986. Reclamation of mine spoils for environmental amelioration. Indian forester, 112 (7) : 621-622.

Wilson, G.W.J. and Martneh, D.C. 1998. Interspecific variation in plant responses to mycorrhizal colonization in tall grass prairie. American Journal of Botany 85 (12): 1732-1738.

ENVIRONMENTAL BIOTECHNOLOGY

Edited By : Professor (Dr.) Arvind Kumar

Published By : DAYA PUBLISHING HOUSE

48

Forest Alternative for Utilization of Effluent and Solid Wastes on Productivity of Forest Tree Saplings

• *M. Elayarajan and P. Jothimani*

Department of Soil Science and Agricultural Chemistry
*Department of Environmental Sciences
Tamil Nadu Agricultural University, Coimbatore

Abstract

Apart from liquid wastes that are discharged into the land and rivers a huge amount of solid wastes such as fly ash and sludge are being dumped by industries leading to accumulation of heavy metals. Total quantity of various types of solid wastes generated from Bilt-IPCL factory is about 300 tonnes per day. Land application of industrial effluent and sludge for the crop production is an effective method of waste disposal wherein the valuable nutreints are recycled back into the ecosystem. Nursery experiment with the fallowing forest tree saplings was carried out at Bilt-IPCL, Thekkampatti, Mettupalayam to study the effect of paper factory effluent with solid wastes on seedling growth of Leguminous *viz. A auriculiformis* and non-leguminous *viz. A. indica, E tereticornis*. Germination percentage, root length, shoot length and vigour index (VI) were measured as described by the standard procedures. Better shoot length, root length, dry matter production, VI and volume index were observed in tree seedlings under FYM+fly ash irrigated with effluent. The seedlings completely withered in case of *A. auriculiformis* due to sludge application. It is clearly inferred that non-leguminous tree seedlings withstand and grow better under treated effluent and solid wastes as amendments.

Key Words : Fly ash, sludge, effluent, germination per cent, root length, shoot length and vigour index (VI)

Introduction

In India, there are about 371 paper and paper board units with a installed capacity of 37,14,043 tonnes per annum. In Tamil Nadu, there are 24 units with an installed capacity of 2,37,872 tonnes per annum (Gupta, 1995). The wastewater generation is in the range of 214-350 m^3 t^{-1} of paper produced. Apart from liquid wastes that are discharged into the land and rivers a huge amount of solid wastes such as fly ash and sludge are being dumped by industries leading to accumulation of heavy metals. Total quantity of various types of solid wastes generated from Bilt-IPCL factory is about 300 tonnes per day. These wastes are reported to promote the crop growth if added below toxic level (Prasanthrajan, 2001). Land application of industrial effluent and sludge for the crop production is an effective method of waste disposal wherein the valuable nutrients are recycled back into the ecosystem.

Materials and Methods

Nursery experiment with the fallowing forest tree saplings was carried out at Bilt-IPCL, Thekkampatti, Mettupalayam to study the effect of paper factory effluent with solid wastes on seedling growth. Leguminous : *A auriculiformis* non-leguminous : *A. indica* and *E. tereticornis.* Thirty day old test tree saplings were transplanted in poly bags (20×10cm) containing nursery mix as per treatments above. The biometric observations were collected at 60 and 120 days after transplanting (DAT). Germination percentage, root length, shoot length and vigour index (VI) were measured as described by the standard procedures. Collar diameter was measured slightly above the root collar region of the seedlings with screw gauge and expressed in millimetres (mm). The normal seedlings were first dried under shade for a week and then dried in a hot air oven maintained at 85°C for 24 hrs then cooled in a desiccator and weighed as Dry Matter Production (DMP). Valume Index (VOI) = Diametre (mm)^2 × Height (cm).

Results and Discussion

Acacia auriculiformis is the predominant leguminous tree species and *Azadirachta indica,* and *Eucalyptus tereticornis* are the predominant non-leguminous tree species of the Indian peninsula having multiple utilities. These are mainly used in afforestation of waste lands wherein growth conditions are not favourable for other species. The principle behind the selection of these three hardy species except *A. auriculiformis* is that the other two species are predominantly used as raw material for paper and pulp production in paper making industries. The wood of the *A. auriculiformis* has the problem of high lignin content which ultimately affects the quality of paper. Investigations are under way to reduce the lignin content in the wood so that *A. auriculiformis* could also be used as raw material for pulp and paper production in future.

Root Length

Significant differences were observed on *Acacia auriculiformis* root length due to amendments and irrigation. The interaction effect was also significant. Among the amendments, A_4 (FYM + fly ash) was significantly superior over A_1 (FYM), A_2 (fly ash) and A_5 (FYM + sludge) which were significant with each other. Sludge (A_2) application caused complete drying of seedlings within 20 to 30 days after transplanting (DAT). Compared to river water, effluent irrigation significantly increased the root length recording 34.2 and 44.3 cm at 60 and 120 DAT respectively. The interaction effect at 60 and 120 DAT showed that A_4 (FYM + fly ash) irrigated with effluent performed better followed by A_4(FYM + fly ash) with river water, A_5 (FYM + sludge) and A_2 (fly ash) with effluent. The other treatment combinations were inferior to the above treatment combinations. The data on root length of the seedlings are presented in Table 48.1.

The root length differed significantly due to amendments and irrigation at both the stages of seedling growth of *Azadirachta indica*. But the interaction effect was not significant at both the stages. Among the amendments, A_4 (FYM + fly ash) recorded the longest root length of 22.6, 36.0 cm at 60 and 120 DAT respectively, which was significant with rest of the treatments at both the stages of seedling growth. The shortest root length was observed in the A_3 (fly ash + sludge) and the values were 18.3 and 27.6 cm at 60 and 120 DAT respectively. Effluent irrigation recorded the longest root length of 21.2 and 33.0 cm at 60

Table 48.1. Impact of Effluent and Amendments on Root Length (cm) of Tree Nursery Seedlings

| Amend- | *Acacia auriculiformis* | | | | | | *Acacia auriculiformis* | | | | | | *Acacia auriculiformis* | | | | | |
| | 60DAT | | | 120DAT | | | 60DAT | | | 120DAT | | | 60DAT | | | 120DAT | | |
ments	I_1	I_2	Mean	I_1	I_2	Mean	I_1	I_2	Mean	I_1	I_2	Mean	I_1	I_2	Mean	I_1	I_2	Mean
A_1	32.2	35.3	33.7	38.5	47.9	43.2	21.2	22.6	21.9	32.0	35.2	33.6	28.3	30.0	29.1	33.5	37.6	33.5
A_2	33.0	33.7	33.3	38.3	43.7	41.0	19.5	20.8	20.1	29.6	32.5	31.0	24.5	25.9	25.2	28.3	31.9	30.1
A_3	*	*	*	*	*	*	17.7	18.8	18.3	26.3	28.9	27.6	16.6	17.7	17.2	17.6	18.5	18.1
A_4	35.9	40.3	38.1	44.4	51.3	47.9	21.9	23.3	22.6	34.3	37.7	36.0	28.4	30.7	29.5	36.2	39.3	37.7
A_5	32.5	33.8	33.1	39.8	43.8	41.8	20.4	21.7	21.0	30.4	33.4	31.9	28.3	28.7	28.5	32.9	35.0	33.9
A_6	26.6	27.9	27.2	32.4	34.6	33.5	18.8	19.9	19.4	27.5	30.2	28.8	22.3	23.9	23.1	29.3	31.5	30.4
Mean	32.0	34.2	33.1	38.7	44.3	41.5	19.9	21.2	20.5	30.0	33.0	31.5	24.7	26.1	25.4	29.6	32.3	31.0
	SEd	CD(0.05)		SEd	CD(0.05)		SEd	CD(0.05)		SEd	CD(0.05)		SEd	CD(0.05)		SEd	CD(0.05)	
I	0.15	0.31		0.20	0.40		0.10	0.20		0.16	0.32		0.13	0.27		0.16	0.33	
A	0.27	0.54		0.34	0.69		0.17	0.35		0.27	0.56		0.23	0.47		0.28	0.57	
IxA	0.37	0.76		0.48	0.97		0.30	NS		0.39	NS		0.32	0.66		0.39	0.80	

Table 48.2 : Impact of Effluent and Amendments on Shoot Length (cm) of Tree Nursery Seedlings

| Amend- | *Acacia auriculiformis* | | | | | | *Acacia auriculiformis* | | | | | | *Acacia auriculiformis* | | | | | |
| | 60DAT | | | 120DAT | | | 60DAT | | | 120DAT | | | 60DAT | | | 120DAT | | |
ments	I_1	I_2	Mean	I_1	I_2	Mean	I_1	I_2	Mean	I_1	I_2	Mean	I_1	I_2	Mean	I_1	I_2	Mean
A_1	34.6	38.5	36.6	73.4	83.5	78.5	17.9	19.4	18.7	30.9	33.6	32.2	31.5	32.9	32.2	68.6	72.8	70.7
A_2	28.2	33.7	31.0	60.4	64.8	62.6	16.9	18.3	17.6	26.9	29.3	28.1	26.9	28.0	27.4	52.3	56.4	54.4
A_3	*	*	*	*	*	*	14.9	16.2	15.6	25.5	27.8	26.7	18.6	18.8	18.7	21.0	21.7	21.3
A_4	38.2	40.3	39.2	81.5	90.4	85.9	21.4	23.3	22.3	35.5	38.6	37.1	35.3	37.5	36.4	75.4	81.7	78.5
A_5	31.1	36.2	33.7	61.9	70.4	66.2	17.2	18.7	18.0	28.9	31.4	30.1	27.3	28.4	27.8	55.6	65.4	60.5
A_6	25.4	30.2	27.8	50.2	60.5	55.4	16.1	17.5	16.8	26.3	28.6	27.4	22.5	24.7	23.6	38.7	40.5	39.6
Mean	31.5	35.8	33.6	65.5	73.9	69.7	17.4	18.9	18.2	29.0	31.5	30.3	27.0	28.4	27.7	51.9	56.4	54.2
	SEd	CD(0.05)		SEd	CD(0.05)		SEd	CD(0.05)		SEd	CD(0.05)		SEd	CD(0.05)		SEd	CD(0.05)	
I	0.16	0.32		0.33	0.66		0.09	0.20		0.15	0.31		0.14	0.28		0.29	0.59	
A	0.27	0.55		0.56	1.15		0.16	0.33		0.26	0.54		0.24	0.49		0.51	1.03	
IxA	0.38	0.78		0.80	1.62		0.22	NS		0.37	NS		0.34	0.69		0.71	1.45	

A - Amendments; A_1-FYM; A_2-FA(Fly ash); A_3-S(Sludge); A_4-FYM+FA; A_5-FYM+S; A_6-FA+S;

I-Irrigation; I_1-River water I_2-Effluent.

and 120 DAT respectively as compared to river water irrigation. The interaction effect was non-significant at both the stages of seedling growth.

The same trend was noticed in *Eucalyptus tereticornis* as that of *Azadirachta indica*. Root length differed significantly due to amendments and irrigation at both the stages of seedling growth. The interaction effect was significant at 60 and 120 DAT. Among the amendments, A_4 (FYM + fly ash) recorded the longest root of 29.5 and 37.7 cm at 60 and 120 DAT respectively which was on par with A_1 (FYM) at 60 DAT and significantly differ at 120 DAT. The shortest root was recorded in A_3 (sludge) and the values were 17.2 and 18.1 cm at 60 and 120 DAT respectively. Effluent irrigation recorded the longest root of 26.1 and 32.3 cm at 60 and 120 DAT respectively as compared to river water irrigation. The interaction effect at 60 and 120 DAT revealed that A_4 (FYM + fly ash) with effluent irrigation recorded significantly higher root length followed by A_1 (FYM) effluent, A_5 (FYM + sludge) with effluent and the least was found in A_3 (sludge) with river water at both the stages of seedling growth.

In the present study, Bilt-IPCL solid wastes except sludge with effluent irrigation had no deleterious effect on these tree seedlings. Among the three seedlings tried, *A. auriculiformis* did not thrive under sludge application whereas the *A. indica* and *E. tereticornis* could withstand the adverse effect of sludge. It is clearly understood that non leguminous tree seedlings withstand the effluent and solid wastes as amendments. Kannapiran (1995) observed better seed germination and growth of seedling at higher concentration of raw paper mill effluent (75 and 100%) for *Acacia leucopholea*, *Albizia lebbeck* and *Leuceana leucocephala*. Under drought or stress conditions, the tree species having maximum root length are termed as 'survival of the fittest'. The root length is taken as an index for evaluating the drought tolerance of a tree species.

In the present investigation, FYM + fly ash, FYM and FYM + sludge recorded longer roots than other solid wastes. This might be due to contribution of more readily available nutrients from these sources. In *E. tereticornis*, sludge inhibited the root growth due to toxicity caused by alkaline pH and higher soluble salt concentration. *Acacia auriculiformis* was sensitive to high alkalinity caused by sludge, which further aggravated the transplanting shock, and thereby the establishment of the seedling was total failure. Very hard surface could be formed due to sludge addition, which reduces the infiltration and root penetration. Sludge exhibited podzolanic properties i.e., reaction with water to form cemented soil layer, resulting in reduced infiltration and root penetration. Such an effect, besides high soluble salts and alkalinity might have been the cause for poor root length in *A. indica* and *E. tereticornis* and total elimination of *A. auriculiformis*. The appreciable level of root length under FYM + fly ash might be due to high moisture holding capacity which enhanced the growth and penetration of the roots. Therefore, the fly ash combined with FYM could safely be used as nursery mix.

Shoot Length

Significant differences were observed on shoot length of *Acacia auriculiformis* due to amendments and irrigation in both 60 DAT and 120 DAT. The interaction effect was also significant. Among the amendments, A_4 (FYM+fly ash) was found superior over A_1 (FYM), A_2 (Fly ash) and A_5 (FYM + sludge) rest of the treatments was significantly different from each other. Sludge application caused drying of seedling with 20 to 30 DAT. Effluent irrigation recorded longer shoot length as compared to river water. The increase being 35.8 to 73.9 cm at 60 to 120 DAT respectively. The data on shoot length of tree seedlings are presented in Table 48.2. The interaction between amendments and irrigation was significant at both 60 DAT and 120 DAT of seedling growth. Effluent with A_4 (FYM + FA) with effluent performed better than rest of the amendment combinations.

Significant differences were observed on *Azadirachta indica* shoot length due to amendments and irrigation at both 60 and 120 DAT. Among the amendments, A_4 (FYM + fly ash) recorded the longest shoot length of 22.3 and 37.1 cm at 60 and 120 DAT respectively followed by A_1 (FYM), A_5 (FYM + sludge) and A_2 (fly ash), which were significanty different from each other. The shortest shoot length of 15.6 and 26.7 cm was recorded in sludge at 60 and 120 DAT respectively. Effluent irrigation recorded the longest shoot length of 18.9 and 31.5 cm respectively as compared to river water at both the stages of seedling growth. The interaction effect was non-significant at both the stages of seeding growth.

A similar trend as above in *Azadirachta indica* was noticed for *E. tereticornis* species also. Significant differences were observed on shoot length due to amendments and irrigation at both the stages of tree seedlings growth. The interaction effect was significant on 60 and 120 DAT. Among the amendments, A_4 (FYM+fly ash) recorded the longest shoot of 36.4 and 78.5 cm at 60 and 120 DAT respectively followed by FYM (A_1), FYM combination with sludge (A_5) and fly ash (A_2). The shortest shoot was recorded in the sludge recording 18.7 and 21.3 cm at 60 and 120 DAT respectively. Effluent irrigation recorded the longest shoot length of 28.4 and 56.4 cm resepctively as compared to river water at both the stages of seedling growth. The interaction effect at 60 and 120 DAT showed that FYM + Fly ash (A_4) with effluent as well as river water combination performed better than the rest of amendment combinations and the sludge (A_3) with river water was found to be inferior at both the stages of seedling growth.

Plant height is an important tool to analyse the site quality, acclimatization of tree species to the existing environment, growth pattern of species, and plant environment interaction. The present study revealed significant differences in shoot length due to effluent irrigation at all growth stages of tree seedlings. Neelay and Dhondizal (1985) reported that pulp and paper mill effluent when used for irrigation tree species, resulted in an increase in height of *Eucalyptus camandulensis*, *Pongamia pinneta*, *Acacia auriculiformis*, *Leuceana eucocephala* and *Dendrocalamus strictus*. Among the amendments, FYM+fly ash recorded the longest shoot. The shoot length under FYM was comparable to that under FYM+fly ash. The reason for this could be attributed to the fact that both FYM+fly ash and FYM were rich in essential plant nutrients. In *E. tereticornis*, the shortest shoot length was recorded in sludge. This might be due to the alkaline nature of sludge and also due to higher proportion of soluble salt in the root zone. Singh and Singh (1971) observed a negative correlation between EC and growth of tree seedlings. The salts present in sludge inturn might affect the aeration essential for survival and growth of tree seedlings. Dolar *et al.* (1972) also were of the opinion that the reduction in growth system was due to the toxic effects of salts and heavy metals on soil structure, aeration and porosity.

Farm yard manure + fly ash with effluent recorded the highest shoot followed by FYM with effluent irrigation than with river water combination. This might be due to the

Table 48.3 : Impact of Effluent and Amendments on Dry Matter Production (g plant^{-1}) of Tree Nursery Seedlings

| Amend- | Acacia auriculiformis | | | | | | Acacia auriculiformis | | | | | | Acacia auriculiformis | | | | | |
| | 60DAT | | | 120DAT | | | 60DAT | | | 120DAT | | | 60DAT | | | 120DAT | | |
ments	I_1	I_2	Mean	I_1	I_2	Mean	I_1	I_2	Mean	I_1	I_2	Mean	I_1	I_2	Mean	I_1	I_2	Mean
A_1	0.65	0.76	0.71	3.66	6.50	5.08	0.78	0.83	0.81	1.03	1.13	1.08	1.18	1.46	1.32	4.66	5.84	5.25
A_2	0.48	0.65	0.57	1.83	3.27	2.55	0.76	0.80	0.78	0.86	0.93	0.90	1.15	1.32	1.24	2.24	3.97	3.10
A_3	*	*	*	*	*	*	0.69	0.72	0.71	0.75	0.83	0.79	0.10	0.83	0.47	0.98	1.25	1.12
A_4	0.93	1.32	1.13	6.07	7.67	6.87	0.89	0.93	0.91	1.30	1.42	1.36	1.49	1.53	1.51	7.04	8.36	7.70
A_5	0.59	0.78	0.69	3.13	4.74	3.93	0.78	0.81	0.80	0.87	0.98	0.93	1.24	1.40	1.32	3.48	4.33	3.90
A_6	0.34	0.48	0.41	0.98	1.75	1.37	0.72	0.76	0.74	0.82	0.90	0.86	1.12	1.28	1.20	1.85	2.21	2.03
Mean	0.60	0.80	0.70	3.13	4.79	3.96	0.77	0.81	0.79	0.94	1.03	0.99	1.05	1.31	1.18	3.38	4.33	3.85
	SEd	CD(0.05)		SEd	CD(0.05)		SEd	CD(0.05)		SEd	CD(0.05)		SEd	CD(0.05)		SEd	CD(0.05)	
I	0.00	0.01		0.02	0.04		0.01	0.01		0.01	0.01		0.01	0.01		0.02	0.05	
A	0.01	0.01		0.04	0.07		0.01	0.02		0.01	0.02		0.01	0.02		0.04	0.08	
IxA	0.01	0.02		0.05	0.10		0.01	NS		0.01	NS		0.02	0.03		0.06	0.01	

Table 48.4 : Impact of Effluent and Amendments on Vigour Index of Tree Nursey Seedlings

| Amend- | Acacia auriculiformis | | | | | | Acacia indica | | | | | | Eucalyptus tereticornis | | | | | |
| | 60DAT | | | 120DAT | | | 60DAT | | | 120DAT | | | 60DAT | | | 120DAT | | |
ments	I_1	I_2	Mean	I_1	I_2	Mean	I_1	I_2	Mean	I_1	I_2	Mean	I_1	I_2	Mean	I_1	I_2	Mean
A_1	4413	5155	4784	5692	6767	6230	2412	2566	2489	2760	2967	2864	2739	2872	2805	4020	4088	4054
A_2	3243	3689	3466	5365	5931	5648	1913	2035	1974	2205	2371	2288	2312	2364	2338	3306	3481	3393
A_3	*	*	*	*	*	*	1708	1817	1762	1933	2078	2005	1837	1879	1858	2577	2699	2638
A_4	4972	6133	5553	8467	10765	9616	2978	3168	3073	3417	3674	3546	3402	3481	3442	6090	6266	6178
A_5	3751	4359	4055	5028	6107	5567	2285	2431	2358	2637	2835	2736	1928	1975	1951	3890	3931	3910
A_6	2370	2448	2409	3924	4241	4082	1814	1930	1872	2053	2208	2130	1726	1752	1739	3272	3296	3284
Mean	3750	4357	4053	5695	6762	6229	2185	2324	2255	2501	2689	2595	2324	2387	2356	3859	3960	3910
	SEd	CD(0.05)		SEd	CD(0.05)		SEd	CD(0.05)		SEd	CD(0.05)		SEd	CD(0.05)		SEd	CD(0.05)	
I	19.31	39.29		29.64	60.31		11.40	23.19		13.15	26.76		11.87	24.16		20.41	41.52	
A	33.45	68.05		51.34	104.5		19.75	40.17		22.78	46.35		20.57	41.84		35.35	71.92	
IxA	47.30	96.23		72.61	147.7		27.35	NS		32.50	NS		29.80	NS		49.55	NS	

A-Amendments; A_1-FYM; A_2-FA(Fly ash); A_3-S(Sludge); A_4-FYM+FA; A_5-FYM+S; A_6-FA+S;

I-Irrigtion; I_1-River water I_2-Effluent.

enrichment of soil by the addition of the above amendments, which contained appreciable amount of a plant nutrients and also due to the ameliorative effect of these amendments on the effluent. Combined use of the effluent along with amendments might have provided enough nutrients with better physical and microbial environment, thus improving the soil fertility. The above findings are in line with those of Kladivko and Nelson (1979), Reddy *et al.* (1981) and Pushpavalli (1990). Kannapiran (1995) also observed better shoot length at higher concentration of untreated effluent in *Acacia nilotica*, *Albeizia lebbeck* and *Leuceana leucocephala.*

Dry Matter Production (DMP)

The data on dry matter production of tree seedlings are presented in Table 48.3. *Acacia auriculiformis* showed significant differences for DMP due to amendment, irrigation and interaction at both the stages of seedling growth. Among the amendments, (A_4) FYM + fly ash performed better than other amendments at both the stages of seedling growth recording 1.13 and 6.87 g seedling^{-1} respectively. The effluent irrigation recorded significantly higher DMP at both the stages than river water irrigation. The interaction between the amendments and irrigation was significant at both the stages of seedling growth. The FYM+fly ash (A_4) with effluent combination recorded sigificantly higher DMP followed by A_4 with river water, FYM (A_1) and FYM + sludge (A_5) with effluent, and the other treatment combinations were inferior to the above treatment combinations.

Significant differences were observed for DMP of *Azadirachta indica* due to amendment, irrigation at both the stages of seedling growth. The interaction effect was non-significant at 60 DAT and 120 DAT. Among the amendments, A_4 (FYM + fly ash) recorded higher DMP of 0.91 and. 1.36 g seedling^{-1} at 60 and 120 DAT respectively which was followed by A_1 (FYM) and A_5 (FYM + sludge) and the later two were at par with each other at 60 DAT and significantly differ at 120 DAT. The DMP was lowest in sludge (A_3) at both the stages of seedlig growth. The effluent irrigation recorded significantly higher DMP at both stages than river water irrigation. The least was found in sludge with river water at all stages of seedling growth.

Eucalyptus tereticornis species also showed the same trend as the above species. Significant differences were observed for DMP due to amendment, irrigation and interaction at both the stages of seedling growth. Among the amendments, A_4 (FYM + fly ash) recorded higher DMP of 1.51 and 7.70 g seedling^{-1} at 60 and 120 DAT respectively which was followed by A_1 and A_5 and the latter two were on par with each other only at 60 DAT and significant difference at 120 DAT. The DMP was the lowest in A_5 at both stages of seedling growth. Compared to river water, effluent recorded higher DMP of 1.31 and 4.33 g seedling^{-1} at 60 and 120 DAT respectively. The interaction effect was significant at both the stages. The treatment FYM + fly ash (A_4) with effluent recorded higher DMP followed by FYM+fly ash (A_4) with river water and FYM (A_1) with effluent recorded higher DMP followed by FYM+fly ash (A_4) with river water and FYM (A_1) with effluent irrigation. The later two were being on par themselves and performed better than the rest of the treatment combinations. The sludge with river water recorded lower values with respect to DMP.

The dry matter production (DMP) increased as the seedling growth advanced. At 120 DAT, the DMP was higher by three times in all the three species. Among the amendments,

FYM + fly ash recorded higher DMP followed by FYM and FYM + sludge. The DMP was very high in *A. auriculiformis* probably due to the genetic nature of the species having broad leaves and thick stem. In non-leguminous, the DMP was very high in *E. tereticornis* than *A. indica*. Effluent irrigation recorded higher DMP compared to river water for the reason that it could provide higher amount of essential nutrients. FYM + fly ash and FYM along with effluent increased the DMP than the rest of treatment combintions (Fig. 48.1). This might be due to the ameliorative effect of these amendments, which could alleviate the ill effects of effluent irrigation resulting in higher DMP. Similar findigns were reported by Sandana (1995) in tomato, Pushpavalli (1990) in sugarcane and Hameed and Udayasoorian (1998) in *A. nilotica, C. equisetifolia* and *E. tereticornis*.

Vigour index (VI)

The data on vigour index of tree seedlings are presented in Table 48.4. Significant differences were noticed for *Acacia auriculiformis* VI due to amendments, effluent irrigation and their interaction at both the stages of seedling growth. The VI was not calculated due to withering of seedlings under sludge. Among the amendments, A_4 was superior over A_1, A_5 and A_2. There was a significant difference between VI of well water and effluent irrigated plants. Effluent irrigation performed better than the river water in both the stages. The interaction was significant at both the stages of the seedling growth and it showed that A_4 (FYM + fly ash) in combination with effluent performed better followed by A_1 (FYM) with effluent. Rest of the treatment combinations were found inferior to them.

The VI of *Azadirachta indica* significantly differed due to amendments and irrigation at both the stages of crop growth but the interaction effect was not significant. Among the amendments, A_4 recorded the highest value of 3073 and 3546 at 60 and 120 DAT respectively which was followed by A_1, A_5 and A_2. The effluent irrigation performed better than the river water irrigation. There was no significant interaction observed at both the stages of seedling growth.

A very similar trend as above was observed for this *Eucalyptus tereticornis* species also. The VI differed significantly due to amendments and irrigation at both the stages of crop growth, while the interaction effect was not significant. Among the amendments, A_4 recorded the highest vigour index of 3442 and 6178 at 60 and 120 DAT respectively which was followed by A_1, A_5 and A_2. The least VI was recorded in sludge (A_3). The effluent irrigation performed better than river water invariably at both the stages. The interaction effect between irrigation sources and amendments was not significant with respect to VI.

Farmyard manure + fly ash recorded the highest VI in all tree species at both the stages of seedlig growth due to enhanced shoot and root growth. The next best amendments were FYM, FYM + sludge and fly ash. The least was recorded in sludge. Kannapiran (1995) observed higher VI in FYM amended soils. The appreciable VI in FYM than FYM + sludge was due to its high moisture retention capacity and high nutrient status which were reponsible for higher germination. The effluent irrigation recorded higher VI when compared to river water. This might be due to the richness of plant nutrients in treated Bilt-IPCL effluent.

The data on volume index of tree seedlings are presented in Table 48.5. Significant differences were noticed in *Acacia auriculiformis* due to amendments, effluent irrigation and

Table 48.5 : Impact of effluent and amendments on volume of index of tree nursery seedlings

| Amend- | *Acacia auriculiformis* | | | | | | *Acacia auriculiformis* | | | | | | *Acacia auriculiformis* | | | | | |
| | 60DAT | | | 120DAT | | | 60DAT | | | 120DAT | | | 60DAT | | | 120DAT | | |
ments	I_1	I_2	Mean	I_1	I_2	Mean	I_1	I_2	Mean	I_1	I_2	Mean	I_1	I_2	Mean	I_1	I_2	Mean
A_1	75.3	99.2	87.2	1094.6	1502.7	1298.6	87.7	112.7	100.2	333.5	437.7	385.6	119.1	124.5	121.8	1537.4	1794.7	1666.0
A_2	45.0	64.6	54.8	665.2	902.0	783.6	76.0	97.6	86.8	274.7	360.6	317.7	91.5	99.4	95.4	633.7	841.3	737.5
A_3	*	*	*	*	*	*	60.1	72.0	66.1	231.2	286.0	258.6	56.8	59.4	58.1	106.7	138.4	122.5
A_4	78.6	119.8	99.2	1818.8	2164.4	1991.6	163.9	210.5	187.2	521.6	684.7	603.2	149.1	191.1	170.1	1835.4	2246.1	2040.8
A_5	58.8	80.8	69.8	498.8	1105.0	801.9	83.9	99.7	91.8	294.5	386.6	340.5	95.9	107.4	101.6	496.4	1030.8	763.6
A_6	42.5	53.7	48.1	158.3	219.5	188.9	66.5	85.4	75.9	254.0	312.9	283.5	74.8	83.1	78.9	222.3	266.9	244.6
Mean	60.0	83.6	71.8	847.1	1178.7	1012.9	89.7	113.0	101.3	318.3	411.4	364.8	97.9	110.8	104.3	805.3	1053.0	929.2
	SEd	CD(0.05)		SEd	CD(0.05)		SEd	CD(0.05)		SEd	CD(0.05)		SEd	CD(0.05)		SEd	CD(0.05)	
I	1.03	2.1		16.03	32.6		1.6	3.3		5.7	11.6		1.6	3.3		17.5	35.6	
A	1.78	3.6		27.76	56.5		2.8	5.7		9.8	20.0		2.8	5.8		30.3	61.6	
IxA	2.52	5.1		39.26	79.9		4.0	8.1		13.9	28.3		4.0	8.2		42.8	87.1	

A-Amendments; A_1-FYM; A_2-FA(Fly ash); A_3-S(Sludge); A_4-FYM-FA; A_5-FYM+S; A_6-FA+S;

I-Irrigation; I_1-River water I_2-Effluent.

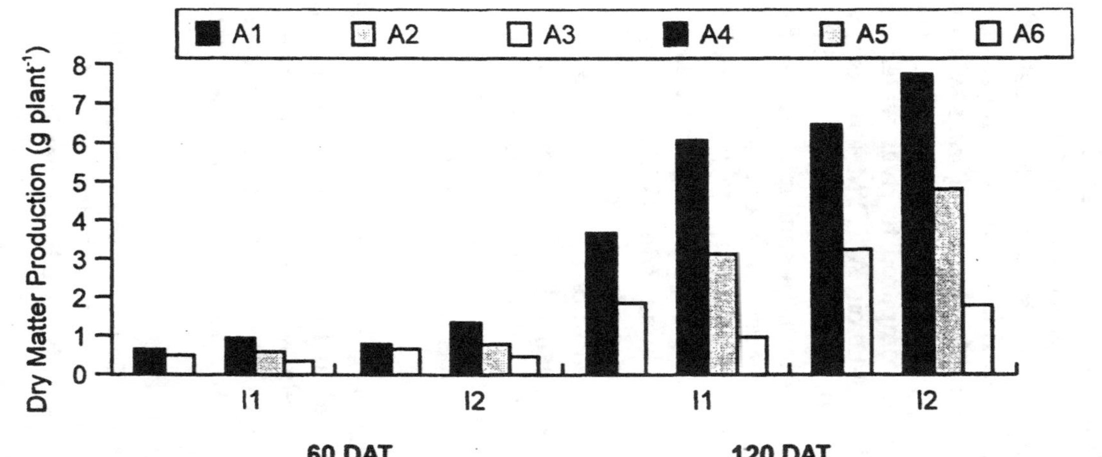

Fig. 48.1 : Effect of effluent and amendments on dry matter production of *A. auriculiformis*.

their interaction at both the stages of seedling growth. Among the amendments, A_4 (FYM + fly ash) performed better with volume index (VOI) values of 99.2 and 1991.6 at 60 and 120 DAT resepctively. The sludge (A_3) recorded the lowest VOI values at both the stages of seedling growth. Effluent performed better than river water with VOI values of 83.6 and 1178 at 60 and 120 DAT respectively. Significant interaction effect at both the stages of seedling growth revealed that A_4 (FYM + fly ash) wth effluent combinaton significantly increased VOI as compared to rest of the treatment combinations. Among Bilt-IPCL solid wastes, the performance of fly ash with effluent was comparable with that of FYM with river water. This showed that fly ash could possibly be used in place of FYM. A very similar trend as in *Acacia auriculiformis* was observed for *Azadirachta indica* species alo except the withering of seedling under sludge. The effluent irrigation performed better with high VOI values as compared to river water irrigation. Significant interaction observed at 60 and 120 DAT wherein A_4 (FYM + fly ash) with effluent irrigation performed better followed by A_1 (FYM), A_5 (FYM + sludge) with effluent. Rest of the treatment combinations was inferior to them. Almost a similar trend as that of *Acacia auriculiformis* and *Azadirachta indica* was noticed for *Eucalyptus tereticornis* species also. The interaction effect was significant at both the stages of 60 and 120 DAT, which showed that A_4 (FYM + fly ash) with effluent irrigation was better than the rest of the treatment combinations.

The volume index (VOI) is a good indicator of quality and ability of the seedlings to survive under adverse conditios. Farm yard manure + fly ash with effluent had higher VOI as compared to rest of the treatment combinations. Among the Bilt-IPCL solid wastes used as amendments, FYM + sludge, and fly ash with effluent was comparable with that of FYM with river water. This clearly showed that sludge and fly ash could be used in place of FYM in the preparation of nursery mix. Maximum VOI was recorded under normal nursery mix (FYM) as in *Acacia nilotica*, *Albezzia lebbeck* and *Pongamia pinnata* by Kannapiran (1995).

Conclusion

Better shoot length, root length, dry matter production, VI and volume index were observed in tree seedlings under FYM + fly ash irrigated with effluent. The seedlings completely withered in case of *A. auriculiformis* due to sludge application. It is clearly inferred that non-leguminous tree seedlings withstand and grow better under treated effluent and solid wastes as amendments.

References

Dolar, S.S., Boyle, J.R. and Kennedy, A.D. 1972. Paper mill sludge disposal on soils : Effects on the yield and mineral nutrition of oats (*Avena sativa* L). J. Environ. Qual., 1 : 405-409.

Gupta, R.C. 1995. Indian Paper Industry–Macro Statistical data. K.K. Kothari Paper Publication, Calcutta, p. 46.

Hameed, S.M. and Udayasoorian, C. 1998. Changes in soil chemical properties and plant nutrient content under Eucalyptus irrigated with paper mill effluent and sludges in forest nurseries. In: Water World 1998. National seminar on applications of treated Effluent for irrigation, Regional Engineering College, Trichi, March 23, pp : 38-49.

Kannapiran, S. 1995. Studies on the effect of solid and liquid wastes from paper and pulp industry on forest species. M.Sc. (Env. Sciences) Thesis, Tamil Nadu, Agrl. Univ., Coimbatore.

Kladivko, E.J. and Nelson, O.W. 1979. Changes in soil properties from application of anaerobic sludge. J. Wat. Pollut. Cont. Fed., 51 : 325-332.

Neelay, V.R. and Dhondizal, L.P. 1985. Observation on the possibility of using industrial effluent water for raising forest plantation. J. Trop. Forestry, 2 : 131-139.

Prasanthrajan, M. 2001. Bio-compost from Paper Board Mill solid wastes. M.Sc. (Env. Science) Thesis, Tamil Nadu Agric. Univ., Coimbatore.

Pushpavalli, R. 1990. Studies on the characterization of pulp and paper mill effluent and its effect on soil profile characteristics and on germination, yield and juice quality of sugarcane (Var. CO 63004 and COC 671) M.Sc. (Ag.) Thesis, Tamil Nadu Agrl. Univ., Coimbatore.

Reddy, M.R., Jivedran, S., Jain, S.C. 1981. Paper mills effluents for sugarcane irrigation. IAWPC Technol. Annu., 8 : 129-146.

Sandana, K.M.C. 1995. Studies on the effect of liquid and solid wastes from industries and the growth of certain crops. M.Sc. (Env. Sciences) Thesis, Tamil Nadu Agri. Univ., Coimbatore.

ENVIRONMENTAL BIOTECHNOLOGY

Edited By : **Professor (Dr.) Arvind Kumar**

Published By : **DAYA PUBLISHING HOUSE**

49

ISOLATION AND CHARACTERISATION OF 1-0-DIGALLOYL-5-0-GALLOYLMANNITOL FROM *PTERLOBIUM HEXAPETALLUM*

• *N.S. Nagarajan*, M.G. Sethuraman, C.N. Manoj and R. Priya*

**Department of Chemistry, Gandhigram Rural Institute, Deemed University Gandhigram, Dindigul District, Tamil Nadu, India.*

Abstract

The ethyl acetate extract of defatted flowers of *Pterlobium hexapetallum* gave a trigalloyl ester of mannitol which was identified as 1-O-digalloyl-5-galloylmannitol by spectral studies.

Key Words : Pterolobium hexapetallum, gallic ester, trigalloyl derivative of mannitol.

Introduction

Pterolobium hexapetallum (Roth) (Mathew, 1981) (family : *Leguminosae*, sub-family : *Caesalpinaceae*) called "Karuindu" in Tamil is locally used as an antiseptic. A perusal of literature revealed that the earlier phytochemical examination of *P. hexapetallum* has yielded methyl gallate, methyl-tri-O-methylgallate, phenanthrene, pterostilbene resveratol and tri-o-methyl reservatol (Jagdish, 1988). Phenolic compounds have been isolated and characterized from *P. stellatum* (Irene, 1987).

Gallic acid is normally encountered in plants in ester form, in which the acid is bound not only in aliphatic ester linkages but also in depside forms (Haslam, 1964). Gallic acid esters with anhydro and branched chain sugars and with quinic acid have been recorded. This acid occurs in fresh plant tissues in the ester form as an important structural unit of ellagitannins (Nonaka, 1991).

Present Work

The present work involves the isolation and characterization of a trigalloyl ester of mannitol from the flowers of *P. hexapetallum*. The air-dried flowers of *P. hexapetallum* were defatted with petroleum ether and then extracted with ethyl acetate and alcohol in hot condition (6 × 6 hours). The ethyl acetate extract was concentrated and left in a refrigerator overnight and the solid separated was filtered, washed with petroleum ether and dried. It was found to be homogenous on TLC.

Characterisation of the Compound

The compound, crystallized from ethyl acetate (m.p. 111-112°C) was found to be homogenous on TLC. It gave blue colour with both neutral ferric chloride and Folin-Ciocalteu reagent (Harbourne, 1984) indicating the nature of the compound to be a derivative of gallic acid.

PMR Spectral Study of the Compound

The PMR spectrum of the compound showed a six-proton multiplet between δ 6.90 and 7.10 ppm, indicating the presence of six aromatic protons. The proton signals between δ 3.45 and 5.2 ppm showed the presence of aliphatic protons. Absence of any anomeric proton in the PMR spectrum of the compound ruled out the presence of any glycosidic linkage.

^{13}C-NMR Spectral Study of the Compound

The ^{13}C-NMR spectrum of the compound showed the presence of the three carbonyl carbons at δ 165.44, 165.15 and 168.52 ppm. The four downfield signals at δ 145.69, 138.76 and 137.93 ppm were characteristic of the oxygenated aromatic carbons. The carbon signals at δ 108.95 and 108.77 ppm were due to the unsubstituted aromatic carbons. The lack of anomeric carbon signals between 59.75 and 70.53 ppm, not only indicated the absence of aldehyde sugar units, but also showed the presence of sugar alcohol unit(s) in the compound.

Hydrolysis of the compound

Compound in methanol, when hydrolysed using 2N HCl for 12 hrs., gave gallic acid and mannitol, both identified by TLC comparison with authentic samples of them.

From the IR, PMR, ^{13}C-NMR spectral studies and the hydrolysis of compound, it was characterised to be a trigalloyl ester of mannitol. From the ^{13}C-NMR spectrum, it was found that the carbon signal due to methylene carbon (C-6) of mannitol resonated at δ 59.75 ppm while the other methylene carbon (C-1) of mannitol resonated at δ 65.55 ppm indicating the galloyl substitution at C-1. It was also concluded that the relative downfield shift of C-6 was due to the galloyl substitution at C-5. The chemical shifts ofthe other carbons of the mannitol unit suggested their unsubstituted nature. Hence, it was identified that the third gallic acid moiety was attached to one of the gallic acid units in C-1 or C-5 in a depside linkage. The highly downfield shift of C-1 carbon at δ 65.55 ppm confirmed the presence of a digalloyl function attached to the C-1 of mannitol (Tsutomu, 1990) further confirming the depside linkage.

The PMR and ^{13}C-NMR spectra showed that all the gallic acid residues were symmetrical suggesting the depside linkage between the hydroxyl group at C-4 of galloyl moiety at C-1 of mannitol and the acid group of the third gallic acid unit.

From the UV, IR, PMR, ^{13}C-NMR spectral studies and chemical evidences, the compound was characterised to be 1-O-digalloyl-5-O-galloyl mannitol. This is the first report of the natural occurrence of a gallotannin from *P.hexapetallum*. Literature survey revealed the isolation of a similar gallotannin, viz. 1, 5-anhydro-D-glucitol having three galloyl groups (Bock, 1980) from plant sources.

Experimental

The flowers of *P. hexapetallum* were identified by the Department of Applied Biology, Gandhigram Rural Institute, Gandhigram and were collected from the foot hills of Sirumalai hills, near Gandhigram in Dindigul district of Tamil Nadu, South India in the month of May. The air dried flowers (860 g) were extracted successively with petroleum ether, ethyl acetate and alcohol in hot condition for 6×6 hours. The ethyl acetate extract was concentrated and left overnight in the refrigerator. The solid that separated was filtered, washed with petroleum ether and dried. It was taken up for characterisation.

Characterisation of the Compound

The compound (m.p. 111-112°C) was crystallised from ethyl acetate as a yellow solid. It was found to be homogenous on TLC using (a) BAW 4:1:5, R_f=0.93, (b) PhOH:H_2O, R_f=0.21 (c) 100% MeOH, R_f=0.77, (d) MeOH:EtOAc:H_2O, 63:12:15, R_f=0.15, (e) C_6H_6 : MeOH:AcOH, 45:10:4, R_f=0.28 as solvent systems. It gave dark blue colour with both ferric chloride and Folin-Ciocalteu reagent.

UV Spectral Data of the Compound : λ_{max} MeOH

216 nm (band I) and 268 nm (band II)

IR Spectral Data of the Compound : V_{KBr}

3156, 3475 (OH Groups); 1618, 1714 (ester carbonyl)

PMR Spectral Data of the Compound : δ, DMSO-d_6, 200 MHz

6.90-7.10 (6H, m, H-2 and H-6 of gallic acid) and 3.45-5.20 (6H, m, protons of mannitol) ppm.

^{13}C-NMR Spectral Data of the Compound δ, DMSO-d_6, 200 MHz

165.44, 165.15 and 168.52 (ester carbonyl of galloyl groups); 145.69 (C-4 of galloyl groups); 138.76 and 137.93 (C-3 and C-5 of galloyl group); 108.95 and 108.77 (C-2 and C-6 of galloyl group); 70.51 (C-5 of mannitol); 69.41 (C-4); 67.37 (C-2 of mannitol); 66.84 (C-3); 65.55 (C-1 of mannitol) and 59.75 (C-6 of mannitol) ppm.

Hydrolysis of the Compound

The compound (5 mg) in MeOH (5 ml) was refluxed with an equal volume of 2 N HCl for 12 hours. The mixture was cooled to room temperature and extracted with EtOAc (10 ml × 3 times). The EtOAc layer was washed with water to neutral pH, dried over anhydrous Na_2SO_4 and concentrated to 2 ml. This was compared with gallic acid, protacatechuic acid, p-hydroxy benzoic acid and salcylic acid, on TLC using HOAc:$CHCl_3$ (1:9) and EtOAc:C_6H_6 (1:1) as the solvent systems. From the TLC comparison and all the spectral data, the compound was found to contain gallic acid. The aqueous layer was evaporated to dryness and dissolved in 2 ml water. It was compared with authentic samples of polyalcohols like, glucitol, mannitol and sorbitol using n.BuOH:HOAc:ether: H_2O (9:6:3:1) as solvent systems and the spots were visualised using alkaline $AgNO_3$. From the TLC behaviour and NMR spectral data, the polyalcohol was identified to be mannitol.

$$\overset{1}{CH_2}-\overset{2}{CH}-\overset{3}{CH}-\overset{4}{CH}-\overset{5}{CH}-\overset{6}{CH_2OH}$$

R = Galloyl

References

Bock, K. *et al.*, 1980. *Phytochemistry*, 19, 2033.

Gen-ichiro Nonaka *et al.*, 1991. *Chem. Pharm, Bull.*, 39, 884.

Harbourne, J.B., 1984. *"Phytochemical methods"* Chapman and Hall, New York, 42.

Haslam, E. *et al.*, 1964. *Progr. Org. Chem*, 6, 1.

Kumar, R. Jagdish *et al.*, 1988. *Phytochemistry*, 27, 3625.

Mathew, K.M. 1981. *"The Flora of the Tamil Nadu Carnatic"*, Part-One : Polypetalae, 515. Mueller-Harvey Irene et al, 1987. *J. Sci. Food Agric.*, 39, 1.

Tsutomu Hatano *et al.*, 1990. *Chem. Pharm. Bull.*, 38, 1902.

ENVIRONMENTAL BIOTECHNOLOGY

Edited By :   **Professor (Dr.) Arvind Kumar**

Published By :   **DAYA PUBLISHING HOUSE**

50

CYTOLOGICAL EFFECT OF SEWAGE WATER ON ROOT MERISTEM OF *ALLIUM CEPA*

• *M. Sakthivel and S.Muthalagi*

P.G. Department of Zoology, Kamaraj College, Thoothukudi.

Abstract

Cytotoxicity of sewage was studied by treating Allium cepa roots with different concentrations of sewage for different durations along with control. The high mitotic index and lower percentage of abnormality was very much evident in roots treated with diluted sewage. The mitotic index showed decreasing tendency with parallel increase in sewage concentration. Binucleates, sticky anaphase, prophase erosion are some of the common abnormalities observed in all treatments. Both duration and concentration of the treatment influenced the cell division. The types and percentage of abnormalities observed confirmed chronic impact of sewage on crop and other biota in the agro-ecosystem.

Key Words : Sewage, Allium cepa, Chromosomal abnormalities.

Introduction

Rapid urbanization combined with industrialization have led to the generation and disposal of enormous amount of wastewater which apart from containing toxic materials are also rich in several plant nutrients. Reclaimed wastewater is most commonly used for irrigation of agricultural crops, while the planned effluent re-use for agricultural irrigation in the United States is less than 1%, whereas other countries such as India, Israel and South Africa are using 20%-25% of wastewater effluents for agricultural purposes (Rose, 1986). Treatment problem combined with water scarcity in many areas of our country have made it obligatory for the farmers living in the low-lying areas of the city to depend on this valuable resource as a source of water for irrigation. Further, many studies on the effect of industrial effluent on crop plants have revealed their beneficial effects on over-all plant growth at lower concentration. But the higher concentrations are likely to affect metabolic and cellular activities because of the presence of several toxic components in them.

Higher plants provide valuable genetic assay system for screening and monitoring environmental pollutants. *Allium cepa* has been considered as an excellent material for the assay of chromosome aberrations following chemical treatment. Their root meristem represents a normal proliferating plant cell population that is sensitive to changes in environmental conditions (Kihlman, 1975). In this present investigation, an attempt has been made to find out the possible toxicity of sewage on the chromosomes of *Allium cepa*, which is used as vegetable.

Materials and Methods

Govinadammal Aditanar College is situated in the skirts of Tiruchendur Taluk, Thoothukudi District. The sewage was collected from the college hostel, at the point of discharge and analyzed for physico-chemical parameters using standard methods (APHA, 1985). The sewage was then used to carry out the cytological study in *A. cepa*. L.

Healthy onion bulbs were selected and the bases of the bulbs were scraped to expose the fresh root primordia. The bulbs were placed on the test tubes with distilled water in such a way that the base of the bulbs were in contact with the water surface. When the bulbs germinated and produced roots of about 1-2 cm length, they were taken out and placed on the test tubes containing different concentrations of sewage viz. 10. 25, 50, 75 and raw sewage (100%) and treated for 24, 48, 72 and 96 hrs. The different concentrations of sewage were prepared using tap water.

The control with dechlorinated tap water was maintained in all the cases. The experiment was conducted at room temperature. After treatment the root tips were excised and fixed in freshly prepared 1:3 mixture of acetic acid and ethanol for 24 hr. Thereafter root tips were stored in 70% alcohol in refrigerator. The root tips were stained in aceto-carmine to study the frequency of cell division and chromosomal abnormalities in different treatments.

Toxicity to mitotic apparatus was calculated using the following formula.

$$\text{Mitotic Index} = \frac{\text{No. of cells in division}}{\text{Total no. of cells.}} \times 100$$

Karyotoxicity is the toxicity caused to a specific component of mitotic apparatus which ultimately lead to the constitutional alteration of the chromosomal component. This was calculated by using the following formula.

$$\text{Percentage of observed cells} = \frac{\text{No. of aberrated cells}}{\text{Total no. of cells.}} \times 100$$

Results and Discussion

The physico-chemical analysis of sewage is presented in Table 50.1. The cytological effects of sewage on somatic cells were estimated on the basis of changes in mitotic index and other induced abnormalities. The inhibitory effect of sewage on *A. cepa* is evident from data given in Tables 50.2, 50.3, 50.4 and 50.5. A strong concentration impact is obvious in terms of decline in the mitotic index with the increase in the concentration (Fig 50.1). Nonetheless the percentage of abnormalities showed a reverse trend (Fig 50.2).

The percentage cell division and abnormalities induced were found to vary in direct proportion to the sewage concentration and duration of exposure to 10% sewage produced

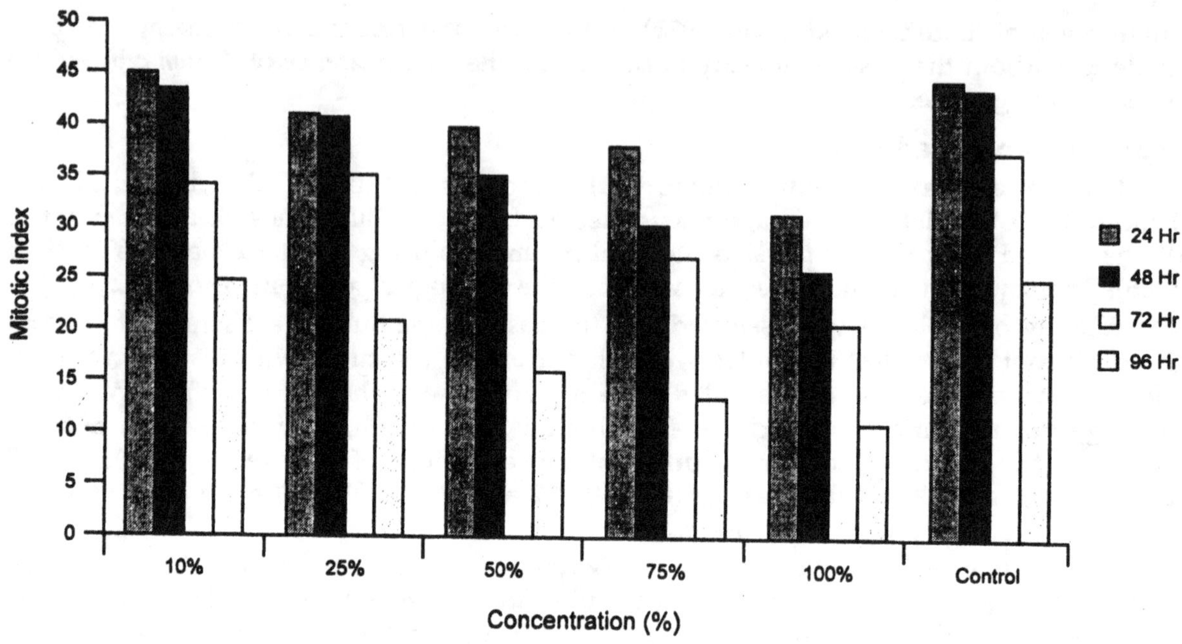

Fig. 50.1 : Effect of sewage on the mitotic index of *Allium cepa* for 24, 48, 72 & 96 hrs.

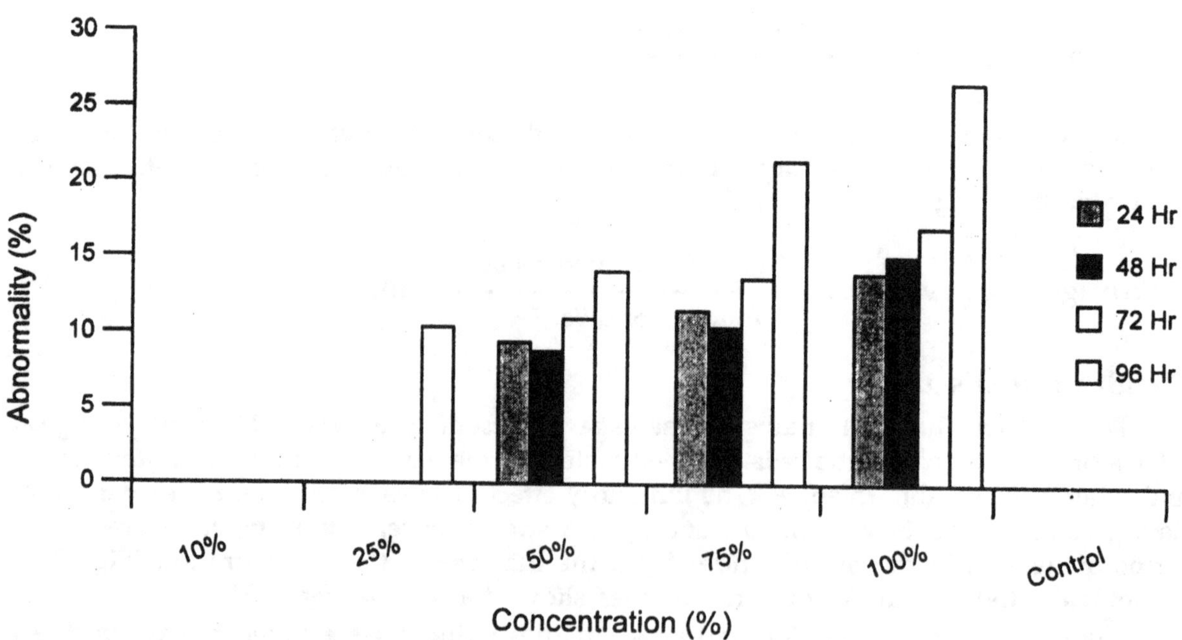

Fig. 50.2 : Effect of sewage on the percentage of abnormalities of *Allium cepa* for 24, 48, 72 & 96 hrs.

Table 50.1 : Physico-chemical and Biological Parameters of Raw Sewage at the Inlet Point (Average of Weekly Samplings for One Year)

Parameters		Raw sewage
pH		
Temperature		8.25
Total dis. solids	°C	29
Total sus. solids	mg/l	865
Dissolved oxygen	mg/l	162
Dissolved carbon-di-oxide	mg/l	1.02
BOD_5 (as O_2)	mg/l	62.75
Ammonia nitrogen	mg/l	285
Nitrite nitrogen	mg/l	1.89
Nitrate nitrogen	mg/l	0.0025
Total nitrogen	mg/l	0.57
Total phosphorus (as P)	mg/l	2.68
Diss. phosphorus (as P)	mg/l	4.35
Zn	mg/l	2.67
Pb	mg/l	3.52
Cu	mg/l	0.21
Cr	mg/l	0.64
Total bacteria	$\times 10^6$/ml	0.12
Coliforms	$\times 10^4$/ml	6.42
Salmonella	$\times 10^3$/ml	32.00
Faecal streptococci	$\times 10^2$/ml	5.72

least impact and 100% sewage caused maximum adverse effect. Both the concentration and duration of treatment showed negative correlation with mitotic index. The root tips treated for 24 hrs induced lowest percentage of abnormalities and maximum occurred following 96 hrs treatment. Srinath Rao *et al.* (1994) reported that higher concentration and longer duration of treatment of the Atrazine compound depressed the mitotic index greatly. The sewage concentration in the range of 75-100% induced higher percentage of abnormalities and lower mitotic index. The root tip treated with lower concentration of sewage, however, showed higher mitotic index and lower percentage of abnormalities. Similar suppressing effect of industrial effluent on mitotic activities has been reported by Somashekar *et al.* (1985).

The different types of chromosomal abnormalities observed are chromosomal disintegration, fragmentation, bi-nucleate condition and sticky anaphase. These abnormalities are probably the outcome of the impact of sewage on spindle apparatus. Kabarity *et al.* (1979) and Soheir *et al.* (1989) reported frequent C- metaphase anamolies in *A. cepa* treated with chemicals and insecticides. Dixit and Nerle (1985) and Joshi and Singh (1979) observed the induction of chromosomal abnormalities in the form of breakage, bridges and pycnotic nuclei in *A. cepa* exposed to industrial effluents.

Table 50.2 : Effect of Sewage of *Allium cepa* – 24 hrs Treatment Percentage of Abnormalities and Mitotic Index

Cont.	Total cells	No. of cells in divison	Pro Phase	Meta Phase	Ana Phase	Telo Phase	Abnormality	Miotic Index
10%	1000	450	298	64	89	09	NIL	45.0
25%	1000	412	318	43	39	12	NIL	41.2
50%	1000	403	302	29	54	18	9.32	40.2
75%	1000	386	292	54	27	13	11.57	38.6
100%	1000	321	298	11	12	NIL	14.02	32.1
CONT	1000	452	312	73	56	11	NIL	45.2

Table 50.3 : Effect of Sewage of *Allium cepa* – 48 hrs Treatment Percentage of Abnormalities and mitotic index

Cont.	Total cells	No. of cells in divison	Pro Phase	Meta Phase	Ana Phase	Telo Phase	Abnormality	Miotic Index
10%	1000	434	387	33	14	NIL	NIL	43.4
25%	1000	410	371	27	12	NIL	NIL	41.0
50%	1000	356	289	20	05	02	8.76	35.6
75%	1000	310	284	12	10	04	10.52	31.0
100%	1000	265	258	05	02	NIL	15.19	26.5
CONT	1000	443	374	45	24	NIL	NIL	44.3

Table 50.4 : Effect of Sewage of *Allium cepa* – 72 hrs Treatment Percentage of Abnormalities and Mitotic Index

Cont.	Total cells	No. of cells in divison	Pro Phase	Meta Phase	Ana Phase	Telo Phase	Abnormality	Miotic Index
10%	1000	342	307	23	12	NIL	NIL	34.2
25%	1000	354	319	14	21	NIL	NIL	35.4
50%	1000	316	284	23	06	03	11.07	31.6
75%	1000	278	261	12	04	01	13.72	27.8
100%	1000	214	198	10	04	02	17.14	21.4
CONT	1000	380	326	32	22	NIL	NIL	38.4

Table 50.5 : Effect of Sewage of *Allium cepa* – 96 hrs Treatment Percentage of Abnormalities and Mitotic Index

Cont.	Total cells	No. of cells in divison	Pro Phase	Meta Phase	Ana Phase	Telo Phase	Abnormality	Miotic Index
10%	1000	248	211	22	08	07	NIL	24.8
25%	1000	212	185	17	06	04	10.37	21.2
50%	1000	164	134	10	19	01	14.22	16.4
75%	1000	140	126	13	NIL	01	21.50	14.0
100%	1000	117	104	12	NIL	NIL	26.79	11.7
CONT	1000	260	234	11	12	03	NIL	26

Chauhan *et al.* (1986) and Thangapandian *et al.* (1995) with respect to industrial effluent also observed bi-nucleate condition as common anomaly in fishes exposed to insecticide and industrial effluent respectively. Onfelt and Klasterska (1983) considered mitotic abnormalities as insignificant from pollution point of view, unless polyploid cells are induced. Moretton *et al.* (1992) assessed the genotypic potential of tannery wastewater by induction of gene conversion and point mutation in *Sacharomyces cerevisae* (d7) strain. This study supports that the induction of C-mitosis and bi-nucleate cells can be considered as reliable parameters for assessing effluent toxicity in *A. cepa*. The different types of abnormalities observed can be attributed to the action of higher concentration of chlorides, sodium, copper, zinc and cadmium ions present in sewage (Gabriel Bitton, 1992). The effluent could be either synergistic or individualistic or both.

In general, the mitotic index of the sewage treated root tips showed gradual decreasing tendency in proportion to the increase in concentration of sewage. Further very high percentages of abnormalities were observed in raw sewage treated roots for longer duration.

References

APHA, AWWA and WPCE. 1985. Standard methods for estimation of water and wastewater 16h *ed.* NewYork,

Chauhan, L.K.S., Dikshith, T.S.S and Sundaraman, V. 1986. Effect of delta methrin on plant cells I cytological effect on the root meristems of *Allium cepa*. Mut. Res. 171, 25 - 30.

Dixit, G.B and Nerle, S.K. 1985. Cytotoxic effect of industrial effluents on *Allium cepa*. L., Geobios 12, 237 - 240,

Gabriel Bitton. 1992, Wastewater Microbiology. John. Wiley and Sons. New York, 371 - 373.

Joshi, B.D. and Singh, T.P. 1989. Cellular and chromosomal effect of tannery effluent in *A. cepa* meristem. XVI Annual Conference EMS I. Abstr. 29.

Kabarity, A., El-Bayoumi, A. and Habib, A.A. 1979. Mitodepressive effect and stanthmoknetic action of Pantopan Hydrochloride Mut. Res 66. 143-148.

Kihlman, B.A. 1975. Root tips of Vicia faba for the study of induction of Chromosomal aberrations Mut. Res. 31. 401-412.

Moretton, J. Barop and Aquino, M. 1992. Genotoxic potential of wastewater from a leather industry. Water, Air and Soil Pollution 63: 81-85.

Onfelt, A. and Klasterska, I. 1983. Spindle disturbance in mammalian cells and multiple chromatid exchanges after treatment of V. 79 Chinese Hamster cells with carbaryl. Mut. Res. 119: 319-330.

Rose, J.B. 1986. Microbial aspects of wastewater reuse for irrigation. Crit. Rev. Environ. Control. 16: 231-256.

Soheir E.L., Khodary, Antoinette Habib and Atif Haliem. 1989. Cytological effect of the Herbicide Gartlon on root mitosis of *A. cepa* cytologia. 54 : 465-472.

Srinath Rao *et al.* 1994. Effect of Atrazine herbicide on *Allium cepa*. L. Root tips. Proc. Conf. On. Cytol. Genet. 4: 63-64.

Somasekar, R.K., Gurudev, M.R and Siddaramaiah. 1985. Somatic cell abnormalities induced by dye manufacturing industry wastewater cytologia 50: 129-134.

Thangapandian, V., Sophia, M. and Swaminanthan, K. 1995. Cytological effect of tannery effluent on root meristem of *Allium cepa*. Linn. Test System. J.Environ Biol 16 : 67-70.

ENVIRONMENTAL BIOTECHNOLOGY

Edited By :   Professor (Dr.) Arvind Kumar

Published By :   DAYA PUBLISHING HOUSE

51

OVICIDAL AND LARVICIDAL ACTIVITY OF CERTAIN PLANT EXTRACTS AGAINST THE TOBACCO ARMYWORM, *SPODOPTERA LITURA* (FAB.) (LEPIDOPTERA : NOCTUIDAE)

• *K. Elumalai, A. Jeyasankar, N. Raja and S. Ignacimuthu*
Entomology Research Institute, Loyola College, Chennai, India.

Abstracts

Organic solvents and aqueous extracts of *Artemesia nilagrica, Acorus calamus, Lobellia leschenaultians, Cassia auriculata. Holoptelia integrifolia, Terrena asiatica Anisomelus malabaricus, Percularia daemia* and *Wedalia calandulacea* were tested against the polyphagous pest, *Spodoptera litura* for their ovicidal and larvicidal activity. Among them highest ovicidal activity was observed in hexane extract of *A. calamus* root (65.66%). Methanol and ethyl acetate-extracts of *A. nilagirica* showed the maximum larvicidal activity (41.91 and 40.24% respectively) against the fourth instar larvae of *S. litura*.

Key Words: Plant extracts, ovicidal, larvicidal, Spodoptera litura.

Introduction

As a result of indiscriminate use of pesticides over a long period, majority of the pests have developed resistance to almost all commercially available chemical pesticides. Worldwide attention now focuses towards alternative methods to control the pests by plant derived bioactive secondary metabolites (biopesticides), which are non-toxic, low cost and biodegradable and also safer to environment (Isman *et al.*, 1990). In total, nearly 2000 plant species are known to possess insecticidal activity, anti-feedant activity, insect growth regulators and repellent activity, (Murthy *et al.*, 1989, Simmonds *et al.*, 1989, McLeod *et al.*, 1990, Isman 1993, Datta and Saxena, 1997, Sajayaraj and Paulraj, 1998, Opender- Koul *et al.*, 2000 and Kannaiyan, 2002). But the reports on the plant extracts against Lepidoptera in general and Noctuidae in particular are scanty and hence the present study is aimed to check the ovicidal and insecticidal activity of various solvent extracts of ten plants against *Spodoptera litura*, which is a major polyphagous pest, infest more than 100 plant species, including economically important crops such as, cotton, groundnut, chilly, tobacco, castor and pulses etc.

Materials and Methods

Artemesia nilagrica (Compositae), *Acorus calamus* (Araceae), *Lobellia leschenaultians* (Lobeliaceae) were collected from Palani hills (Kodaikanal) and *Cassia auriculata*

(Caesalpiniaceae), *Holoptelia intefrifolia* (Ulmaceae), *Terrena asiatica* (Rubiaceae) were collected from the villages of Dindugaï district and *Anisomelus malabaricus* (Lamiaceae), *Percularia daemia* (Asclepiadaceae) and *Wedalia calandulacea* (Compositae) were collected in and around Chennai, They were brought to the laboratory and shade dried under room temperature (27±2°C), The dried plant materials were powdered using electric blender. The powdered plant tissues (100 gm) were loaded in a glass column and extracted with 500 ml of different solvents with increasing polarities i.e., hexane, dietylether, dichloromethane, ethylacetate, methanol and water. The solvent from the crude extracts was evaporated using rotary evaporator and the bioassays were carried out on S, lituta.

Ovicidal activity : Egg masses of *S. litura* were dipped in 5% concentration of various solvent extracts of the selected plants for a few minutes and the egg masses were allowed to dry the solvents. Egg masses were dipped in water was used as control. Treated and control egg masses were kept separately in a plastic container covered with muslin cloth. Number of eggs hatched in control and treated was recorded. Five replications were maintained for all the experiments and the percentage of ovicidal activity was calculated using the following formula

$$= \frac{\text{\% of eggs hatched in control} - \text{\% of eggs hatched in treated}}{\text{\% of eggs hatched in control}} \times 100$$

Insecticidal activity : Fresh castor leaves were treated with 5% concentration of crude extract and the petiole was tied with cotton and dipped in water to avoid drying. The control was treated as that of anti-feedant studies, Treated leaves were placed inside the plastic trough and 20 fourth instar larvae of *S. litura* were introduced in each trough and covered with muslin cloth. Five replicates were maintained and the number of dead larvae were recorded after 48 hrs. Percentage of larval and adult mortality was noted and corrected by Abbott's formula

$$= \frac{\text{\% mortality in treated} - \text{\% mortality in control}}{100 - \text{\% mortality in control}} \times 100$$

The data collected from the experiments were corrected with the respective formula as mentioned earlier. The corrected data were angular transformed before subjected to analysis of variance (two-way ANOVA) and Least Significance Difference (LSD) was calculated to separate mean significant difference within the solvent extracts.

Results and Discussion

Ovicidal activity of different solvent extracts of various plants at 5% concentration against 4th instar larvae of *S. litura* is presented in table 51.1. It was noted that the hexane crude extract of *A. calamus* root showed significant ovicidal activity. In diethyl ether extract, maximum ovicidal activity was observed in *A. malabaricus* followed by *A. calamus* root and *W. calandulacea*. Dichloromethane extract of *L. leschenaultians* showed maximum ovicidal activity followed by *T. assiatica* and *A. nilagrica* with marginal variation among them. The maximum of 47.11% of ovicidal activity was observed in ethyl acetate extract of *A. nilagrica*.

Methanol extract of *A. calamus* leaf and root were found to have maximum ovicidal activity among the other plant extracts. *A. malabaricus* aqueous extract showed significant result than the other plants. Among the six solvent extracts of ten plants, highest ovicidal activity was observed in hexane extract of *A. calamus* root (65.66%).

Table 51.1 : Ovicidal Activity of Different Plant Extract at 5% Concentration Against the Eggs of *Spodoptera litura*

Plants used	Hexane	Diethylether	Dichloro methane	Ethyl acetate	Methanol	Water
Artemesia nilagrica	51.62 ± 4.06 (45.92)	32.11 ± 4.21 (34.51)	37.15 ± 3.64 (37.52)	47.11 ± 3.14 (43.34)	45.68 ± 3.96 (42.48)	35.41 ± 2.54 (36.51)
Acorus calamus (leaf)	47.33 ± 3.99 (43.45)	38.98 ± 4.21 (28.59)	35.68 ± 3.76 (36.63)	41.98 ± 3.98 (40.34)	64.59 ± 3.99 (53.43)	42.06 ± 6.01 (40.4)
Acorus calamus (root)	65.66 ± 3.54 (54.04)	40.66 ± 5.22 (39.58)	29.92 ± 2.67 (33.15)	44.54 ± 4.64 (41.84)	55.85 ± 4.56 (48.33)	43.96 ± 4.13 (41.5)
Anisomelus malabaricus	57.19 ± 3.54 (49.08)	41.33 ± 5.93 (39.99)	35.06 ± 2.62 (36.27)	45.43 ± 6.01 (42.36)	29.32 ± 4.71 (32.77)	46.55 ± 4.89 (42.99)
Cassia auriculata	47.31 ± 3.98 (43.45)	40.55 ± 5.10 (39.52)	29.80 ± 2.87 (33.09)	44.06 ± 5.12 (41.55)	38.52 ± 3.99 (38.35)	41.84 ± 4.16 (40.28)
Holoptelia integrifolia	36.36 ± 3.79 (37.05)	35.87 ± 3.98 (36.75)	32.55 ± 3.41 (34.14)	41.88 ± 4.64 (40.28)	34.91 ± 5.01 (36.21)	36.30 ± 4.15 (37.05)
Lobellia leschenaultians	42.09 ± 3.46 (40.4)	34.07 ± 4.29 (35.67)	41.33 ± 3.46 (39.99)	42.01 ± 4.97 (40.4)	43.36 ± 2.87 (41.15)	27.84 ± 4.09 (32.46)
Terrena asiatica	39.20 ± 3.78 (38.76)	37.91 ± 3.99 (38)	37.88 ± 3.79 (37.94)	44.29 ± 4.09 (41.67)	37.86 ± 3.98 (37.94)	37.26 ± 4.79 (37.58)
Percularia daemia	44.11 ± 5.97 (41.61)	45.21 ± 4.55 (42.25)	29.07 ± 3.21 (32.58)	37.76 ± 3.98 (37.88)	40.61 ± 4.97 (39.58)	39.36 ± 3.88 (38.82)
Wedalia calandulacea	41.88 ± 6.01 (40.28)	40.66 ± 5.41 (39.58)	33.23 ± 2.98 (35.18)	45.76 ± 3.71 (42.53)	40.28 ± 4.44 (39.35)	37.60 ± 3.41 (37.82)
LSD at 5% level p<0.05	1.97	2.81	1.87	1.67	3.9	2.39

Values presented are mean of five replications ± S.D. Figures in parentheses are angular transformed. LSD for comparison of any two means for statistical significance.

The larvicidal activity of various solvent extracts of different plants is shown in Table 51.2. Though less than 50% larvicidal activity were obtained in all the solvent extracts of different plants, among them the methanol and ethyl acetate extracts of *A. nilagrica* showed the maximum larvicidal activity (41.91 and 40.24% respectively) against the 4 instar larvae of *S. litura*.

In the present study the various plant extracts tested, hexane extract of *A. calamus* root showed the maximum ovicidal activity. The hexane extract of this plant may have one or more chemical substances, which may block the microphyle region of the egg, thereby, prevent the exchange of gases that will ultimately kill the developing embryo in the egg itself, which was noticed as the eggs with black spot stage. Slama (1974) earlier reported, similar findings and the results were supported for the present work to study the ovicidal activity of the plants. Another possible reason for ovicidal activity is that the unequal penetration of extracts through the egg chorion to different parts of egg at different times of the sensitive period could also be associated with observations on variability of morphological effects and also resulted incomplete blastokinesis and abnormal breakage of extra embryonic membranes in the embryo (Sahayaraj and Paulraj, 2000).

Table 51.2 : Larvicidal Activity of Different Plant Extract at 5% Concentration Against Fourth linstar Larvae *Spodoptera litura*

Plants used	Hexane	Diethylether	Dichloro methane	Ethyl acetate	Methanol	Water
Artemesia nilagrica	20.12 ± 1.45 (26.64)	28.50 ± 1.00 (32.27)	35.21 ± 2.45 (36.39)	40.24 ± 2.83 (39.35)	41.91 ± 4.28 (40.34)	18.44 ± 2.74 (25.4)
Acorus calamus (leaf)	15.02 ± 1.36 (22.79)	18.44 ± 1.49 (25.4)	31.85 ± 2.54 (34.33)	31.85 ± 2.82 (34.33)	35.21 ± 2.35 (36.39)	11.73 ± 1.77 (20.00)
Acorus calamus (root)	28.05 ± 1.01 (32.27)	31.85 ± 0.98 (34.33)	33.53 ± 2.52 (35.37)	33.53 ± 2.45 (35.37)	33.53 ± 2.11 (35.37)	23.47 ± 1.89 (28.93)
Anisomelus malabaricus	13.41 ± 1.00 (21.47)	21.79 ± 1.09 (27.76)	28.50 ± 1.52 (32.27)	28.50 ± 1.75 (32.27)	30.18 ± 2.00 (33.27)	13.41 ± 1.64 (21.47)
Cassia auriculata	15.09 ± 2.45 (22.79)	16.76 ± 1.00 (24.12)	20.12 ± 1.27 (26.64)	20.12 ± 1.69 (26.64)	21.79 ± 2.56 (27.76)	13.41 ± 2.48 (21.47)
Holoptelia integrifolia	16.77 ± 2.01 (24.12)	18.44 ± 1.87 (25.4)	26.82 ± 3.00 (31.18)	26.82 ± 1.89 (31.18)	28.50 ± 2.46 (32.27)	15.09 ± 2.01 (22.79)
Lobellia leschenaultians	8.38 ± 1.93 (16.74)	11.73 ± 1.64 (20.0)	26.82 ± 2.89 (31.18)	26.82 ± 2.11 (31.18)	26.82 ± 3.66 (31.18)	11.73 ± 1.22 (20.00)
Terrena asiatica	6.71 ± 0.97 (15.0)	15.09 ± 1.48 (22.79)	23.47 ± 2.41 (28.93)	23.47 ± 1.99 (28.93)	25.15 ± 2.48 (30.07)	13.41 ± 1.00 (21.47)
Percularia daemia	15.09 ± 2.07 (22.79)	21.79 ± 1.89 (27.76)	26.82 ± 2.07 (31.18)	26.82 ± 2.11 (31.18)	26.82 ± 2.10 (31.18)	18.44 ± 1.64 (25.4)
Wedalia calandulacea	16.76 ± 1.87 (24.12)	15.09 ± 1.45 (22.79)	26.82 ± 2.00 (31.18)	26.82 ± 2.88 (31.18)	28.50 ± 1.99 (32.27)	18.44 ± 1.09 (25.4)
LSD at 5% level p<0.05	1.52	1.96	1.53	1.42	1.68	1.59

Values presented are mean of five replications ± S.D. Figures in parentheses are angular transformed. LSD for comparison of any two means for statistical significance.

Methanol and ethyl acetate extract of *A. nilagrica* were found to have significant larvicidal activity. It may be possible that the insecticidal compounds present in the selected plant extracts may arrest the various metabolic activities of the larva, ultimately the larvae were failed to feed further which led to larval mortality. In the present study the feeding rate of the larva was decreased even after feeding the larval development was arrested in various instar stages. Some of the final instar larvae were dead with the symptoms of larval-pupal intermediate stage and if the larva goes to pupal stage the newly emerged adult from the pupa showed various malformations in its morphology such as wings deformities. Earlier Sahayaraj and Paulraj (2000) also reported similar findings in *Tridax procumbens* leaf extract on *S. litura*. Further studies of Chen *et al.* (2000) strongly support the present findings. The various activities of the selected crude extracts may due to the presence of steroid, glycosides, phenol and flavonoids. The crude extracts showed five fractions in the TLC study (unpublished data). These groups of chemicals may have either individual or cumulative effect against *S. litura*.

References

Chen, L.I., Xu-Hanhon, G., Zhao-Shanhuan, Chen, L. and Zhao, S.H. 2000. Studies on anti-feeding activity of *Daphne tangutica* against tobacco cutworm. *Spodoptera litura*. Journal of South China Agricultural University. 211: 44-46.

Datta, S. and Saxena, D.B. 1997. Parthenin and azadirachtin-A as antifeedants against. *Spodoptera litura* Fab. Pesticide Research Journal. 92: 263-266.

Isman, M. B., Koul, O, Luczynski, A. and Kaminski, J. 1990. Insecticidal and anti-feedant, bioactives of Neem oils and their relationship to azadirachtin content. J. Agric. Food Cherm. 38: 1406 - 1411.

Isman, M. B. 1993. Growth inhibitory and anti-feedant effects of azadirachtin on six noctuids of regional, economic importance. Pesticide Science. 381: 57-63.

Kannaiyan, S. 2000. Environmental Impact of Pesticides in Agro-ecosystem In: Environmental Impact of Pesticides in Agro-ecosystem – Assessment and Abatement. Eds. Santharam, G., Jayakumar, R., Kuttalam, S., Chandrasekaran, S. and Manoharan, T. pp. 3 - 9.

McLeod, J.K., Moeller, P.D.R., Molinski, T.F and Koul, O. 1990. Anti-feedant activity against. *Spodoptera litura* larvae and [13C] NMR spectral assignments of the meliatoxins. Journal of Chemical Ecology. 168: 2511-2518.

Murthy, P.S.N., Ramaprasad, G., Sitaramaiah, S., Rao, B.V.K., Prabhu, S.R and Rao, S.N. 1989. Studies on the enrichment of neem fractions and their relative efficacy against tobacco caterpillar *Spodoptera litura* F. on Lanka and FCV tobacco. Tobacco Research. 151: 65-70.

Koul, Opender. Jain, M.P. and Sharma, V.K. 2000. Growth inhibitory and anti-feedant activity of extracts from *Melia dubia* to *Spodoptera litura* and *Helicoverpa armigera* larvae. Indian Journal of Experimental Biology. 381: 63-68.

Sahayaraj, K. and Paulraj, M.G. 1998. Screening the relative toxicity of some plant extracts to *Spodoptera litura* Fab. Insecta: Lepidoptera: Noctuidae of groundnut. Fresenius Environmental Bulletin. 7: 557-560.

Sahayaraj, K. and Paulraj, M.G. 2000. Impact of *Tridax procumbens* leaf extract on *Spodoptera litura* Fab. behaviour, development and juvenometry. Insect Environment. 54: 149-150.

Slama, K. 1974. Physiological and biochemical effects of juvenoids. In: Insect hormones and Bio-analogues. Springer - Verlag, New York, pp. 217 - 281.

ENVIRONMENTAL BIOTECHNOLOGY

Edited By :   **Professor (Dr.) Arvind Kumar**

Published By :   **DAYA PUBLISHING HOUSE**

52

REACTION OF TAMATO GERMPLASM TO TAMATO LEAF CURL VIRUS

• *Dharmesh Gupta, S.C. Chowfla, P.D. Thakur, Anil Handa and Bhupesh Gupta*

Department of Mycology and Plant Pathology,
Dr. Y.S. Parmar University of Horticulture and Forestry,
Nauni, Solan, (H.P.) India

ABSTRACT

Eighty-seven entries (varieties/collections/lines) were evaluated both under glasshouse and field conditions to see their reaction against tomato leaf curl virus (TCLV). Only one collection EC-104395 was found to be resistant to the virus both under field and glasshouse conditions. Seven varieties/lines viz. Ramaya, Meghana. Abhimaan, CLN-2001, MDT 9, Russian-II and FT-I were moderately susceptible both under field and glasshouse conditions. Remaining lines possessed varying degree of susceptibility to the virus.

Key Words: Lycopersicon esculentum, tomato leaf curl virus, resistance, whitefly.

Introduction

Tomato (*Lycopersicon esculentum* Mill.) is one of the most widely cultivated annual herbs of the family Solanaceae. The cultivation and production of tomatoes in India is seriously hampered by the tomato leaf curl virus (TCLV). This virus disease has assumed alarming propositions in the plains of India and is nearly becoming a limiting factor in tomato cultivation in many parts of the country. In the recent past tomato leaf curl disease has been recorded in different tomato growing regions of Himachal Pradesh. In these areas the infestation of whitefly (*Besmisia tabaci*) vector has also been recorded. Due to versatility of uses, commercial expediency and a lucrative business enterprise there is a great upsurge in its cultivation within the fanning community.

Studies on the control of this disease has been carried out by several workers using insecticides, oils, antibiotics and nylon net coverings with a view to reduce the vector population *Bemisia tabaci*. These strategies check the disease to varying degree, but clearly, none is beneficial as growing of TCLV resistant cultivars. No commercially grown cultivars are resistant to TCLV. Thus, there is a need to develop TCLV resistant cultivars for effective management of the disease. Chenuarayappa *et al.* (1992) screened more than 1200 tomato leaf curl virus in Mangalore, India and reported all accessions susceptible to the virus.

Materials and Methods

Eighty-seven tomato entries (varieties/collections/lines) were evaluated both under glasshouse and field conditions during summer season of 1997 and 1998 at the main campus of Dr. Y.S. Parmar University of Horticulture and Forestry, Nauni, Solan, under glasshouse conditions. Some 3-week old seedlings of each line were challenged with viruliferous whiteflies. For the whitefly (*Bemisia tabaci*) transmission tests, adult whiteflies were collected from healthy stock colonies maintained on tomato and okra plants and transferred to TLCV infected tomato plants for 24 hours acquisition period. The viruliferous whiteflies were then enclosed with healthy tomato seedlings covered with glass chimney. The end of the chimney was covered with muslin cloth to permit free exchange of air. The whiteflies were allowed to feed for 2 days after which these were sprayed with insecticide to kill the whiteflies. The plants were subsequently monitored for symptom development over a period of three months.

Table 52.1 : Reaction of Tamato Leaf Curl Virus Under Glasshouse/field Conditions

Reaction	Varieties/lines/collection	
	Glasshouse	Field
HR		
R	EC - 104395	EC - 104395
MS	Ramya, Meenakshi, Meghana, Abhimaan, CLN-2001, MDT-9, Russian - II, FT - 1	Ramya, Meghana, Abhimaan, CLN - 2001, MDT - 9, Russian - II, FT - 1
S	Vaishali, Rajni, Sioux, Rohini, Rupali, UHF-II, CLN-2026, Momotaro, BL-737, BL-444, Azad Kirti, Arka Vikas, Pearson, Srinagar, Trucker, B7-10, LA-1388, M-1, M-7, T-11564, Sioux X EC-9101652, NA-101, NA-501, NA-601, NTH-81	Vaishali, Meenakshi, Sioux, Rohini, Rupali, BL-737, BL-444, LA-1388, M-1, T-1564, NA-601
HS	Rashmi, Sultan, Rotilic, Naveen, Sultan Roma, Comato-101, Money Maker, CLN-1466-H, CLN-1466-J, Solan Vajer, V-16, Ratana, Lerica, MTH-1, BSS-40, ARTH-128, Punjab Chhuhara, Cal Ace, Homestead, Hisarlalima, Selection-2, Selection-16, TERI, LA-2912X EC 174057, LA-1401, LA-1467, LA-1468, Marglobe, HS-101, K-10, Punjab Kesari, Pusa Ruby, Rupali, S-28, S-29, Sheetal, Sonali, IAHS-883, EC-56351, EC-95152, EC-103595, EC-129156, Rutger, Red Cherry, Sel-30, SGP8M2, Marglobe supreme, NDT-96665, KT-4	Rajni, Rashmi, Sultan Rotilic, Naveen, Sultan Roma, Camato-101, Money Maker, CLN-2026, CLN-1466-H, CLN-1466-J, Solan Vajer, Momotaro, V-16, Ratana, Lerica, MTH-1, BSS-40, ARTH-128, Punjab Chhuhara, Azadkirti, Arka Vikas, Cal Ace, Homestead, Hisarlalima, Pearson, Srinagar, Selection-2, Selection-16, Trucker, BT-10, M-7, TERI, LA-2912 X EC 174057, LA-1401, LA-1467, LA-1468, Marglobe, HS-101, K-10, Punjab Kesari, Pusa Kranti, Pusa Ruby, Rupali, S-28, S-29, Sheetal, Sonali, IAHS-883, EC-56351, EC-95152, EC-103595, EC-129155, EC-129156, EC-251665, EC-910652, Rutger, Red Cherry, Sel-30, NA-101, NA-501, NTH-81, SGP8M2, Marglobe supreme, NDT-96665, KT-4.

Under field conditions

Twenty plants of each variety were followed for at least three months period. All screening was done between May and August when epiphytotic conditions for the occurrence of leaf curl virus in field was best. For screening disease reaction, the following scale was adopted.

Results and Discussion

It is apparent from the data presented in Table 52.1 that out of 87 entries, only one collection EC-104395 was found to be resistant to the tomato leaf curl virus both under glasshouse and field conditions. However, eight entries viz. Ramaya, Meenakshi, Meghana, Abhimaan, CLN-2001, MDT 9, Russian -II and FT - I were moderately susceptible and 29 lines/ varieties were found to be susceptible under glasshouse conditions and rest were found to be highly susceptible. Under field conditions, seven entries viz. Ramaya, Meghana, Abhimaan, CLN-2001, MDT -9, Russian -II and FT - I were moderately susceptible and eleven entries were found highly susceptible. Varma *et al.* (1980) also reported the line EC-104395 resistant to tomato leaf curl virus. Banerjee and Kallo (1991) recorded 83.3 per cent leaf curl disease incidence in the cultivar HS-101 and 100 per cent in Punjab Chhuhara and Red Cherry under field conditions. In the present studies also a high incidence of the disease was recorded in the cultivars HS-101 and Red Cherry. However, Punjab Chhuhara recorded a low incidence of the disease which might be due to different environmental conditions. The variation in disease incidence in different lines/selections may be attributed to their different genetic makeup.

Percent infection	Rating
0 – 5	Highly resistant (HR)
6 – 10	Resistant (R)
11 – 20	Moderately Susceptible (MS)
21 – 30	Susceptible (S)
Above 30	Highly Susceptible (HS)

References

Banerjee, M.K. and Kallo, G. 1991. Evaluation of tomato breeding material against tomato leaf curl virus: technique of inoculation. Veg. Sci.18:109-112.

Chenuarayappa, Shivshankar, G., Muniayappa, V. and Frist, R.H. 1992. Resistance of *Lycopersicon esculentum* species to *Bemisia tabaci*, a tomato leaf curl virus vector. *Can. J Bot.* 70:2184-2192.

Varma, J.P., Hayati, J. and Poonam. 1980. Resistance in *Lycopersicon* species to tomato leaf curl disease in India. Zeitschrift fur Pflanzenkrankheiten and Pflanzenschutz. 87: 137-144.

ENVIRONMENTAL BIOTECHNOLOGY

Edited By :   **Professor (Dr.) Arvind Kumar**

Published By : **DAYA PUBLISHING HOUSE**

53

USE OF BIOPESTICIDES AND BIOCONTROL AGENTS FOR THE MANAGEMENT OF COLLAR ROT PATHOGEN *PHYTOPHTHORA CACTORUM*

• *Bhupesh Gupta, L.N. Bhardwaj, Anil Handa and Usha Sharma*

Department of Mycology and Plant Pathology,
Dr. Y.S. Parmar University of Horticulture and Forestry, Nauni, Solan (H.P.)

Abstract

Collar rot of apple caused by *Phytophthora cactorum*, a major soil borne disease in Himachal Pradesh is presently managed by application of copper oxychloride or metalaxyl or mancozeb. Non-judicious use of fungicides has many negative effects. Therefore, biopesticides and biocontrol agents were evaluated in vitro for their effectiveness against the pathogen. Among different biopesticides clove oil resulted in maximum inhibition (97.59%) followed by Ovis (93.93%) though statistically at par with each other. Among biocontrol agents tested *Trichoderma longibrachiatum* resulted in maximum inhibition (70.00%) of mycelial growth followed by *T. viride* (68.33%) and *T. harzianum* (68.17 %). All these treatments were statistically at par with each other

Key Words: Apple, collar rot, biopesticides.

Introduction

Apple (*Malus domestica Borkh*) is one of the important temperate fruit crops grown throughout the world wherever agro-climatic conditions are met with for its cultivation. In India, it is commercially grown in the States of Himachal Pradesh, Jammu and Kashmir, Kumaon hills of Uttaranchal and parts of Arunachal Pradesh and Sikkim on an area of 2,31,000 ha with an annual production of 13,80,000 MT (FAO, 2001), however, area under apple cultivation in Himachal Pradesh is about 88,673 ha with an annual production of 49,129 MT (FAO, 2001). Apple suffers heavily due to variety of diseases which cause serious damage to crop under favourable conditions, thus affecting fruit quality. Of various diseases attacking apples, collar rot is the most important soil-borne disease following white root rot, prevalent throughout the apple growing areas. Presently, collar rot caused by *Phytophthora cactorum* is managed by application of copper oxychloride or metalaxyl or mancozeb and use of resistant rootstocks like M2, M4, M9 etc., however, non-judicious use of chemical fungicides has many negative effects. Therefore, search for an alternative means of disease control is of utmost importance. Biopesticides and biocontrol agents offer best environment friendly answer to this problem.

Materials and Methods

Evaluation of some biopesticides viz. Ovis, Nimbicidine, clove oil, ginger oil, Neem Jeevan Spray oil, mentha oil and eucalyptus oil were evaluated for their inhibitory effect on mycelial growth of *P. cactorum* by poisoned food technique (Falck, 1907) under *in vitro* conditions. Double strength oat meal agar medium prepared in distilled water was sterilized at 15 p.s.i. for 20 minutes. Simultaneously, double concentrations of biopesticides were also prepared in sterilized water. The suspensions were added separately into equal quantities of double strength agar medium aseptically before pouring into petriplates. These were then inoculated with mycelial discs of 7 mm diameter of vigorously growing fungal culture. Each treatment was replicated thrice. Data on diametric growth of fungus was recorded ten days after incubation at 25°±1°C when control plates were fully covered with fungal mycelium. Per cent inhibition is calculated as Vincent (1947)

$$I = \frac{C - T}{C} \times 100$$

Where,

I = Per cent growth inhibition

C = Growth of fungus in control

T = Growth of fungus in treatment

Five fungal antagonists namely, *Trichoderma viride, T. harzianum, T. hamatum, T. longibrachiatum* and *T. virens* were evaluated for their inhibitory effect against *P. cactorum* under *in vitro* conditions by dual culture method. Mycelial discs (7 mm diameter) of all the antagonists and test pathogen were taken from the margins of actively growing cultures (8 days old) and transferred to petriplates containing oat meal agar. A check having only mycelial discs of test pathogen was also kept for comparison. All the treatments were replicated thrice. Observations on diametric growth were taken after ten days of incubation at 25°±1°C. Per cent inhibition was also calculated as per Vincent (1947). Four bacterial antagonists namely, *Pseudomonas fluorescens* IISR-8, *P. fluorescens* T$_1$R$_2$K$_4$, *P. fluorescens*-81 and *Bacillus subtilis* were screened for their inhibitory effect against *P. cactorum* causing collar rot of apple. A loopful of each bacterial culture (48 hours old) was streaked in the centre of the petriplates containing oat meal agar medium. Mycelial discs (7mm) of test fungus cut from 8 days old culture was placed simultaneously on either sides of the Petriplates where streaking was done with antagonists at the centre. Each treatment was replicated thrice and kept at 25°±1°C for ten days. Per cent inhibition was calculated as described earlier in the text.

Result and Discussion

Biopesticides are the safest means of plant disease control and are supposed to have inhibitory effect on vegetative growth of plant pathogens. Thus the inhibitory effect of some biopesticides and biocontrol agents were studied separately under present investigations. Data presented in Table 53.1 revealed that all the biopesticides tested at different concentrations inhibited the mycelial growth of *P. cactorum*, however, clove oil resulted in maximum inhibition of the test fungus. This was followed by Ovis (a formulated product from *Lantana camera*) which gave 9393 per cent inhibition, though statistically at par with clove oil. Mentha oil followed by Neem Jeevan Spray oil were the next best giving 58.10

and 52.59 per cent inhibition, respective y. However, minimum inhibition (7.08%) was recorded in eucalyptus oil. The concentrations of different fungicides tested show that higher concentrations of fungicides were more effective than lower concentrations, however, maximum mycelial inhibition was achieved with 1000 ppm (63.49%) followed by 750 ppm (55.08%) and 500 ppm (50.08%), while the least inhibition was recorded with 250 ppm (37.94%) concentration. The interaction studies showed that the increase in concentration of fungicides from 250 to 1000 ppm, the percentage of mycelial inhibition also increased. Ovis and clove oil gave cent per cent mycelial inhibition at 500 ppm, however, the same amount of inhibition was achieved with mentha oil at 1000 ppm. Clove oil at 250 ppm was found significantly superior to all other biopesticides tested at this concentration, this was followed by Ovis which gave 75.74 per cent inhibition, however, least inhibition was recorded with eucalyptus oil. These findings corroborate the observations of Sharma (1998) and Gupta (2001) who also reported Ovis, clove oil and mentha oil against *P. cactorum* causing leather rot of strawberry.

Table 53.1 : *In vitro* Evaluation of Bio-pesticides and Plant Oils Against *Phytophthora cactorum*

Treatment	Mycelial inhibition (%)				Overall Mean
	250 ppm	500 ppm	750 ppm	1000 ppm	
Clove Oil	90.37 (72.00)	100.00 (90.00)	100.00 (90.00)	100.00 (90.00)	97.59 (85.50)
Neem Jeevan Spray Oil	18.52 (25.39)	20.93 (27.05)	21.30 (27.26)	31.67 (34.22)	23.10 (28.48)
Ginger Oil	44.63 (41.92)	52.96 (46.71)	53.33 (46.93)	59.44 (50.46)	52.59 (46.50)
Nimbicidine	19.07 (25.74)	25.37 (30.08)	34.45 (35.93)	43.70 (41.38)	30.65 (33.28)
Ovis	75.74 (60.67)	100.00 (90.00)	100.00 (90.00)	100.00 (90.00)	93.93 (82.67)
Mentha Oil	13.33 (19.22)	44.26 (40.14)	74.81 (60.08)	100.00 (90.00)	58.10 (52.53)
Eucalyptus Oil	3.89 (9.25)	7.04 (15.38)	7.78 (16.11)	9.66 (18.06)	7.08 (14.70)
Overall Mean	37.94 (36.41)	50.08 (49.47)	55.96 (52.33)	63.49 (59.15)	

Figures in parentheses are arcsin transformed values.

$CD_{0.05}$

Treatment	=	4.50
Concentration	=	3.40
Treatment × concentration	=	6.37

Five fungal and four bacterial antagonists were evaluated against collar rot pathogen *Phytophthora cactorum* under *in vitro* conditions by dual culture method. The per cent mycelial inhibition is presented in Table 53.2. A perusal of the data in Table 53.2 depicts that all the

Table 53.2 : *In vitro* Evaluation of Antagonists Against *Phytophthora cactorum*

Antagonist	Growth inhibition (%)
Trichoderma virens	63.33
Trichoderma viride	68.33
Trichoderma harzianum	68.17
Trichoderma hamatum	65.83
Trichoderma longibrachiatum	70.00
Bacillus subtilis	59.83
Pseudomonas fluorescens 81	55.00
Pseudomonas fluorescens IISR-8	64.67
Pseudomonas fluorescens $T_1R_2T_4$	58.50
$CD_{0.05}$	3.05

fungal as well as bacterial antagonists, in general, inhibited the vegetative growth of the test fungus, *P. cactorum*, however, the percentage of inhibition varied with the antagonists. *Trichoderma longibrachiatum* resulted in maximum growth inhibition (70.00%) followed by *T. viride* (68.33%) and *T. harzianum* (68.17%), though satistically at par with each other. Amongst the bacterial antagonists *Pseudomonas fluorescens* IISR-8 was significantly superior in inhibiting the vegetative growth of *P. cactorum* followed by *Bacillus subtils* and *P. fluorescnes* $T_1 R_2K_4$ though statistically at par with each other. *P. fluorescens*-81 was the least effective antagonist in inhibiting the mycelial growth of the test fungus. Inhibitory effect of *Trichoderma sp.* was also reported by *Chambers* and *Scott* (1995) against *Phytophthora cinnamomi*. Recently, Rajan *et al.* (2002) reported the management of foot rot disease (*P. cacpcisi*) of black pepper with *Trichoderma spp.* Several workers have also reported *Psuedomonas* species to be inhibitory to vegetative growth of different species of *Phytophthora* includig *P cactorum* (Utkhede and Smith, 1991, Sauer and Zeller, 1992, Grosch and Grote, 1998, Gulati *et al.*, 1999). *Bacillus subtilis* and other species of *Bacillus* have been found effective biocontrol agents against *P cactorum*, collar rot pathogen of apple (Gupta and Utkhede, 1986, Utkhede and Smith, 2000).

References

Chambers, S.M. and Scott, E.S. 1995. In vitro antagonism of *Phytophthora cinnamomi* and *P. citricola* by isolates of *Trichoderma spp.* and *Gliocladium virens*. J. Phytopath. 143: 471-477

Falck, R. 1907 Wachstomgesetze, Wachstm-taktoren und temperature Wertder holzesterenden. Mycelien 1: 43- 154.

F.A.O. 2001. FAO Web site, www.fao.org.

Grosch. R. and Grote, D. 1998. Suppression of *Phytophthora nicotianae* by application of *Bacillus subtilis* in closed soilless culture of tomato plants. Gartenbauwissenschaft. 63: 103-109.

Gulati, M.K., Koch, E., Zeller, W. and Sisler, H.D. 1999. Isolation and identification of anti-fungal metabolites produced by fluorescent Pseudomonas, antagonist of red core disease of strawberry. In: 12th International Reinhardsbrunn Symposium, Germany. 24th-29th May 1998. pp 437-444.

Gupta, M. 2001. Studies on *Phytophthora* diseases of strawberry. Ph. D. Thesis, Dr. Y.S. Parmar UHF, Nauni, Solan (H.P.). p. 83.

Gupta, V.K. and Utkhede, R.S. 1986. Factors affecting the production of anti-fungal compounds by *Enterobacter aerogenes* and *Bacillus subtilis*, antagonists of *Phytophthora cactorum*. J. Phytopathol. 117: 9-16.

Rajan, P. P. and Sarma, Y. R. 1997. Compatibility of potassium phosphonate (Akomin-40) with different species of *Trichoderma* and *Gliocladium virens*. In: Proceedings of the National Seminar on Biotechnology of Spices and Aromatic Plants, Calicut, India. 24-25 April, 1996. pp. 150-155.

Sauer, H. and Zeller, W. 1992. Biological control of collar rot on apple (*Phytophthora cactorum* (Leb. et Cohr.) Schroet.) by bacteria of soil and bark. Acta Phytopathologica er - Entomologica Hungarica. 27: 1-14

Sharma, A. 1998. Studies on *Phytophthora* rot of strawberry. M.Sc. Thesis. Dr Y.S. Parmar University of Horticulture and Forestry, Nauni, Solan, H.P 53p.

Utkhede, R.S. and Smith, E.M. 1991. Biological and chemical treatments for control of *Phytophthora* crown and root rot caused by *Phytophthora cactorum* in a high-density apple orchard. Can. J. Plant Path. 13: 267-270.

Utkhede, R.S and Smith, E.M. 2000. Impact of chemical, biological and cultural treatments on growth and yield of apple in replant-disease soil. Aus Plant Path. 29: 129-136.

Vincent, J.M. 1947. Distortion of fungal hyphae in the presence of certain inhibitors. Nature 150:850.

ENVIRONMENTAL BIOTECHNOLOGY

Edited By :   Professor (Dr.) Arvind Kumar

Published By :   DAYA PUBLISHING HOUSE

54

IN VITRO SCREENING FOR POTENT VIRUS INHIBITORS OF PLANT ORIGIN

• *Bhupesh Gupta, Anil Handa and Dharmesh Gupta*

Plant Virus Laboratory, Department of Mycology and Plant Pathology,
Dr. Y.S. Parmar University of Horticulture and Forestry, Nauni, Solan (H.P.)

Abstract

Four wild growing plants namely *Mirabilis jalapa, Vitex negundo, Asparagus adscendens* and *Symphytum perigrinum* were screened for possessing anti-viral substances using acetone, alcohol and cold water as organic solvents. Cold water fraction of *M. jalapa* roots and *S. perigrinum* roots, acetone fraction of *A. adscendens* roots and alcohol fraction of *V. negundo* leaves were found to contain potent antiviral substances against tomato mosaic tobamovirus. In an *in vitro* experiment the alcohol fraction of *Vitex negundo* leaves resulted in maximum inhibition (91.63%) of tomato mosaic tobamovirus at 7 ppm.

Key Words: *Tomato mosaic tobamovirus, virus inhibitors, tomato*

Introduction

Tomato (*Lycopersicon esculentum* Mill.) an important off-season vegetable crop is grown in mid hills of Himachal Pradesh and is sold in the States of Punjab, Haryana and Delhi where it fetches remunerative prices. Unfortunately, such an important cash crop falls prey to viruses, which otherwise take a heavy toll of the crop in terms of economic losses and in many cases render its cultivation an unproductive venture. Tomato mosaic tobamovirus is one such virus that causes a severe mosaic disease of tomato, besides many strains of cucumuber mosaic *cucumovirus* and potato *Y potyvirus* which cause tremendous losses to this highly remunerative crop. Keeping in view the economic importance of the crop and seriousness of the disease, effects of partially purified fractions of *Mirabilis jalapa* roots (cold water fraction), *Asparagus adscendens* roots (acetone fraction), *Vitex negundo* leaves (alcohol fraction) and *Symphytum perigrinum* roots (cold water fraction) were observed on the inhibition of tomato mosaic *tobamovirus*.

Materials and Methods

Cold water fractions of *Mirabilis jalapa* and *Symphytum perigrinum* roots, acetone fraction of *Asparagus adscendens* roots and alcohol fraction of *Vitex negundo* leaves, found to contain virus inhibitor(s) in a primary experiment, were applied at five concentrations viz., 1, 3, 5,

7 and 10 ppm. The desired concentration of anti-viral fractions was mixed with virus inoculum in equal quantity (V/V) and applied to the bioassay plant after about two hours of mixing. The virus concentration was assayed on *Chenopodium amaranticolor*, the bioassay plant. Six leaves of uniform size were inoculated with the treated inoculum on the right hand side by leaf rub method. The other half of the leaves were inoculated with untreated inoculum. Rest of the leaves were clipped off. The inoculum was diluted to 1:100 with distilled water, prior to inoculation on bioassay plant, because it produced countable number of local lesions. Concentration of the virus was calculated by local lesion method (Holmes, 1929). The inhibition of virus was calculated with the help of half leaf assay method (Samuel and Bald, 1933) described as under:

$$\text{Inhibition percentage} = \frac{a - b}{a} \times 100$$

Where,

a = number of local lesions in control

b = number of local lesions in treatment

Results and Discussion

The results presented in Table 54.1 revealed that cold water fraction of *Mirabilis jalapa* roots inhibited virus significantly at 3 ppm concentration as compared to inhibition at 1 ppm. Later the inhibition started declining with the increase in concentration. These findings are in line with those observed by Kubo *et al.* (1990). Acetone fraction of *Asparagus adscendens* roots was found to be effective in inhibiting the virus significantly at 1 ppm, closely followed by concentration of 7 ppm. The same extract used by Garg (1995) was found to be effective in eliminating the carnation latent *carlavirus* at 5 and 7 ppm concentrations when combined with meristem tip culture. The differences in result could be attributed to the difference in stage of extraction. Inhibition of tomato mosaic *tobamovirus* by alcohol fraction of *Vitex negundo* leaves increased with increase in concentration upto 7 ppm, however, it reduced later but very close to the maximum i.e. inhibition achieved at

Table 54.1 : Effect of Concentration of Effective Fractions for Antiviral Activity Against Tomato Mosaic Tobamovirus

Fraction	Percent inhibition					Overall Mean
	1 ppm	3 ppm	5 ppm	7 ppm	10 ppm	
M. jalapa roots cold water extraction	63.38 (52.76)	87.57 (69.37)	83.38 (65.95)	62.59 (52.29)	58.56 (49.93)	71.10 (58.06)
A. adscendens roots acetone fraction	89.30 (70.93)	60.81 (51.24)	57.47 (49.30)	84.97 (67.20)	73.59 (59.08)	73.23 (59.55)
V. negundo leaves alcohol fraction	59.50 (50.48)	72.55 (58.41)	81.49 (64.52)	91.63 (73.22)	89.95 (71.53)	79.02 (63.63)
S. perigrinum roots cold water fraction	42.22 (40.52)	57.44 (49.28)	65.09 (53.79)	77.41 (61.63)	80.49 (63.79)	64.53 (53.80)
Overall Mean	63.60 (57.67)	69.59 (57.08)	71.86 (58.39)	79.15 (61.08)	76.65 (63.59)	

7 ppm. The cold water fraction of *Symphytum perigrinum* roots was quite effective at 10 ppm concentration. The percentage inhibition of tomato mosaic *tobamovirus* increased with an increase in concentration of the extract.

References

Garg, M. 1995. *In vitro* multiplication of virus tested carnations. M.Sc. Thesis, Dr. Y.S. Parmar University of Horticulture and Forestry, Nauni, Solan, India.

Holmes, F.O. 1929. Local lesions in tobacco mosaic. Bot. Gaz. 87:39.

Kubo, S., Ikeda, T., Imaizumi, S. Takayami, Y. and Mikami, Y. 1990. A potent virus inhibitor found in *Mirabilis jalapa* L. Annals of Phytopathological Society of Japan 56: 481-487.

Samuel, G. and Bald, J.G. 1933. On the use of primary lesions in quantitative work with two plant viruses. Annals of Applied Biology. 20: 70-99.

ENVIRONMENTAL BIOTECHNOLOGY

Edited By :   **Professor (Dr.) Arvind Kumar**

Published By :   **DAYA PUBLISHING HOUSE**

55

EXTRACTION OF ANTI-VIRAL SUBSTANCES AS INFLUENCED BY DIFFERENT ORGANIC SOLVENTS

• *Bhupesh Gupta, Anil Handa and R.C. Rana**

Plant Virus Laboratory, Department of Mycology and Plant Pathology,
* Department of Forest Products and Utilization,
Dr. Y.S. Parmar University of Horticulture and Forestry, Nauni, Solan (H.P.)

Abstract

Four wild growing plants namely *Mirabilis jalapa, Vitex negundo, Asparagus adscendens* and *Symphytum perigrinum* were screened for possessing anti-viral substances using acetone, alcohol and cold water as organic solvents. Cold water fraction of *M. jalapa* roots and *S. perigrinum* roots, acetone fraction of *A. adscendens* roots and alohol fraction of *V. negundo* leaves were found to contain potent anti-viral substances against tomato mosaic *tobamovirus*.

Key Words: antiviral substances, tomato mosaic tobamovirus, organic solvents

Introduction

The search for efficient anti-viral chemotherapeutic substances are still being continued for the control of plant viruses. There is a great demand for the general purpose anti-viral chemicals. The occurrence of anti-viral agents often referred to as inhibitory substances in plant extracts was first noticed when sap transmissible viruses could not be transmitted from certain host plant to other hosts. Moreover, the use of plant extracts as a means of anti-viral chemotherapy has long been recognized as an area of investigation. Only a few inhibitory substances from some plants like *Abutilon striatum, Boerhavia diffusa, Clerodendron* sp. *Chenopodium* sp. and *Mirabilis jalapa* have been satisfactorily purified and characterized (Verma, 1986). Inhibitors in plants largely occur in leaves but other plant parts like seeds, flowers, fruits, roots, rhizomes, bark also contain strong inhibitors.

Materials and Methods

Four plants namely, *Mirabilis jalapa, Vitex negundo, Asparagus adscendens* and *Symphytum perigrinum* were selected for the study. The plant material was washed thoroughly with tap water to remove dust/debris and soil from the surface. The leaves were air dried, whereas, roots after removal of water from surface was cut into small pieces and fruits were oven dried at 50°C temperature till constant weight was obtained. The dried plant materials were grounded to form fine powder. Cold extraction of the inhibitors was carried out by taking 50 g of dried powder of each and sufficient amount of acetone was added to cover the material. This was kept as such for 6-8 hours and then filtered through filter paper. To the residue obtained so, a sufficient amount of alcohol was added then filtered after keeping it undisturbed for 6-8 hours. Residue obtained after second filtration was covered with cold water for 6-8 hours and then filtered. All the filtrates were collected in separate flasks. The solvents were evaporated completely by keeping the flasks containing filtrates in oven at 50°C. These fractions were completely dried in vacuum and utilized as starting material for further experimentation. For convenience sake, stock solutions of 100 ppm of each fraction were prepared. All the fractions were evaluated for the presence of inhibitor(s) by mixing virus inoculum and fraction in equal quantity (V/V) at 5 ppm concentration. About 4-5 hours of mixing, these were inoculated on bioassay plant (*Chenopodium amaranticolor*). For assaying virus concentration six leaves of uniform size were inoculated with the treated inoculum on the right hand side by leaf rub method. The other half of the leaves were inoculated with untreated inoculum. Rest of the leaves were clipped off. The inoculum was diluted to 1 :100 with distilled water, prior to inoculation on bioassay plant, because it produced countable number of local lesions. Concentration of the virus was calculated by local lesion method (Holmes, 1929). The inhibition of virus was calculated with the help of half leaf assay method (Samuel and Bald, 1933) described as under:

$$\text{Inhibition percentage} = \frac{a-b}{a} \times 100$$

Where,

    a = number of local lesions in control

    b = number of local lesions in treatment

Results and Discussion

Data presented in Table 55.1 revealed that cold water fraction of *Mirabilis jalapa* leaves and roots contained potent inhibitor(s) as compared to acetone and alcohol fractions. However, the extract of *M. jalapa* roots was more effective as compared to the extract obtained from leaves. Cold water fraction of *M. jalapa* roots, caused maximum inhibition of tomato mosaic *tobamovirus* (84.45%) followed by cold water fraction of *M. jalapa* leaves. The results are in conformity with Kubo *et al.* (1990),

Table 55.1. Effect of Different Fractions and Plant Parts of *Mirabilis jalapa* on Infectivity of Tomato Mosaic *tobamovirus*

Plant/part	Percent inhibition			Overall Mean
	Acetone	Alcohol	Cold water	
Leaves	15.67 (22.32)	16.28 (23.79)	72.62 (57.45)	34.86 (35.19)
Roots	18.24 (25.28)	18.02 (25.18)	84.45 (66.57)	40.15 (38.99)
Overall Mean	16.95 (24.30)	17.15 (24.46)	78.40 (38.99)	

Figures in parentheses are arcsin transformed values.

$CO_{0.05}$

Plant parts	= 0.58
Fractions	= 0.71
Plant parts x Fractions	= 1.00

who reported roots of *Mirabilis jalapa* to be most effective in yielding inhibitory substance(s) than any other part of the plant. In the present studies cold water fraction has been found to be more effective inhibitory substance(s), thereby suggesting that inhibitory substance(s) to be mostly polar in nature. A number of workers have suggested that *Mirabilis jalapa* contains potent inhibitor(s) of the viruses (Verma and Kumar, 1980, Kubo *et al.*, 1990, Takayami *et al.*, 1990).

Table 55.2. Effect of Different Fractions and Plant Parts of *Asparagus adscendens* on Infectivity of Tomato Mosaic *tobamovirus*

Plant/part	Percent inhibition			Overall Mean
	Acetone	Alcohol	Cold water	
Fruits	63.03 (52.56)	19.05 (26.24)	18.76 (25.67)	33.78 (34.82)
Roots	65.47 (54.01)	21.20 (27.42)	19.94 (26.52)	35.54 (35.98)
Overall Mean	64.25 (53.28)	20.38 (26.83)	19.35 (26.09)	

Figures in parentheses are arcsin transformed values.

$CD_{0.05}$

Plant parts	= 0.48
Fractions	= 0.59
Plant parts x Fractions	= 0.83

Table 55.3 : Effect of Different Fractions and Plant Parts of *Vitex negundo* and *Symphytum perigrinum* on Infectivity of Tomato Mosaic *tobamovirus*

Plant/part	Percent inhibition			$CD_{0.05}$
	Acetone	Alcohol	Cold water	
Vitex negundo leaves	14.26 (22.18)	80.19 (63.58)	12.74 (20.91)	1.15
Symphytum perignum Roots	15.91 (23.51)	14.81 (22.63)	65.72 (54.16)	0.64

Figures in parentheses are arcsin transformed values.

Similarly the results presented in Table 55.2 indicate that the acetone fraction of *A. adscendens* contained the potent inhibitor. Root extract was found to be more effective over fruit extract in yielding the inhibitor(s). Acetone fraction of roots resulted in maximum inhibition (65.47%) followed by the acetone fraction of *A. adscendens* fruits thereby suggesting that inhibitory substance(s) present in *A. adscendens* is moderately polar in nature. Similar results were obtained by Garg (1995) who compared acetone extract with a combination of 'acetone and methanol extract' against carnation latent *carlavirus*.

The results outlined in Table 55.3 indicated that alcohol fraction of *Vitex negundo* leaves and cold water fraction of *Symphytum perigrinum* contained potent inhibitor(s) indicated by 80.19 per cent and 65.72 per cent inhibition of tomato mosaic *tobamovirus*, respectively. In case of *V. negundo* cold water fraction (12.74%) was least effective, while in case of *S. perigrinum* alcohol fraction (14.81%) was least effective in inhibiting tomato mosaic *tobamovirus*. This clearly indicates that inhibitor(s) present in *V. negundo* is sufficiently polar whereas that present in *S. perigrinum* is mostly polar in nature.

References

Garg, M. 1995. *In vitro* multiplication of virus tested carnations. M.Sc. Thesis, Dr. Y.S. Parmar University of Horticulture and Forestry, Nauni, Solan, India.

Holmes, F.G. 1929. Local lesions in tobacco mosaic. Bot. Gaz. 87:39.

Kubo, S., Ikeda, T., Imaizumi, S., Takayami, Y. and Mikami, Y. 1990. A potent virus inhibitor found in *Mirabilis jalapa* L. Annals of Phytopathological Society of Japan. 56: 481-487.

Samuel, G. and Bald, J.G. 1933. On the use of primary lesions in quantitative work with two plant viruses. Annals of Applied Biology. 20: 70-99.

Takayami, Y., Kuwata, S., Ikeda, T. and Kubo, S. 1990. Purification and characterization of anti-plant viral protein from *Mirabilis jalapa* L. Annals of Phytopathological Society of Japan. 56: 488-494.

Verma, H.N. 1986 Interferon like anti-viral agents from plants. In: Vistas in Plant Pathology, pp. 481 Eds Verma, A. and Verma, J.P. Malhotra Publishing House, New Delhi - 64. India.

Verma, H.N. and Kumar, V. 1980. Presentation of plant virus diseases by *Mirabilis jalapa* leaf extract. New Botanist 7: 87-91.

ENVIRONMENTAL BIOTECHNOLOGY

Edited By :   **Professor (Dr.) Arvind Kumar**

Published By :   **DAYA PUBLISHING HOUSE**

56

HAEMATO-BIOCHEMICAL VARIATIONS INDUCED BY MONOCROTOPHOS IN *CYPRINUS CARPIO* DURING THE EXPOSURE AND RECOVERY PERIOD

• *C. Maruthanayagam. and G. Sharmila*

A.V.C. College (Autonomous), Department of Zoology Mannampandal,
Myiladuthurai

Abstract

In the haemato-biochemical parameters, the blood glucose was found to be increased during exposure period. This may be due to increased glucogenesis as well as due to the inhibition of glucogenolysis and glyconeogenolysis. The increased level of glucose may also enable the animal to withstand the toxicant imposed stress conditions. However, the serum protein and serum cholesterol level decreased in fishes during exposure period. The decrease may be attributed to the inhibition of RNA synthesis or increased proteolytic activity and retardation of fat or lipid metabolism by pesticides respectively. During recovery period, the restoration of glucose, protein and cholesterol level was found to be slow and gradual which might be due to slow elimination of the toxicant from the tissues and reduced proteolysis.

Introduction

Generally aquatic organisms are susceptible to pollution effects by pesticides as well as by industrial effluents. However, an organism tries to adapt itself to these changes by changing their metabolic activities. But at higher concentrations, these, pollutants can cause damage to the physiological systems by affecting the organisms either at organ and cellular level or even at molecular level, which in turn cause changes in the biochemical composition and can be used to study the different protective mechanisms of the body to resist toxic substances and detoxification. The pesticides cause a number of subsidiary problems like affecting the ecosystem, affecting the growth, reproduction and behaviour by causing pathological and physiological changes (Meenakshi, 1993, Holden, 1973) and alterations in biochemical constituents of fishes (Anon, 1962).

Pesticides alter the nutritional value of fishes as well as their biochemical constituents (Korschagen and Murthy, 1967). It is well known that protein, carbohydrates and lipid play a major role as energy precursors for fishes under stress conditions (Idler and Clemens, 1959, Umminger, 1970). The tissue total proteins are important substrates required for building process of the body. They are also energy sources for fishes during thermal stress and muscular exercise which have been demonstrated by several investigators (Fontaine and Hatlay, 1953, Idler and Clemens, 1959). The fishes when exposed to an

Organophosphorous insecticide showed elevated levels of cholesterol and decrease in total lipid, free fatty acids and total proteins of liver and other tissues (Gill *et al.*, 1989). The enzymes play significant role in food utilization and metabolism. The proteolytic enzymes participate in the breakdown of protein molecule into amino acids (Ramalingam and Ramalingam, 1982). The pollutants also produce metabolic changes at cellular levels by way of influencing enzyme systems.

Similarly, the fishes when exposed to pesticide also showed a shift in the metabolic pathway and change in glycogen mobilization in the tissues (Vasanthy and Ramaswamy, 1987). Baskaran (1991) reported that the glycogen level decreased in fishes exposed to pollutant containing media.

Lipids are normally used by animals, whenever the carbohydrates are not sufficient to meet the energy demands. Many reports are there to show the altered lipid levels when the fish was exposed to toxicants (Bakthavathsalam and Srinivasa Reddy, 1981, Pandey and Pandey, 1985 and Palanichamy *et al.*, 1986). Adibi (1980) has found out that the quality and quantity of free amino acids can be considered as best diagnostic tool in understanding the physiological state of the cell. However, studies on the impact of toxicants on tissue energy sources are relatively very few, though considerable information is available dealing with determination of acute toxic levels of several pollutants.

Besides this, the pesticides also bring some changes in the blood parameters. The metabolic status in an organism is very much reflected in its milieu interior (Lu, 1985). Blood being the medium of intercellular and intracellular transport which comes in direct contact with various organs and tissues of the body, thus the physiological state of an animal at a particular time is reflected in its blood. Pesticides rapidly bind to the blood proteins and induced haematological changes such as changes in blood glucose, serum protein and serum cholesterol levels are of some value in assessing the impact of exposure under natural conditions and may also serve as tools for biological monitoring.

Materials and Methods

Haemato-biochemical analysis

The fishes were collected from the control and experimental tanks at the end of 15 days of exposure period and 15 days of recovery period and were subjected to the haemato-biochemical analysis individually.

Preparation of the Sample

By severing the caudal region of the fish, 0.2 ml of blood was collected from each and every fish with the help of a micropipette pre-rinsed with cold water. Then the blood collected was released into the serum tubes containing 1.8 ml of 80% ethanol and centrifuged for 5 minutes at 5000 rpm. The supernatant was then carefully transferred to another test tube. Finally the supernatant was divided into two aliquots of 0.5 ml (for blood glucose estimation) and 1 ml (for serum protein analysis) each. The precipitate was taken for the serum cholesterol analysis.

Estimation of Protein

1 ml of the sample was taken in a centrifuge tube and to this 1 ml of 80% ethanol was added. The mixture was centrifuged at a velocity of 3000 rpm for 10 minutes. After

discarding the supernatant, IN sodium hydroxide was added to dissolve the precipitate in it. From this 1 ml was taken and mixed with 5 ml of alkaline copper solution and kept for 10 minutes. To this mixture, 0.5 ml of folinphenolciocaltue reagent was added. The blue colour of the solution was measured at 720 nm by using spectronic 21. The G.D. value of each sample was recorded. A blank solution of pure IN sodium hydroxide was treated likewise. The amount of protein present in the sample was read from the graph drawn for standard solution.

Estimation of Carbohydrates

1 ml of the sample was taken in a centrifuge tube and to this 1 ml of 80% ethanol was added. The mixture was centrifuged at a velocity of 3000 rpm for 10 minutes. From this, 0.5 ml of the supernatant was taken and 5 ml of the anthrone reagent was added to it. The test tube was kept in boiling water bath for 15 minutes. The intensity of the green colour developed was measured at 620 nm in spectronic 21 and the O.D. value of each sample was recorded. A blank solution of pure 80% ethanol was treated likewise. The amount of carbohydrate present in the sample was read from the graph drawn for standard solution.

Estimation of Lipid

1 ml of sample was taken in a centrifuge tube and 2 ml of chloroform : methanol mixture (2:1) was added to it. The mixture was centrifuged at a velocity of 3000 rpm for 10 minutes. From this 0.5 ml of lipid extract was taken into a test tube and to it 0.5 ml of concentrated sulphuric acid was added. The test tube was placed in boiling water bath for 2 to 5 minutes and cooled at room temperature. From this, 0.2 ml was taken in a separate test tube and to this 5 ml of phosphovanillin reagent was added. The test tube was allowed to stand for 30 minutes. The pink colour developed was measured at 520 nm using spectronic 21. The O.D. value of each sample was recorded. A blank sample of pure chloroform: methanol mixture (2:1) was treated likewise. The amount of lipid present in the sample was read from the graph constructed for standard solution.

Results

The fish *Cyprinus carpio* exposed to sublethal concentration of Monocrotophos for 15 days exposure period showed the following alterations in the blood parameters.

Alterations in the blood parameters during exposure period

Impact of the pesticide, Monocrotophos on blood glucose level as a function of period of exposure is presented in Table 56.1.

The blood glucose level increased significantly in the sublethal concentration treated fishes. The initial blood glucose level before the experiment was 34.00 mg/100ml whereas after 15 days of exposure the level of blood glucose increased significantly to 40.00 mg/100 ml.

The serum protein level in the fishes before the commencement of the experiment was 514.13 mg/100 ml. During the experimental period, the protein level gradually decreased to 463.14 mg/100 ml at the end of exposure period of 15 days of exposure in fishes treated with the sublethal concentration.

The initial serum cholesterol level in the blood of the test fishes before the start of the experiment was found to be 280.13 mg/100 ml. The cholesterol level increased significantly in the sublethal concentration treated fishes and attained the value of 314.14 mg/100 ml.

Alterations in the blood parameters during recovery period

The blood glucose level was 32.14 mg/100 ml at the end of the recovery period in fishes pretreated with sublethal concentration against the recovery normal fishes having 36.34 mg/100 ml of glucose.

The serum protein level gradually increased to a level of 480.13 mg/100 ml in the sublethal concentration pretreated fishes in the recovery period against the recovery control fishes with 523.14 mg/100 ml of serum protein.

The cholesterol level was found to be 290.13 mg/100 ml at the end of the recovery period in fishes pretreated with sublethal concentration which has gradually increased as compared to the recovery control fishes having 285.14 mg/100 ml of serum cholesterol. The percentage alterations over control for the haemato-biochemical parameters are depicted in the Table 56.2 and Fig 56.1.

Discussion

Changes in the blood glucose level

After exposure to sublethal concentration of Monocrotophos, the *Cyprinus carpio* showed alterations in the blood parameters. It was observed that in the blood of the fish treated with sublethal concentration of pesticide, the glucose level was increased by 17.6% than the control fishes. The elevation of blood glucose in the *Cyprinus carpio* exposed to the toxicant may be attributed to physiological stress caused by the Monocrotophos. The elevation in the blood glucose level gains support from the works of Metelev *et al.* (1981) and Amudha (1986). These authors have suggested that such an increase is due to inhibition of glucogenesis and glyconeogenolysis. They may also be attributed to the utilization of glucose for energy during stress. In the toxicant medium the fish absorbs little oxygen from the environment, their respiratory metabolism being depressed hence the stored intracellular glycogen is utilized. Under such conditions, the hyperglycemic hormone is released for the degradation of glycogen. This glucose leaks into the blood causing

Table 56.1 : Changes in Selected Haemato-biochemical Parameters in *Cyprinus carpio* During exposure and recovery periods. Each value is Mean ± S.E (n=5)

Haemato-biochemical parameters	Nature of tissues	Exposure period (15 days)	Recovery period (15 days)
Blood Glucose (mg/100 ml)	Control	34.00 ± 1.3	36.34 ± 2.4
	Treated	40.00 ± 1.18** (+17.6)	32.14 ± 1.3** (−11.5)
Serum Protein (mg/100 ml)	Control	514.13 ± 1.43	523.14 ± 1.34
	Treated	463.14 ± 1.03* (−9.9)	480.13 ± 1.43* (−8.2)
Serum Cholesterol (mg/100 ml)	Control	280.13 ± 1.03	285.14 ± 0.34
	Treated	314.14 ± 1.14* (+12.14)	290.13 ± .045** (+1.75)

(−) Value in parenthesis indicates percent decrease over control.

(+) Indicates percent increase over control.

** Values are statistically significant at p<0.5 (Student's test).

* Values are satistically significant at p<5.

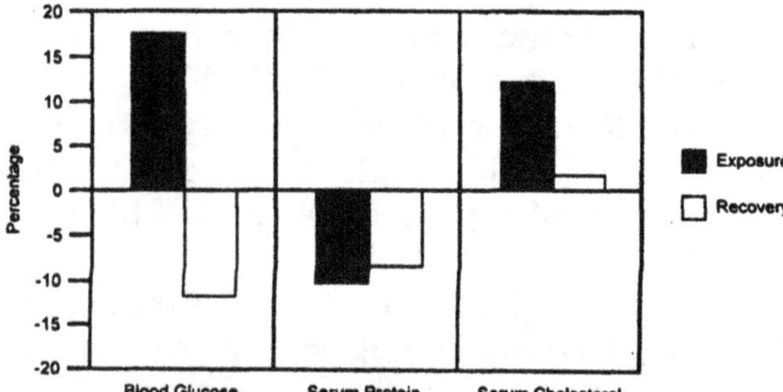

Fig. 56.1 : Percentage change over control in the selected Haemato-biochemical parameters of *Cyprinus carpio* during exposure and recovery periods.

hyperglycemia (Bhattacharya, *et al.*, 1975). Similar observations were made for the rainbow trout *S. gairdneri* (Grant and Mehrle, 1972). The present investigation therefore gives indirect confirmation that hyperglycemia is a good indicator for assessing general metabolic responses of fish to minor environmental changes.

Vijaymohanan and Achuthan Nair (2000) reported that glycogen, being the chief source of energy for the fish is the first metabolite to be affected by any stress. Further they stated that the exposure of flounders to the titanium effluent significantly enhances the levels of blood glucose due to the enhanced breakdown of liver glycogen (Glucogenesis).

Glucose is the major fuel for most of the organisms and it can be mobilized from glycogen at the time of demand and also to meet day-to-day energy requirements. A significant level of glucose content was evident in the blood in the present investigation. Increased level of glucose could be to withstand the toxicant imposed stress in the fishes (Keller and Andrew, 1973).

Changes in the serum protein level

Protein is the most important biochemical constituent present in the animal body. It performs the vital functions that are necessary for the sustenance of life. The enzymes and hormones which are considered essential for body functions are not just passive structures but they are active lipoprotein complexes at cellular and sub-cellular levels.

The blood of *Cyprinus carpio* treated with sublethal concentration of Monocrotophos showed a decline in protein content (-9.9%) when compared with that of the protein content of -8.2% of recovered fishes, thereby indicating the occurrence of proteolytic activity in the tissues. Anitha's (1984) observation on the effect of Phorate on *Mystus vittatus* and the observation of Rajamanickam (1985) on the effect of Lindane on the levels of blood protein in the freshwater fish *Tilapia mossambica*, lend support to the findings in *Cyprinus carpio*.

In general, any toxic substance like pesticides or heavy metals are known to depress blood protein in fishes (Jana and Bandyopadhyaya, 1987). In the present study also, the decrease in the protein content in blood was observed. The urea and ammonical compounds toxicity increase the value of protein. But several similar findings have reported the depletion in the level of plasma protein (Goel and Guptha, 1985). The pesticides may inhibit the RNA synthesis or increase the proteolytic activity resulting in low protein content or it may somehow affect the uptake of amino acids in the polypeptide chain. It is possible that pollutant stress influence the conversion of tissue protein into soluble fractions reaching the blood for utilization. The depletion in protein level as observed in *Channa striatus* (Baskaran, 1980) and *Barbus stigma* (Manoharan and Subbaiah, 1982) may be due to the diversification of energy to meet the impending energy demand caused by toxic stress. Gray and Mckenzie (1970) have stated that protein pattern in fishes is not influenced by the inherent factors like sex, stage of maturation and geographical location. Therefore, the alterations in the protein patterns that are noticed in the present study may be due to the influence of exogenous factors like toxic environment as has been suggested by Castell *et al.* (1970). Hence, it is likely that observed reduction in the protein level of *Cyprinus carpio* may therefore be direct consequences of stress imposed by Monocrotophos.

Changes in the serum cholesterol level

The cholesterol level of the blood was found increased to 12.14% in *Cyprinus carpio* on sublethal treatment of Monocrotophos. Increase in cholesterol in Monocrotophos exposed fishes, indicates increased lipid content in blood and retardation of fat metabolism. Such an increase in lipid content on sublethal concentration may be related to the heavy stress imposed by the pesticides. Bakthavathsalam (1980) reported the occurrence of a similar trend in *Anabas testudineus* when treated with Lindane. The significant increase of cholesterol indicates lipid profile of blood, hyper lipid anaemia may be due to abnormal lipid metabolism which is probably the result of hepatic dysfunction and chronic hypoxic condition related to the result of Goel and Guptha (1985). Gargi Sen (1992) also obtained similar results while working on the stress of zinc on *Heteropneustes fossilis*. The cholesterol level increased during exposure period whereas it decreases during recovery period. This corroborates the earlier finding of Larsson and Leward (1973) and Dave *et al.* (1975). The significant decrease of cholesterol level in recovery period depicted the hypocholes-terolamia which frequently results in anaemia.

Moreover, the decreased protein level is compensated by the enhancement of cholesterol which may help to withstand the sublethal toxic stress. Similar finding was made by Patil and Dhande (2000) who reported that the cholesterol level is significantly high when compared to control during $HgCl_2$ and $CdCl_2$ induced medium of fresh water fish, *Channa punctatus*. Thus it can be concluded that decreased protein content, increased cholesterol and carbohydrate (hyper glycemia) can be used as an index of pesticide (Monocrotophos) pollution.

During recovery period (free from Monocrotophos medium), it was found that after short-term exposure of (15 days), restoration of protein, glucose and cholesterol level to their normal level was found to be slow and gradual which might be due to, slow elimination of the toxicant from the tissues and reduced proteolysis.

Acknowledgement

We express our sincere thanks to Mr. S. Swetharanyam, Head of the Department, Department of Zoology, Dr. V. Radhakrishnan. Principal, A.V.C. College for giving constant support and encouragement and providing us the necessary laboratory facilities in doing this research work.

References

Adibi, S.A. 1980. Roles of amino acid in metabolic regulation. J. Lab. Clin. Med., 95 : 475.

Amudha, R. Christy. 1986. Effects of Malathion on blood glucose level and its histopathology on the endocrine adrenal gland of *Anabas testudineus*. M. Phil. Thesis, Bharathidasan University, Trichy.

Anitha, J.P. 1984. Effect of Phorate on the bimodal respiration and changes in the blood glucose, lactic acid and amino acids in the fresh water fish, *Mystus vittatus*. M. Phil, Thesis, Annamalai University.

Annon, M. 1962. The wealth of India. Raw materials of Vol. IV supplement fish and fisheries. Council of Scientific and Industrial Research. New Delhi. 132.

Bakthavathsalam, R. 1980. Toxicity and physiological impact of selected pesticides on air breathing fish, *Anabas testudineus*. Ph.D. Thesis, Annamalai University.

Bakthavathsalam, R. and Srinivasa Reddy, Y. 1982. Lipid kinetics in relation to the toxicity of three pesticides in the climbing perch, *Anabas testudineus* (Bloch). Proc. Indian Natu. Sci. Acad. B. 47 (5): 670-676.

Baskaran, R. 1980. Biological studies on a chosen thermoconformer, *Channa Striatus*. Ph.D. Thesis, Madurai Kamaraj University.

Baskaran, P. 1991. Use of biochemical parameters in biomonitoring of pesticide pollution in some fresh water fishes. Ecotoxicol. Environ. Monit., 9(2): 104-109.

Bhattacharya, S., Mukherjee, and Bhattacharya, S. 1975. Toxic effects of Endrin on Hepatopancreas of the teleost fish *Clarius batrachus*. Indian J. Exp. Biol., 13: 185-186.

Castell, C.H., Smith, B. and Neal, W. 1970. Effects of transition metal ions on the extractable proteins of fish muscles. J. Fish. Res. Bd. Can., 27.701-714. .

Dave, G., Boman, A., Johanson-sjobeck, M.L and Lindman, U. 1975. Metabolic effects of PCB in the European eel *Anguilla anguilla*. In: Sublethal effects of toxic chemicals on aquatic animal Eds Koeman, J.H and Strik, J.J. TWA (Netherlands: E.sevier scientific Pub. Co.).

Fontaine, M and Hatlay, T. 1953. Contributional elude du metabolism glucidique du salmon salmo salr La liver sestapes deson development etduses migrations; Physiol. Compo Et. Oecologia 3:37-52.

Gargi sen, Behara, Milan Kumar and Patel, Premananda. 1992. Effect of Zinc on haemato-biochemical parameters of *Channa punctatus*. J. Ecotoxicol. Environ. Monit. 2(2): 89-92.

Gill, S.T., Pant, J.C and Tiwari, H. 1989. Sublethal effects of an Organophosphorous insecticide on certain metabolic levels in a fresh water fish, *Puntius conchonius*. (Hamilton). Indian Bull. Environ. Contam, Toxicol., 290-299.

Goel, K.A and Guptha, Premananda. 1985. Haematobiological characteristics of *Heteropneustes fossils* under the stress of Zinc. Ind. J. Fish, 32: 256-259.

Grant, B.F. and Mehrle, Premananda. 1972. Endrin toxicosis in rainbow trout *Salmo gairdneri*. J. Fish Res. Bol. Canada. J. Fish. Res. Bd. Can. 27, 2109-2112.

Holden, A.V. 1973. Effects of pesticides on fish. In : Environmental pollution by pesticide (Ed.C.A.Edwards). Plenum Press, Ny. 213-253.

Idler, D.R and Clemens, Premananda. 1959. The energy expenditures of fresh water river sockeye salmon during spawning migration to chicks and stuar Lakes. Inct. Pacific salmon fish. Comm.. Prog. Rep. Jackson. Printing, New Westminster B.C : 80.S

Jana, S. and Bandopadhyaya, N. 1987. Effect of heavy metals on some biochemical parameters in the fresh water fish, *Channa punctatus*. Environ, Col. 5(3): 488-493.

Keller R. and Andrew, E.M. 1973. The site of action of the crustacean hyperglycemic hormone. Gen. Comp. Endocrinol. 20: 572-578.

Korschagen, L.T. and Murthy, D.A. 1967. Missouri federal aid project progress report.

Larsson and Leward. 1973. Biochemical and haematological effect of Titanium dioxide Industrial effluent on fish. But Environ. Contam. Taxicol., 25: 427-435.

Lu, F.C. 1985. Basic toxicology. Hemisphere Publishing Corporation, Washington, USA : 184-195.

Manoharan, T. and Subbaiah, G.N. 1982. Toxic and Sublethal effects of Endosulfan on *Barbus stigma*. Proc. Ind. And. Acad. Anim. Sci. 91: 523-532.

Meenakshi, V. 1993. Histological and haematological studies on the effect of pesticide Phosphamidon in an air breathing fresh water fish, *Anabas testudineus*, M.Phil, Thesis, Annamalai University.

Metelev, V.V., Kanaev, A.I and Dzaskhova, N.G. 1981. Water Toxicology. Amerind Publ. Company. New Delhi. : 174-175.

Palanichamy, S., Malliga Devi, T. and Arunachalam, S. 1986. Sublethal effects of Thiodon and Ekalux on food utilization, growth and conversion efficiency in the fish, *Lepido cephalichthys*. U.P.J. Zool. 6: 58-63.

Pandey, R.K and Pandey, S.K. 1985. Seasonal fluctuations in lipid and cholesterol content of ovary and testis in relation to annual sexual cycle in *Mystus vittatus* under the stress of the agrochemical NPK. Proc. 72nd Ind. Sci. Cong., Part II No. 280.

Patil. G.P. and Dhande, R.P. 2000. Effect of $HgCl_2$ and $CdCl_2$ on Haematobiochemical parameters of the fresh water fish *Channa punctatus* (Bloch). J. Ecotoxicol. Environ. Monit. 10(3): 177-181.

Rajamanickam, C. 1985. Impact of lindane (r-BHC) on the levels of glycogen, glucose, lactic acid, protein and free amino acid in the fresh water fish *Tilapia mossambica* M. Phil. Thesis. Annamalai University.

Ramalingam, K. and Ramalingam, K. 1982. Effects of sublethal levels of DDT, Malathion and mercury on tissue protein of *Sarothrerodon mossambicus* (Peters). Proc. Indian Acad. Sci. (Animal Sci.) 90: 501-505.

Umminger, B.L. 1970. Physiological studies on super cooled hill fish, *Fundulus heteroclitus* III carbohydrate metabolism and survival at sub. Zero temperature. J. Exp. Zool., 173: 159-174.

Vasanthi, M and Ramaswamy, M. 1987. A shift in the metabolic pathway of *Sarotherodon mossambicus* (Petess) exposed to Thiodon (Endosulfan). Proc. Indian. Acad. Sci. (Arim. Sci). 95(1) : 55-61.

Vijayamohan, T.G and Nair, G. Achuthan. 2000. Impact of Titanium dioxide factory effluent on the biochemical composition of the fresh water fishes *Oreochromis mossambicus* and *Etroplus maculates*. Poll. Res. 19(1): 67-71.

ENVIRONMENTAL BIOTECHNOLOGY

Edited By :   Professor (Dr.) Arvind Kumar

Published By : DAYA PUBLISHING HOUSE

57

EFFECT OF MONOCROTOPHOS ON BIOCHEMICAL MARKERS DNA AND RNA IN *CYPRINUS CARPIO* DURING EXPOSURE AND RECOVERY PERIOD

• *C. Maruthanayagam and G. Sharmila*

A.V.C. College (Autonomous), Department of Zoology Mannampandal,
Mayiladuthurai

Abstract

The toxicity of Monocrotophos on the fresh water fish *Cyprinus carpio* has been studied here. The pesticide leads to several changes in the biochemical markers like DNA and RNA may be due to the increased activity of the enzyme DNA as and the inhibition of RNA polymerase function. But during recovery period, the DNA and RNA level both progressively increased indicating a probable recovery from the disruption of internal organs.

Introduction

Pesticides alter the nutritional value of fishes as well as their biochemical constituents (Korschagen and Murthy, 1967). It is well known that protein, carbohydrates and lipid play a major role as energy precursors for fishes under stress conditions (Idler and Clemens, 1959 and Umminger, 1970). The tissue total proteins are important substrates required for building process of the body. They are also energy sources for fishes during thermal stress, spawning and muscular exercise which have been demonstrated by several investigators (Fontaine and Hatlay, 1953, Idler and Clemens, 1959). The fishes when exposed to an organophosphorous insecticide showed elevated levels of cholesterol and decrease in total lipid, free fatty acids and total proteins of liver and other tissues (Gill *et al.*, 1989). The enzymes play significant role in food utilization and metabolism. The proteolytic enzymes participate in the breakdown of protein molecule into amino acids (Ramalingam and Ramalingam, 1982). The pollutants also produce metabolic changes at cellular levels by way of influencing enzyme systems.

Similarly, the fishes when exposed to pesticide also showed a shift in the metabolic pathway and change in glycogen mobilization in the tissues (Vasanthy and Ramaswamy, 1987) Baskaran (1991) reported that the glycogen level decreased in fishes exposed to pollutant containing media.

Lipids are normally used by animals, whenever the carbohydrates are not sufficient to meet the energy demands. Many reports are there to show the altered lipid levels when the fish was exposed to toxicants (Bakthavathsalam and Reddy, 1981, Pandey and Pandey, 1985 and Palanichamy *et al.*, 1986). Adibi (1980) has found out that the quality and quantity of free amino acids can be considered as best diagnostic tool in understanding the physiological state of the cell. However, toxicants on tissue energy sources are relatively very few, though considerable information is available dealing with determination of acute toxic levels of several pollutants.

Another toxic effect of the pesticide is seen in the alterations it brings in the nucleic acid content of the animal. The role of nucleic acids particularly RNA/DNA and protein/DNA ratios, which are used as an index of protein synthesis and cell size, are considered to be important and form an integral part of the growth and toxicity studies in animals. But such studies in fishes are scanty. The treatment with the pesticides causes variability in the nucleic acid content in different tissues and the degree of variability or extent of alterations caused by the pesticides is found to be dose dependent (Malla Reddy and Bashamohideen, 1988).

Materials and Methods

Extraction of nucleic acids from tissues

Nucleic acids were extracted from the tissues by the method of Schneider (1957). 100 mg of tissues were homogenized in 5 ml of 10% TCA and was added to allow complete precipitation of protein and nucleic acids for 30 minutes. The mixture was centrifuged for 10 minutes at 5000 rpm and the precipitate was washed thrice with 10% ice cold TCA to remove any soluble compounds.

The precipitate was later treated with 95% ethanol to remove the lipid content and centrifuged for 10 minutes. The lipid free tissue residue was suspended in 5 ml of 5% TCA and placed in a water bath maintained at 90°C for 15 minutes with occasional stirring. This facilitated the quantitative separation of nucleic acids from the precipitated protein. This was centrifuged for 10 minutes and the supernatant was used for the estimation of nucleic acids.

Estimation of DNA

The DNA concentration was estimated by the method of Burdon (1957). DNA on hydrolysis yields acids soluble or dialyzable diphenylamine - positive materials. The blue colour was obtained after the reaction of DNA with diphenylamine glacial acetic acid. This method was chosen because it is sensitive, specific and less susceptible to interference by other chromogenic compounds.

One ml of extract was made upto 3.0 ml with 1 N perchloric acid. To this 2 ml of diphenylamine reagent was added. Tubes containing known amount of standard (Calf thymus DNA) and a reagent blank containing 1 N perchloric acid were also run simultaneously. The test tubes were placed in water bath maintained at 90°C for 20 minutes and cooled at room temperature. The intensity of blue colour developed was read at 600 nm against blank. The concentration of DNA in the tissue sample was read from the graph for standard solution. The DNA concentration was expressed as mg/g wet weight of tissues.

Estimation of RNA

The RNA concentration was estimated by the method of Ceriotti (1965). One ml of tissue was made upto 2ml with water and 3ml of Orcinol reagent was added to this. The content were shaken well and heated in a boiling water bath for 20 minutes and then cooled. The intensity of green colour obtained was read at 660 nm against the reagent blank. The standard tube was simultaneously subjected to the same procedure. The amount of RNA present in the sample was read from the graph drawn for standard solution.

Results

Changes in the Nucleic acid content during exposure period

When the fishes were exposed to presumable harmless concentration of 51.57 ppm of Monocrotophos, the tissues of muscle, liver and intestine showed a significant amount of decrease in the nucleic acid content perhaps due to stress in the toxicant medium.

Muscle

The initial RNA content before the experiment was 5.4 mg/g wet weight of tissues. Due to exposure to sublethal concentration of Monocrotophos over a period of 15 days, the RNA content of the fish gradually decreased to the level of 3.4 mg/g wet weight (Table 57.1).

Similarly the DNA content before the experiment was 1.8 mg/g wet weight of tissues which at the end of 15 days of exposure period decreased significantly to 1.4 mg/g wet weight (Table 57.2).

Liver

The initial RNA content of the fish before the commencement of the experiment was 6.3 mg/g wet weight of tissues. Due to exposure to sublethal concentration of Monocrotophos over a period of 15 days, the RNA content of the fish gradually decreased to 4.2 mg/g wet weight.

The same is the case with DNA content also. The initial level of DNA content was found to be 2.9 mg/g in the fishes but in the sublethal concentration treated fishes, the DNA content was found to decrease to a level of 1.9 mg/g at the end of 15 days of exposure period.

Intestine

Table 57.1 : Sublenthal Effect of Pesticide on RNA Content (mg/g) in Different Tissues of *Cyprinus carpio* During Exposure and Recovery periods. Each value is Mean ± S.E. (n=5)

Period of Study	Muscle (mg/g)	Liver (mg/g)	Intestine (mg/g)
Control (0 day)	5.4 ± 0.33	6.3 ± 0.41	4.8 ± 0.26
Exposure (15 days)	3.4 ± 0.24*	4.2 ± 0.31**	3.1 ± 0.18**
	(−37.03)	(−33.3)	(−35.41)
Recovery Control	5.5 ± 0.53	6.8 ± 0.63	4.9 ± 0.16
Recovery (15 days)	5.0 ± 0.32**	5.9 ± 0.33**	4.8 ± 0.09*
	(−9.09)	(−13.2)	(−2.04)

(−) Values in parenthesis indicates percent decrease over control.

** Values are statistically significant at $p<0.5$ (Student's 't' test).

* Values are statiscally significant at $p<5$.

Table 57.2 : Sublenthal Effect of Pesticide on DNA Content (mg/g) in Different Tissues of *Cyprinus Carpio* During Exposure and Recovery Periods. Each value is Mean ± S.E. (n=5)

Period of Study	Muscle (mg/g)	Liver (mg/g)	Intestine (mg/g)
Control (0 day)	1.8 ± 0.07	2.9 ± 0.07	1.4 ± 0.15
Exposure (15 days)	1.4 ± 0.04**	1.9 ± 0.13**	1.0 ± 0.12**
	(−22.2)	(−34.4)	(−28.5)
Recovery Control	1.9 ± 0.09	2.4 ± 0.03	1.5 ± 0.11
Recovery (15 days)	1.5 ± 0.08**	1.98 ± 0.13**	1.3 ± 0.13**
	(−21.0)	(−17.5)	(−13.3)

(−) Values in parenthesis indicates percent decrease over control.

** Values are statistically significant at $p<0.5$ (Student's 't' test).

Generally the nucleic acid content of intestine was lower than the muscle and liver. The initial RNA content before the experiment, was found to be 4.8 mg/g wet weight of tissues which gradually decreased to a level of 3.1 mg/g wet weight in sublethal concentration treated fishes at the end of the 15 days of exposure period.

Similarly the DNA content initially was found to be 1.4 mg/g wet weight of tissues. In the sublethal concentration treated fishes, the DNA content decreased gradually to a level of 1.0 mg/g wet weight at the end of the 15 days of exposure period.

Changes in the Nucleic acid content during recovery period

Muscle

The RNA level was found to be gradually returning to their normal level of 5.0 mg/g wet weight of tissues as compared to the recovery control fishes with the value of 5.5 mg/g wet weight of tissues.

The DNA level was also found to be gradually increased to a level of 1.5 mg/g wet weight of tissues as compared to the recovery control fishes with 1.9 mg/g wet weight of tissues.

Liver

The RNA level was found to increase to a level of 5.9 mg/g wet weight of tissue against the recovery control fishes with 6.8 mg/g wet weight of tissues.

Similarly the DNA level also increased to a level of 1.98 mg/g wet weight of tissues against the recovery normal fishes with 2.4 mg/g wet weight of tissues.

Intestine

At the end of 15 days of recovery period, the sublethal concentration pretreated fishes showed a gradual increase in the RNA content to a level of 4.8 mg/g wet weight of tissues as compared to the recovery normal fishes with the value of 4.9 mg/g wet weight of tissues.

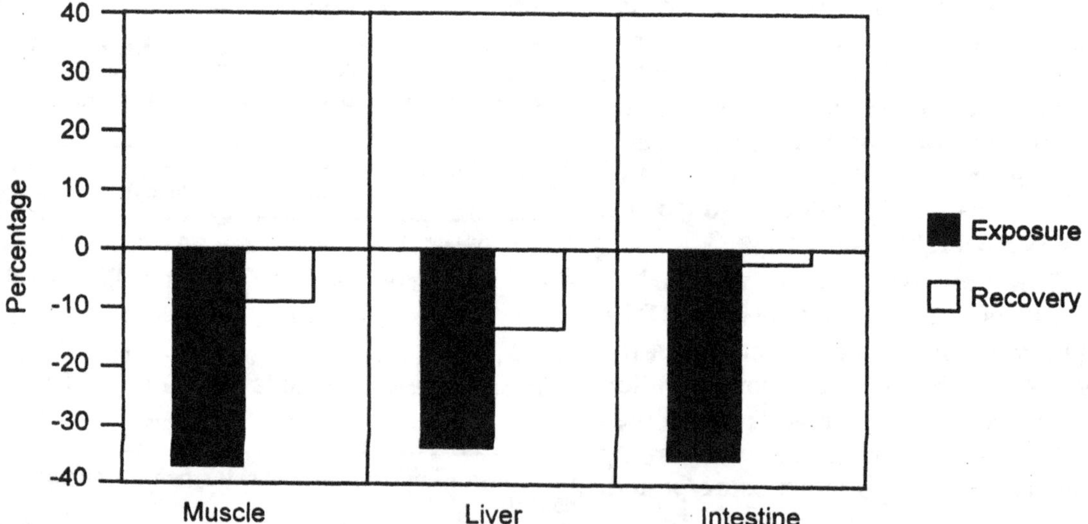

Fig. 57.1 : Percentage change over control in the RNA content of Muscle, Liver and Intestine of *Cyprinus carpio* during exposure and recovery periods.

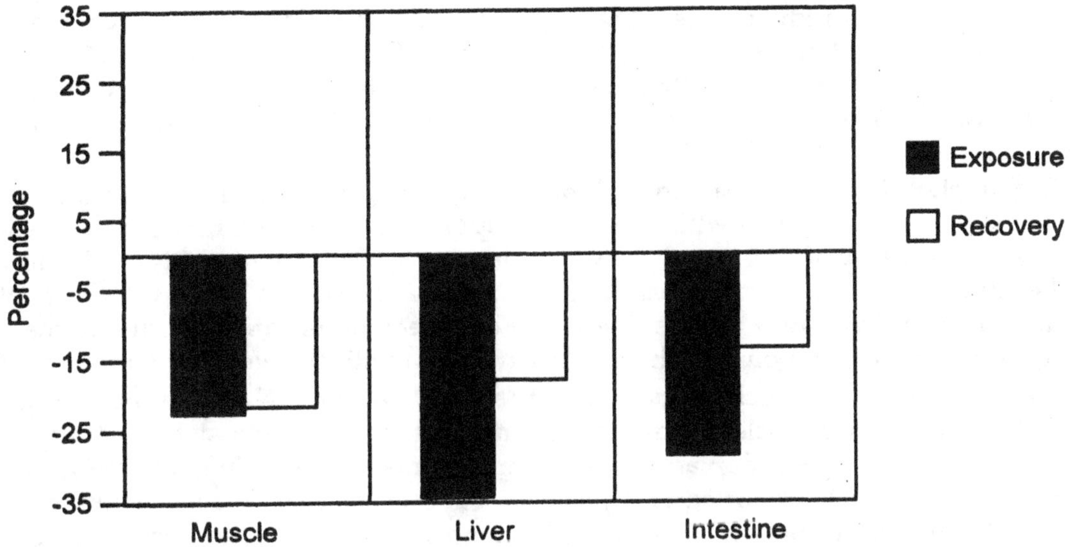

Fig. 57.2 : Percentage change over control in the DNA content of Muscle, Liver and Intestine of *Cyprinus carpio* during exposure and recovery periods.

The DNA content also showed a gradual increase in the sublethal concentration pretreated fishes at the end of 15 days of recovery period and attained the value of 1.3 mg/g against the recovery control fishes with 1.5 mg/g wet weignt of tissues.

The percentage change over control in RNA content as well as DNA content of fishes in exposure and recovery period are depicted in the Fig. 57.1 and Fig. 57.2 respectively.

Discussion

The pesticide leads to several changes in the biochemical markers such as RNA and DNA, the non-target agents of aquatic pollution resulting in secondary responses (Mazeaud *et al.*, 1977). The present investgation is aimed to understand the alteration in biochemical constituents with special reference to genetic materials, RNA and DNA content at both exposure and recovery periods, in tissues like muscle, liver and intestine of fresh water fish *Cyprinus carpio* exposed to Monocrotophos toxicity.

In the present study, the exposure period of fishes to Monocrotophos resulted in the depletion of RNA and DNA. This decrease in nucleic acid suggests the decrease in protein synthesis and further damage to the liver which is the major metabolic organ of drug detoxification (Ramalingam *et al.*, 2000). This decreasing trend is further supported by Durairaj *et al.*, (1997) who opined that RNA and DNA content level decreased significantly in *Oreochromis mossambicus* when exposed to Quniolphos and Phenthoate.

The decrease of RNA may be suggested that the daily addition of pesticides resulting in swelling and chromatolysis of Nissle bodies which are rich in RNA (Watson, 1969). According to Mukhopadhyay and Dehadrai (1980), the decrease of RNA may also be due to interference in the incorporation of precursor in the nucleic acid synthesis or inhibiting the function of RNA polymerase. Since the nucleic acid regulates protein synthesis, the alteration in the RNA level at different intervals leads to the dependent changes in the protein level and its reduction level is higher for Organochlorines when compared to Organophosphates (Durairaj *et al.*, 1997). Dawood (1986) and Benjamin (1990) have suggested that the decrement of RNA may also be due to the non-coding for the process of protein synthesis, thereby decrease in the RNA content which inturn would have reduced the concentration of RNA.

Organophosphorous compounds exhibit strong mutagenic and clastogenic potentiality (Patankar and Vaidya, 1980) which may be responsible of the alteration of DNA level. However the decrease of DNA is not very prominent when compared to RNA. The decrease may be attributed to the increased activity of DNA asd as suggested by Tayyaba *et al.* (1981). The decrement of DNA may also be due to the degeneration hepatic cells due to the effect of Organophosphate compounds which is in agreement with the work of Bachelard (1974).

The decrease in the nucleic acid content (both DNA and RNA) may be presumably due to the metabolism of nucleic acids that accompanies the end of massive protein synthesis and autophagy of rough endoplasmic reticulum (Domodaran Reddy *et al.*, 1994). Sabitha Borah and Yadav (1995) reported that pesticide exposed fishes showed a significant low value in the amount of nucleic acids; i.e. DNA and RNA. The higher decreasing trend is evident with the increase of pesticides in the exposure period.

Since DNA is the chemical estimate of cell number, its decline also suggested some loss of cells during exposure to pesticides. Loss of DNA and RNA content of fishes showed the possible interference in the nucleic acid synthesis.

Usually RNA content indicates the intensity of protein synthesis in a tissue (Brachet, 1995). High level of RNA reflects its involvement in the cellular growth. In view of the significant correlation of RNA and protein, a deficient synthesis of any type of RNA should have its reflection in a corresponding failure of protein synthesis (Borah and Yadav, 1995).

The observed persistent significant decrease in both the protein and nucleic acids would suggest that pollutant impair the process of protein synthesis in the tissues of fishes exposed to pesticides. Since RNA is the biochemical mid-wife in the formation of proteins, the diminished RNA content also affects the cellular protein content. Eichorn (1995) have also suggested that the depression in DNA synthesis is not energy dependent and may be due to the disruption of the replication process.

In the recovery period, the RNA and DNA level progressively increased, indicating a probable recovery from the disruption of internal organ and decrease of hypoxic condition to the fish, as evidenced by the recovery of whole animal activity after their shift from pesticide water to normal water.

Thus in the present investigation, the reduction of nucleic acids can be taken as a meaningful biochemical indices of pesticides toxicity to assess the extent of pollution of the aquatic environment.

Acknowledgement

We express our sincere thanks to Mr. S. Swetharanyam, Head of the Department, Department of Zoology, Dr. V. Radhakrishnan. Principal, A.V.C. College for giving constant support and encouragement and providing us the necessary laboratory facilities in doing this research work.

References

Adibi, S.A. 1980. Role of amino acid in metabolic regulations. J. Lab. Clin. Med., 95:475.

Anon, M. 1962. The wealth of India. Raw materials of vol IV Supplement fish and fisheries. Council of Scientific and Industrial Research, New Delhi. 132.

Bachelard, H.S. 1974. Outline studies in biology. Brain Biochen Chapman and Hall, London: 8.

Bakthavathsalam, R. and Reddy, Y. Srinivasa. 1981. Lipid Kinetics in relation to the toxicity of 3 pesticides in the climbing perch, *Anabas testudineus* (Bloch). Proc. Indian Natu. Sci. Acad. B. 47Z(5): 670-676.

Baskaran, P. 1991. Use of biochemical parameters in biomonitoring of pesticide pollution in some fresh water fishes. Ecotoxicol. Environ. Monit. 1(2) : 104-109.

Benjamin, N. 1990. Biochemical changes due to learning in the brain of rat exposed to an Organophosphate pesticide: M. Phil. Thesis, University of Madras.

Brachet, J. 1995. In: The nucleic acids. Vol. II, Academic Press, New York.

Burdon, R.M. 1957. In: Methods in Enzymology. S.P. Colowick and N.O. Kaplam (Eds). Academic Press, New York. 1 : 267.

Ceriotti, G. 1965. A microchemical determination of deoxyribonucleic acid. J. Biol. Chem., 198: 297-303.

Reddy, Damodaran. K., Chaudhuri, A. and Thangavelu, K. 1994. Effect of L-thyroxic-sodium Pentahydrate on protein and nucleic acid turn over in fat body during fifth larval stage of non-diapausing tasar silk worm, *Antheraca mylitta* Drury. Ind. J. Exp. Biol. Vol. 32, 413-417.

Dawood, N. 1986. Studies on the functional organization of the brain of rabbit; *Oryctolagus cuniculus*. Ph.D. Thesis, University of Madras.

Durairaj, S., Devaraj, P. and Selvarajan, V.R. 1997. Neurochemical responses of fish, *Oreochronis mossambicus* to Quinolphos, Phenthoate and their combination. J. Ecobiol. 9(2) 141-144.

Eichorn, G.L. 1995. Active sites of biological micromolecules and their interaction with Heavy metals. In: Ecological toxicology research: Effects of heavy metals and Organohalogen compound. A.D. McIn tyre and C.F. Mills (Eds). Plenum Press, New York. 123-143.

Fontaine, M and Hatlay, T. 1953. Contributional elude du metabolism glucidique du salmon salmo salr La liver sesetapes deson development etduses migrations; physiol. Comp. Et. Oecologia 3: 37-52.

Gill, S.T., J.C and Tiwari, H. 1989. Sublethal effects of an Organophosphorous insecticide on certain metabolic levels in a fresh water fish, *Puntius Conchonius* (Hamilton). Indian Bull. Environ. Contam, Toxicol., 290-299.

Holden, A.V. 1973. Effects of pesticides on fish. In: Environmental pollution by pesticide (Ed. C.A. Edwards). Plenum Press, NY. 213-253.

Idler, D.R. and Clemens, W.A. 1959. The energy expenditures of fresh water river sockeye salmon during spawning migration to chicks and stuar lakes. Inct. Pacific Salmon fish. Comm. Prog. Rep. Jackson. Printing, New Westminster B.C. 80.

Korschagen, L. T and Murthy, D.A. 1967. Missouri federal aid project progress report.

Malla Reddy, P. and Bashamohideen, M.D. 1988. Toxic impact of Fenvelerate on the Protein metabolism in the branchial tissues of fish *Cyprinus carpio*. Curr. Sci., 57: 211-212.

Mazeaud, M.M., Mazeaud, F. and Donaldson, E.M. 1977. Primary and secondary effects of stress in fishes: some new data with a general review. Trans. Am. Fish. Soc. 106: 201-212.

Meenakshi, V. 1993. Histological and haematological studies on the effect of pesticide Phosphamidon in an air breathing fresh water fish, Anabas testudineus, M. Phil. Thesis, Annamalai University.

Mukhopadhyay, P.K. and Dehadrai, P. V. 1980. Biochemical changes in the air-breathing cat fish, Clarias batrachus (Linn.) exposed to Malathion. Environ. Pollut. A. 22: 149-158.

Palanichamy, S., Malliga Devi, T. and Arunachalam, S. 1986. Sublethal effects of Thiodon and Ekalux on food utilization, growth and conversion efficiency in the fish, *Lepido Cephalichthys* thermalis. U.P.J. Zool. 6: 58-63.

Pandey, R.K. and Pandey, S.K. 1985. Seasonal fluctuations in lipid and cholesterol content of ovary and testis in relation to annual sexual cycle in *Mystus vittatus* under the stress of the agrochemical NPK. Proc. 72nd Ind. Sci. Cong., Part II No. 280.

Patankar, N. and Vaidya, V.G. 1980. Evaluation of genetic toxicity of the insecticide Phosphamidon using invitro mammalian test systems. Ind. J. Exp. Biol. 18: 1145-1147.

Ramalingam, K. and Ramalingam, K. 1982. Effects of sublethal levels of DDT, Malathion and Mercury on tissue protein of *Sarotherodon mossambicus* (Peters). Proc. Indian Acad. Sci. (Animal Sci.). 90: 501-505.

Ramalingam, V., Vimaladevi, Narmadaraji, R. and Prabakaran, P. 2000. Effect of the lead on the haematological and biochemical changes in the fresh water fish, *Cirrhina mrigala*. Poll Res. 19(1): 81-84.

Borah, Sabitha and Yadav, R.N.S. 1995. Alteration in the protein, free amino acids, nucleic acids and carbohydrate contents of freshwater fish Heteropneustes fossilis. J. Pollution Research Vol. 14(1): 99-103.

Schneider, R.P. 1957. Mechanism of inhibition of rat brain (Na^{++}K$^+$) Adenosine triphosphate by 2, 2-bis (P-chlorophyll)-l, 1, 1, Trichloroethane (DDT). Bio chem. Pharmacol., 24: 939-946.

Tayyaba, K., Hasan, M. F. and Khan, N.H. 1981. Organophosphate Metasustox-R, Induced regional alterations in brain nucleic acid metabolism., Indian J. Exp. Biol.19: 688-690.

Umminger, B.L. 1970. Physiological studies on super cooled hill fish, *Fundulus heteroclitus* III Carbohydrate metabolism and survival at sub-zero temperature J. Exp. Zool., 173: 159-174.

Vasanthy, M. and Ramaswamy, M. 1987. A shift in the metabolic pathway of *Sarotherodon mossambicus* (Peters) exposed to Thiodon (Endosulphan). Proc. Indian. Acad. Sci. (Anim. Sci.). 95 (1) : 55-61.

Watson, W.E. 1969. The response of motor neurons to intra-muscular injection of botulinum toxin. J. Physiol. 202: 611-630.

ENVIRONMENTAL BIOTECHNOLOGY

Edited By : **Professor (Dr.) Arvind Kumar**

Published By : **DAYA PUBLISHING HOUSE**

58

THE CHANGING ROLE OF ANALYTICAL INSTRUMENTATION IN ENVIRONMENTAL ANALYSIS

• *G.H. Pandya, V.K. Kondawar*

Senior Scientists,
National Environmental Engineering Research Institute,
Nehru Marg, Nagpur.

Abstract

In the early years the development of analytical instrumentation were mainly devoted to solving the scientific and technological problems of designing and constructing equipment capable of performing analytical procedures in a precise and repeatable manner. Very little thought was directed to the wider problems of a laboratory requiring change over from a manual analysis to one dependent on automation. In recent years a change over to automatic techniques has increased the efficiency but has eroded the manipulative skills and experimental judgment. With the introduction of advanced injection systems in the form of purge and trap, headspace, automated thermal desorption, solid phase micro-extraction, super critical fluid extraction system, the analytical instrumentation has changed the role of analysts for performing better, quicker and reliable analysis. The paper discusses the changing role of analysts as a result of automation in analytical instrumentation.

Introduction

IUPAC has defined automation as "the use of combinations of mechanical and instrumental devices to replace, refine, extend or supplement human effort and facilities in the performance of a given process, in which at least one major operation is controlled without human intervention, by a feedback mechanism".

The major thrust for R&D in automation is

1. To improve the cost effectiveness in discharging large analytical workloads, especially those of repetitive nature, and

2. To improve the method performance, notably precision in these circumstances.

The principal steps involved in chemical analysis include sampling, sample pretreatment, sample measurement, calculation of results, validation of results, and report preparation. Sampling requirements are often not amenable to automation and the task of reporting the result to end user is not clearly defined. Analysts are not involved in either choice of the sample nor in the reporting of the result. The sample measurement step has received the most attention and almost all measurement techniques used in chemical analysis have been automated. The recent trend is to focus attention on post-analysis data processing and reporting the results with better software packages.

The reason for automation depends upon the economics of investment. In case of an industry, the automation will improve the efficiency by reducing the storage of intermediates or the final products and the financial advantage gain would exceed the cost of the automatic equipment.

Education

Many times the commercial suppliers of the analytical instruments offer training needs to the customers as per the manufacturers instruction methods at the place of installation or demonstrate at their premises. However for those large institutions the automatic instrumental facilities is viewed as modification to meet the new needs. Sometimes the laboratory staff is engaged as a team in development of a chemical method. This involves familiarization with chemical analysis procedure, component design, software development, performance testing and data processing. For these people education and training needs are more demanding. Teaching of analytical instrument automation in educational institution is rarely available due to its interdisciplinary nature. It involves chemistry, engineering, electronics, computing facilities and requires demonstration facilities which is quite expensive. Educational institutions are aware of the problems and have made some headway in this direction by developing micro-processor based instruments such as automated titrator which is of value as teaching aid. Some modular experiments are also developed for illustrating the principles of automatic analysis. This may not be considered as adequate for the academic staff as new generation has limited successes and are still not exposed to failures of automation. There is a greater need to provide background and education in automation. A scope of awareness training programme in analytical instrumentation is strongly felt involving the user or experience workers in the field who can pass on not only the practice but also thinking that would generate interest in automated analysis. The programme could cover the various aspects as illustrated in Table 58.1.

The Table 58.1 illustrates that small groups could be formed initially to discuss the strategy for finding a solution to automation of an analytical problem and then combining to form a large group. Each group would present and discuss the development methodology. The various approaches are then discussed with the faculty experts. The participant gains additional information on how to prepare specification of their requirements for better performance of new automated instrument to be purchased and subsequently maintain.

Table 58.1 : Awareness Training on Analytical Automation Instrumentation

Mode of training	Allocation	Area of Expertise
Lectures	Informal	Instruction to be provided by academic, industrial and experience workers in the field
Tutorials	Small Groups to have a strategy to find an automated solution to an analytical problem	Real world problems to be discussed with the experts for a solution
Demonstration	Group	Hands on training with full range of automated system
Conclusion	Small groups to unite	To discuss the developmental strategy.

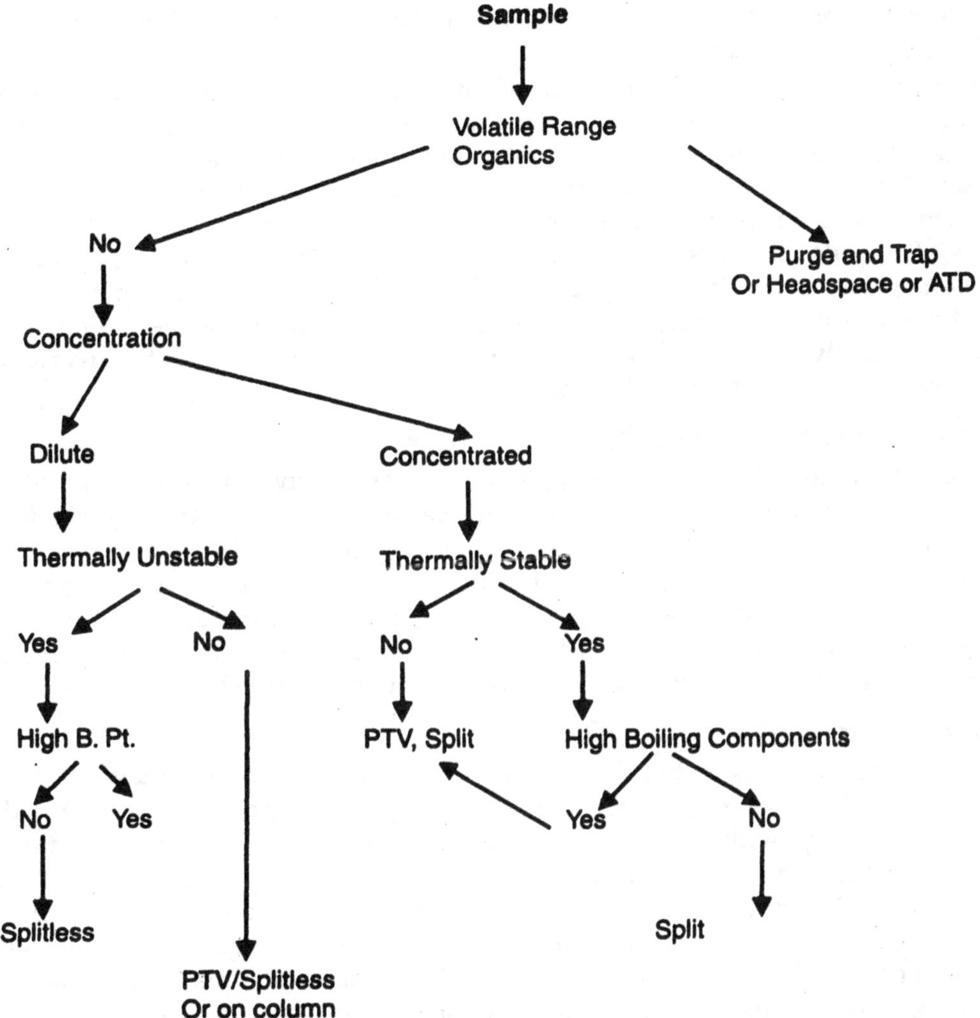

Fig. 58.1 : Flow chart for selecting the injection mode.

Automated Analytical Techniques

Automation of laboratory methods

The conventional bench level methods of analysis involve manual preparation of reagents, heating on hot plates, and colorimetry measurements. Extensive time and chemicals are consumed in analysis. Recently lot of automated methods have been introduced to improve analytical procedures and laboratory efficiency. Flow analysis has been developed to automate most of the wet chemical methods. Automation of manual procedures improves the sample throughput and sample reproducibility of the results. It also decreases the reagents and sample use, human exposure to toxic chemicals and cost per sample analyzed. In environmental analysis the greatest use of flow analyzers is in the analysis of water quality parameters such as nitrate, orthophosphate and ammonium.

Environmental laboratory adoption to such automated analysis saves time, improves sample throughput and cost of reagents.

Elemental analysis for all matrices of environmental samples such as biological, geological, botanical, aqueous, etc require dissolution of the matrix prior to instrumental analysis. Dissolution of these matrices usually involves acid treatment in an open or closed vessel heated by a hot plate or convection oven. The most common technique is the open beaker heating on a hot plate. The convection method is limited by factors like dissolution time, loss of volatile elements, contamination of sample by excessive amounts of reagents and airborne particulates. The changing role of an analyst demands switching over to some advanced measurement and analysis techniques for meeting the challenges of the millennium. Some of the recently developed techniques useful for automation with the instruments like Gas Chromatograph are discussed below.

Selection of Proper Injection Modes

The injection mode in chromatographic techniques depends upon the separation needs and the nature of the specific sample. The sample considerations include sample concentration, the presence of non-volatile contaminants, and the stability (thermal and otherwise) and volatility range of the analytes of interest. Various injection modes are now a days available making the technique automated saving time and chemicals used in extraction and concentration. These injection modes are discussed below. A sample based flow chart for selecting the injection mode is illustrated in Fig 58.1.

Atomatic Thermal Desorption (ATD)

With current emphasis on both environmental and quality control analyses, the modern laboratory faces demands for greatly increased sample throughput. The ATD is designed to meet this challenge with optimum thermal desorption performance and versatile, fully automatic operation. Thermal desorption is a clean, labour saving technique which simplifies a wide range of GC applications. Volatile and semi-volatile components can be desorbed directly from the sample matrices or from sorbent or cryogenic traps without any significant sample preparation. The technique has proved invaluable for monitoring environmental levels of organic pollutants and simplifies many routine determinations of volatile in products such as polymers, pharmaceuticals, foodstuffs and paints.

Head space Analysis

The technique, often referred as static headspace sampling is becoming increasingly popular as it eliminates many of the tedious and error-producing steps necessary in other sample introduction techniques. Since headspace sampling relies totally on volatilization to separate the compounds from the sample matrix, sample extraction, cleanup, and pre-concentration are not necessary. Samples can be sealed at the point of collection to reduce losses in handling and storage. Moreover, headspace sampling is readily automated so that 24-hour unattended operation is possible.

The HS employs a unique sampling technique - a pneumatic pressure-balanced system. The headspace sample is introduced onto the column without resorting to a gas syringe, thus avoiding fractionation due to pressure changes in the syringe. Since the needle is sealed, there is no loss of headspace gas during transfer. When using the pressure-balanced system no multi-port valves are required (multi-port valves are prone to adsorption problems and may introduce dead volumes). The fewer components in contact with a delicate sample the better.

Initially the sampling needle is in the upper position as it is in thermostatting position. Carrier gas passes solenoid valve V1 to the column and at the same time the needle cylinder is flushed. In the second stage the sampling needle moves downward to a lower position. This occurs at the end of the sample thermostatting time. Carrier gas passes through the solenoid valve V1 and branches to the column and the vial. The vial is now pressurized to the column head pressure. For sampling, solenoid valve V1 interrupts the carrier gas supply for the preset injection period. The vial is at the column head pressure and acts as a reservoir of carrier gas and delivers the sample to the column. At the end of the sampling period valve V1 is reopened and the normal carrier gas supply returns. The cycle is completed by the sampling needle returning to the original position.

Purge and Trap Process

'This method is useful for sample preparation and extraction of volatile organic compounds that have boiling points below 200°C and are insoluble or slightly soluble in water.

An inert gas is bubbled through the solution at ambient temperature, and the volatile components are efficiently transferred from the aqueous phase to the vapour phase. The vapour is swept through a sorbent column where the volatile components are adsorbed. After purging is completed, the sorbent column is heated and back flushed with inert gas to desorb the components on to a gas chromatographic column.

The Purge and Trap Device : The device consists of three separate pieces of equipment: the sample purger, the trap, and the desorber.

The purging chamber is designed to accept 5 ml samples with a water column at least 3 cm deep. The gaseous headspace between the water column and the trap must have a total volume of less than 15 ml. The purge gas must pass through the water column as finely divided bubbles with a diameter of less than 3 mm at the origin. The purge gas must be introduced no more than 5 mm from the base of the water column.

The trap must be at least 25 cm long and have an inside diameter of at least 0.105 in. The trap contains the adsorbents : 1/3 of 2,6-diphenylene oxide polymer, 1/3 of silica gel, and 1/3 of coconut charcoal. A 1 cm of methyl silicone-coated packing (OV-1 3% on chromosorb-w) be inserted at the inlet to extend the life of the trap.

Super Critical Fluid Extraction (SFE)

Another sample preparation technique of prominence in the millennium is the Super Critical Fluid Extraction (SFE) technique for preparing difficult samples rapidly, efficiently and without organic solvents. Excellent candidates for SFE include highly viscous, poorly soluble materials and heat sensitive analytes from matrices such as soils, sludges, foods and plastics. SFE employs as a solvent a highly compressed gas that is near to or above its critical temperature and critical pressure. At the critical temperature of a substance the vapour and the liquid have identical densities. A gas cannot be liquefied when it is above its critical temperature regardless of the pressure. Above the critical temperature the substance exists distinctly as a supercritical fluid. Carbon dioxide (CO_2) is the most widely used supercritical fluid. It is superior solvent over conventional liquid hydrocarbons such as hexane, methylene chloride for extraction because it is inert, non-toxic, non-inflammable, non-corrosive and environmental friendly. CO_2 has also shown to possess the excellent solubility characteristics for many target analytes of environmental significance. It will be very useful for extracting PAHs and PCBs from environmental samples. Dissolution of sample is a major time consuming factor, hence a new approach to direct heating of the sample is required. Microwave heating has provided an efficient and rapid procedure for dissolution of sample. It has provided a new capability for exploiting chemical reaction control in laboratory. The application of microwave assisted decomposition and extraction in analysing environmental samples is rapidly increasing. The technique is useful for sediment, sludges, soil and dust samples.

Conclusion

Environmental Analysis is an inter-disciplinary field involving careful observation, measurement, analysis and assessment of contamination of air, water and land components. The laboratory requirements for such a purpose vary from single room to large institutions involving scientific workers, supported by technical and administrative personnel.

The tasks for such environmental laboratories vary with the mandate of the organization. Research laboratories to probe frontiers of knowledge would have an flexibility to adjust itself to changing and emerging areas of environmental significance, whereas organizations primarily responsible for surveyor quality monitoring work to generate database would have different options. The requirements and functions also vary when the environmental laboratories operate as process control laboratories for maintenance of treatment plants or to provide testing or analytical services for developmental projects. The role of analyst has thus changed from using conventional to far advanced automatic chemistry instrumental methods for better results.

Acknowledgement

The authors are thankful to Dr. R.N. Singh, Director, NEERI for the encouragement and permission to publish the paper.

ENVIRONMENTAL BIOTECHNOLOGY

Edited By :   Professor (Dr.) Arvind Kumar

Published By : DAYA PUBLISHING HOUSE

59

MONITORING OF PESTICIDE (METHYL PARATHION) TOXICITY BY USING POLLEN AS INDICATORS - POLLEN OF *CATHARANTHUS ROSEUS*–A CRITICAL REVIEW

• S.A. Salgare and L.A. Yeragi

Department of Botany, Institute of Science, Mumbai, India

Abstract

Present paper deals with the effect of methyl parthion on the pollen germination and tube growth of successive flowers *i.e.*, F and F-24 series of pink and white-flowered cultivars of *Catharanthus roseus*. Except for the lowest concentration (10^{-20} mg/ml) of methyl parathion all the different concentrations (10^{-20}, 10^{-15}, 10^{-10}, 10^{-5}, 10^{-1} mg/ml) of methyl parathion tried proved to be toxic for the pollen germination of F-24 series of pink-flowered cultivar. However the pollen of F-24 series of white-flowered cultivar failed to germinate in sucrose medium supplemented by the highest concentration (10^{-1} mg/ml) of the pesticide. This proves that the pollen of the said series are highly sensitive and act as an ideal indicators of pollution. It should be pointed out that all the different concentrations of the pesticide tried inhibited the pollen germination and tube growth of either cultivars of *C. roseus* throughout the experiment.

Key Words : Pesticides, Palynology, Toxicology, Environmental Sciences.

Introduction

In spite of the very varied approach of study and the extensive work done, the large number of pesticides being developed in industry and used in agriculture stand only in testimony of the necessity of more work in the field.

Materials and Methods

Refer Salgare and Phunguskar (2001).

Results and Discussion

As a rule the percentage of pollen germination is always less than the pollen viability. However, Banerji and Gangulee (1937) and Dharurkar (1971) reported higher percentage of pollen germination than the pollen viability in *Eichhornia crassipes*. The claim of Banerji and Gangulee (1937) and Dharurkar (1971) is challendged by Salgare (2000b) who stated that the observation of Banerji and Gangulee (1937) and Dharurkar (1971) are exaggerating.

Johri and Chowdhury (1957) stated that in *Citrullus colocynthis*, where pollen grains 'mostly remained attached in tetrads' satisfactory germination is observed. Kwack (1965) stated that the pollen of *Mimosa pudica* does not germinate in the absence of calcium in the culture medium. However, Kwack's (1965) argument is challenged by Salgare (2000a). Salgare (2000a) observed the germination of pollen of *M. pudica* in the culture medium of glucose and sucrose in the absence of calcium. This proves the superficial and misleading observations of Kwack (1965).

It should be pointed out that except for the lowest concentration (10^{-20} mg/ml) of methyl parathion all the different concentrations (10^{-20}, 10^{-15}, 10^{-10}, 10^{-5}, 10^{-1} mg/ml) tried proved to be toxic for the pollen germination of F-24 series of pink-flowered cultivar. However the pollen of F-24 series of white-flowered cultivar failed to germinate in sucrose medium supplemented by the highest concentration (10^{-1} mg/ml) of the pesticide (Tables 59.1 and 59.2). This proves that the pollen of the said series are highly sensitive and acts as an ideal indicators of pollution (Tables 59.1 and 59.2). It should be pointed out that Salgare (2001),

Salgare and Phunguskar (2001) have already proved that the pollen is an ideal indicator of pollution. Methyl parathion inhibited the pollen germination and tube growth of either cultivars of *C. roseus* throughout the experiment (Tables 59.1 and 59.2).

It is interesting to note that even the lowest concentration (10^{-20} mg/ml) of methyl parathion tried inhibited the pollen germination as well as tube growth of *C. roseus* throughout the experiment (Tables 59.1 and 59.2). It also inhibited the rate of pollen germination and tube growth throughout the experiment. The pesticide also inhibited the germination of pollen and tube growth (Tables 59.1 and 59.2). This proves that the pollen of the said series are highly sensitive and act as an ideal indicator of pollution. Thus it is confirmed that pollen development and activity are more sensitive indicators of adverse factors in the botanical environment and the use of an entire vascular plant (Berg, 1973, Brandt, 1974, Vick and Bevan 1976, Rasmussan, 1977, Navara, Horvath and Kaleta 1978; Mhatre, 1980, Mhatre, Chaphekar, Ramani Rao, Patil, Haldar, 1980, Shetye, 1982 and Giridhar, 1984) as an indicator of pollution is a very crud method and

Table 59.1 : Effect of Methyl Parathion on Pollen Germinaton and Tube Growth of Successive Flowers of Pink-Flowered Cultivar of *Catharanthus roseus*

(Values given are mean ± SD of 100 for germination and ± SD of 20 for tube growth)

	% Pollen germination		Pollen tube growth in μm	
Conc.	F	F-24	F	F-24
10^{-20}	71.00±3.74	2.00±0.89	346.00±45.45	30.00±4.74
10^{-15}	61.00±8.60	ng	318.00±38.06	ng
10^{-10}	60.00±7.07	ng	305.00±36.81	ng
10^{-5}	58.00±6.78	ng	242.50±29.69	ng
10^{-1}	26.00±3.74	ng	185.00±11.18	ng
C	79.00±4.74	58.00±5.09	431.00±30.47	294.50±26.68

C, pollen germination and tube growth in sucrose medium (control); Conc, concentrations of methyl parthion in mg/ml; ng, no germination.

Table 59.2 : Effect of Methyl Parathion on Pollen Germination and Tube Growth of Successive Flowers of White-Flowered Cultivar of *Catharanthus roseus*

(Values given are mean ± SD of 100 for germination and ± SD of 20 for tube growth)

	% Pollen germination		Pollen tube growth in μm	
Conc.	F	F-24	F	F-24
10^{-20}	56.00±5.83	68.00±5.09	617.00±32.53	270.00±15.96
10^{-15}	39.00±3.74	55.00±7.07	330.00±26.57	204.00±14.72
10^{-10}	31.00±4.24	53.00±6.78	320.00±17.72	144.00±16.07
10^{-5}	19.00±3.74	12.00±4.00	208.00±14.08	126.00±15.32
10^{-1}	13.00±2.44	ng	180.00±14.27	ng
C	69.00±3.74	76.00±8.60	918.00±41.05	299.00±16.93

C, pollen germination and tube growth in sucrose medium (control); Conc, concentrations of methyl parthion in mg/ml; ng, no germination.

rather a wrong choice. There is no evidence of any entire vascular plant exhibiting this much degree of sensitivity. This is very clearly confirmed in the present critical review (Tables 59.1 and 59.2). It should be pointed out that Salgare (2001), Salgare and Phunguskar (2001) have already proved that the pollen is an ideal indicator of pollution.

Sudhakaran (1967) stated that in *Vinca rosea* besides pollen grains which produced single pollen tube, it has also been noticed that tetraploid grains frequently produce more than one pollen tube. Pollen tubes are branched quite frequently. Aberrations of this type in the pollen tube development are not observed in diploid pollen tubes, but quite frequently met with the pollen grains of irradiated plants. Extensive work of Salgare (1979) made it very clear that Sudhakaran (1967) had failed to trace out the branched pollen tubes and polysiphonous condition which is fairly common even in diploid pollen grains. Apart from this Sudhakaran (1967) was not able to report the various types of pollen tube deformities either with diploid or tetraploid grains.

References

Banerji, I. and Gangulee, H.C. 1937. Spermatogenesis in *Eichhornia crassipes* Solms. J. Indian Bot. Soc. 16 : 289-296.

Berg, H. 1973. Plants as indicators of air pollution. Toxicol. 1:79-89.

Brandt, C.C. 1974. Plants as indicators of air quality In : Indicators of Environmental quality. Eds Thomas, W.A. Plenum Press, New York p.p. 101-107.

Brewbaker, J.L. and Kwack, B.H. 1963. The essential role of Ca in pollen germination and pollen tube growth. Amer. J. Bot. 50 : 859-865.

Dharurkar, R.D. 1971. Effect of herbicides on the cytomorphology of *Eichhornia crassipes* Solam (Mart). Ph.D. thesis, Univ. Bombay.

Giridhar, B.A. 1984. Study of interactions between industrial air pollutants and plants. Ph.D. Thesis, Univ. Bombay.

Johri, B.M. and Chowdhury, Chhaya Roy. 1957. A contribution to the embryology of *Citraullus colocynthis* Schare. and *Melothria maderaspatana* Cong. New Phytol. 56 : 51-60.

Kwack, B.H. 1965. The effect of calcium on pollen germination. Proc. Amer. Soc. Hort. Sci. 86 : 818-823.

Mhatre, G.N. 1980. Studies in responses to heavy metals in industrial environment. Ph.D. Thesis, Univ. Bombay.

Mhatre, G.N., Chaphekar, S.B., Ramani Rao, I.V., Patil, M.R., and Haldar, B.C., 1980. Effect of industrial pollution on the Kalu river ecosystem. Environ. Pollut. Series A 23 : 67-78.

Navara, J., Horvath, I. and Kaleta, M. 1978. Contribution to the determination of limiting of values of SO_2 for vegetation in the region of Bratislava. Environ. Pollut. Series A 16 : 249-262.

Rasmussan, L. 1977. Epiphytic bryophytes as indicators of changes in the background levels of air borne metals from 1951 to 1975. Environ. Pollut. Series A 14 : 34-45.

Salgare, S.A. 1979. Reinvestigation of Palynology of *Vinca rosea* Linn. Proc. 66th Session of Indian Sci. Congr. Botany Section, Abstract No. 130.

Salgare, S.A. 2000a. Germination of pollen of the sensitive plant, *Mimosa pudica* Linn. In glucose. Bionotes 2:35.

Salgare, S.A. 2000b. A criticism on the findings of Banerji an Gangulee (1937) and Dharurkar (1971). Him. J. Environ. Zoo. 14:159-160.

Salgare, S.A. 2001. Monitoring of herbicide (nitrofen) toxicity by using pollen as Indicators–A critical review-I. Website www. microbiologyou.com. U.S.A.'s Publication.

Salgare, S.A. and Phunguskar, K.P. 2001. Monitoring of pesticide (endosulphan) toxicity by using pollen as indicators–Pollen of pink–flowered cultivar of *Catharanthus roseus* (L.) G. Don. - A Critical Review. Bioved 12 : 99-102.

Shetye, R.P. 1982. Effect of heavy metals on plants. Ph.D. Thesis, Univ. Bombay.

Sudhakaran, I.V. 1967. Cytogenetic studies in *Vinca rosea* Linn. Ph.D. Thesis, Univ. Bombay.

Trisa Palathingal. 1990. Evaluation of industrial pollution of Bombay by pollen-I. M. Phil. Thesis Univ. Bombay.

ENVIRONMENTAL BIOTECHNOLOGY

Edited By :   **Professor (Dr.) Arvind Kumar**

Published By : **DAYA PUBLISHING HOUSE**

60

EFFECT OF MULTIFLORAL HONEY ON BLOOD GLUCOSE PROFILE OF RABBITS AFTER *INDUCED LIPIDOSIS*

• *Shashikala*, D. Belsare and B. Jayanti*

* Biochemistry Laboratory, Department of Bioscience, Barkatullah University, Bhopal, India

Abstract

The saturated fat fed rabbits develop hypoglycemia and glucose intolerance. The animals tolerated blood glucose load after dietary administration of honey and did not develop hypoglycemia. It is suggested that honey contains glucose tolerance factor (GTF) that restores hypoglycemia to normal blood sugar level in saturated fat fed animals.

Key Words: Saturated Fat Diet, Hypoglycemia, Honey, Glucose Tolerance Factor.

Introduction

The rabbits given hydrogenated saturated fat in diet develop lipidosis (Belsare, 1980) as well as glucose intolerance during fat mobilization by ACTH (Belsare, 1981). The elevated level of plasma free fatty acids (FFA), which occur during it, may be secondary to the increased utilization of lipid, a metabolic condition that diminishes peripheral oxidation of glucose. Treatment with honey prevents hepatocellular damage (Lutomski, 1987) which is due to bioelements (Dobrowolski, 1987) present in honey. More recent studies indicate that sugars present in honey do not enter cells by osmosis, but require 'active carrier' to move them across the membrane barrier which means less of a 'rush' of sugar to the body with honey (Ladas and Rapitis, 1999). In the present investigation attempts have been made to study the effect of honey on blood glucose profile during dietary-induced lipidosis (fatty liver) in rabbits.

Materials and Methods

Albino rabbits of about 1 to 1.5 kg/BW were used in the present experiment. Eighteen animals were divided in 3 groups of 6 each and were used for blood glucose profile. Six animals of group 1 were maintained on standard diet and six of group 2 was given saturated fat diet (casein-vitamin free 20%, starch 10%, sucrose 28%, hydrogenated saturated fat 10%, Brewer's yeast dried 10%, 10 units of vitamin A and 1 unit of vitamin D per g of diet). Six animals were given saturated fat diet with honey (1.5g/kg). The animals were kept in cages individually with free access to water *ad lib*.

The peripheral blood was drawn from marginal ear vein for glucose estimation according to Nelson (1944) and Somogui's (1945) methods. At the end of experiment the animals were fasted overnight and a loading dose of glucose (0.5 g/kg) was administered by intravenous route as reported earlier (Belsare, 1981). The honey used in these experiment was collected from natural beehives of Rock-bee - *Apis dorsata* from the Pachmarhi forest. The saturated fat was purchased from Hindustan Lever Co., India. Administration of quantity of honey in diet of animals (per kg BW) was calculated from dietary requirement for human (100g/60 kg /day). Significance of data was determined by Student's t-test.

Observations

One week after treatment the blood glucose level in control animals was 98 mg/dl and showed marginal increase six weeks after the treatment, but in saturated fat fed animals it showed significant decrease in its level as compared to the initial control one (Table 60.1). The dietary treatment with honey with saturated fat diet for six weeks resulted in restoration of the blood glucose level to its normal one.

After loading oral dose of glucose, the blood glucose level was significantly increased in control as well as in honey treated animals within one hour, but it reached the initial level three hours after treatment in controls (Fig 60.1). However, in the saturated fat-fed animals, it was also increased within one hour, but reached initial level five hours after giving a loading dose of glucose. In honey and saturated fat fed diet, the animals tolerated loading dose of glucose like those, which were maintained on normal diet.

Discussion

There is tendency to hypoglycemia in saturated fat-fed animals which is due to decrease in oxaloacetate and thus inhibits gluconeogenesis (Belsare, 1981). The honey treatment restores this hypoglycemia to normal condition, because honey enters the cells directly with the help of 'active factor'. This factor was postulated as glucose tolerence factor of which trivalent chromium was identified as its active ingredient (Yamamoto *et. al.*, 1989). Lefavi *et al.* (1999) reviewed the literature and concluded that chromium plays a necessary role in the normal physiological system by acting as a "cofactor" for insulin and amplify its action as well as the efficiency of insulin's effects on glucose, amino acid and fatty acid flux into cell with subsequent glycogen, protein and triglyceride synthesis which is dependent on the maintenance of adequate chromium store. Although the exact structure of GTF and its exact location is still unkown (Anderson, 1987), a nicotinic acid-chromium is essential in the GTF complex (Lefavi *et al.*, 1999). Honey contains nicotinic acid, glycine, glutamic acid and cysteine (Zelenski and Kossan, 1985), which are essential for the synthesis of GTF complex. There are several reports that administration of honey in animals and human restores hypoglycemia to normal state (Genter and Ipp, 1994).

References

Anderson, R.A. 1987. In Trace Elements and Human Nutrition. W. Mertz (Ed.) New York, Academic Press., p.225.

Belsare, S.D. 1980. Chronic effects of some drugs on histopathology and lipid content of liver in rabbits. Z. Tierphysiol. *Tierernahrg* u. Futtermittelkde. 45: 245-250.

Belsare, S.D. 1981. Studies on glucose tolerance during fat mobilization: Infleuence of heparin and salicylate. J. Expt. Biol. 19, 88-89.

Dobrowolski, J. W. 1987. Bee made products, environmental protection, prophylaxis of diseases caused by deficiency of elements. Apipol. 8: 4-11.

Genter, P. and Ipp, E. 1994. Metabolism 43: 98

Ladas, S.D. and Rapitis, S.A. 1999. Honey, fructose absorption and the laxative effect. Nutrition 15 (7-8): 591-592.

Lefavi, R.G., Anderson, R.A., Keith, R.F., Wilson, G.D., McMillan, J.L. and Stone, M.H. 1992. Efficacy of chromium supplementation in athletes. Internat J. Sport. Nutrit. 2: 111-122

Lutomski, J. 1987. Opinion about the sense of production of herb honey nourishment. Apipol. 8 : 14-15

Nelson, N. 1944. J. Biol. Chern. 153: 375.

Somogui, M.J. 1945. J. Biol. Chern. 160,42.

Yamamoto, A., Wada, O. and Manabe, S. 1989. Evidance that chromium is an essential factor for biological activity of a low molecular weight, chromium-binding substance. Biochem. Biophys. Res. Comm. 163:189-103

Zalenski, W. and Kossan, R. 1985. Aminoacids and protein content in bee-bread and corbicular pollen collected from several places in Poland. International Symposium Apitherapy. Krakow, Poland; Abstract. pp.46.

ENVIRONMENTAL BIOTECHNOLOGY

Edited By :   Professor (Dr.) Arvind Kumar

Published By :   DAYA PUBLISHING HOUSE

61

BIOSORPTION AND SUBSEQUENT BIOAVAILABILITY OF LEAD AND COPPER THROUGH SOIL COLUMN INOCULATED WITH RHIZOSPHERE MICROBIAL ISOLATES

• *A. Bharani, S. Mahimai Raja and D. Augustine Selvaseelan*
Department of Environmental Sciences, Tamil Nadu Agricultural
University, Coimbatore.

Abstract

The downward movement of Pb and Cu through a soil column under the influence of rhizosphere microbial consortia was evaluated. Mobility of Pb was reduced in the rhizosphere microbial consortia added soil column. A maximum of 72.8% was sorbed in the soil inoculated with microbial consortia which subsequently reduced the mobility of Pb in the leachate collected that was evident from the lower concentration of Pb in the leachate. In the colum that received no microbial consortia, a maximum of 31.8% of Pb in the leachate was only observed. Biosorption potential of Cu was only 6.95% at 50 mg kg^{-1} concentration in the soil inoculated with microbial consortia and the corresponding leachate showed 35.48% of Cu. The mobility of Cu was maximum as evidenced in the leachate with 38.8%. Hence, it is evident that Cu is not sorbed as that of Pb by microbial consortia.

Key Words : Lead Copper, Mobility, Leachate, Biosorption.

Introduction

For most of the heavy metals of concern, the experimental evidence in the field suggests that there is relatively little movement of sludge applied below the surface soil, ever over periods of several decades (Alloway and Jackson, 1991). In general, however, unless soils are coarse textured or very acidic, the mobile fraction of most sludge applied metals is small, at least as long as organic matter from the biosolid persists (McBride, 1995). The mobility of heavy metals is governed by a series of chemical and biological mechanisms. It is therefore essential to study the behaviour of heavy metals in soil to characterize the mechanism of soil pollution and arrive at suitable land disposal system for sewage biosolid. An understanding of distribution of heavy metals between the solid and solution phase in various soils that differ in their physico-chemical characteristics is desirable in establishing loading limits for each soil to ensure the contamination to the food chain is regulated.

Many micro-organisms are known to accumulate metals from their environment. Recently, the research interest has been on the use of biomass as an absorbent material to remove toxic metals for waste. Biomass from a vriety of organisms including fungi, algae, bacteria have been evaluated for this purpose. Continuously growing organisms could be used for the metal removal and control of toxicity of metals (Gunasekharan, 1999). Hence, an investigation was carried out to study the mobility of Pb and Cu under the influence of microbial consortia.

Materials and Methods

The downward movement of Pb and Cu through a soil column under the influence of rhizosphere microbial consortia was evaluated. About 500 g of red soil (previously dried and sieved) was used in the column with 1.5 foot height and 6 cm diameter connected with a polypropylene tube at the bottom. To the column, first coarse gravel was added to the base then the soil and the microbial consortia were added simultaneously such that contents are moistened. The soil columns along with microbial consortia consisting of *Baccilus* sp. @ 10^7 cells per gram of soil and *Aspergillus niger* and *Penicillium* sp. @ 10^4 cells per gram of soil were kept for three days as such for the micro-organisms to stabilise. On the 4th day, the known concentration of Pb and Cu (50 mg l^{-1}) solution were added to columns. This set up was kept for another three days for stabilization. The weight of the column was recorded and maintained at a constant. On the 7th day the column weight was noted and to maintain the original weight water was added. The leachate then was collected from 7th day onwards adding a known volume of water to each column. Leachate was collected for upto 10 days from each column. On the 10th day, the soil from the column were removed and homogenized, dried, digested and analysed for Pb and Cu. The leachate collected were also analysed for Pb and Cu.

Results and Discussion

Biosorption and Mobility of Lead

The column with soil, 50 mg l^{-1} of Pb and rhizosphere microbial consortia (C_1) recorded very low concentrations of Pb in the leachate collected from 7th to 10th day (0.63 to 0.83 mg l^{-1}). Control samples (C_0) recorded very low concentrations of Pb, which received no metal and microbial consortia. In the column C_2, which received only 50 mg l^{-1} of Pb, the concentrations in the leachate were 2.62, 4.18, 11.38 and 15.98 mg l^{-1} on the 7th, 8th, 9th and 10th day respectively. While the residual soil of the column, after the 10th day on analysis recorded 11.62 mg l^{-1} of Pb in column C, which received 50 mg l^{-} of metal concentration and rhizosphere microbial consortia and column that received only 50 mg l^{-1} metal concentration recorded 36.42 mg l^{-1} of Pb.

The maximum biosorption potential of the rhizosphere microbial consortia with 50 mg l^{-1} of Pb at the column showed an increasing trend in Pb concentration from 7th to 10th day. Both the quantity of Pb in the leachate was low as compared to the soil. The reduction in the bioavailability of Pb in the leachate is due to biosorption by rhizosphere microbial consortia enriched in the soil. These organisms were able to biosorb as much Pb as possible in their cells and immobilize them thus reducing the concentration in the leachate. Under the absence of the microbial consortia (at- column C_2), with 50 mg l^{-1} of Pb, the biosorption potential registered 23.2 per cent (Table 61.1) which is lesser than the leachate concentration. This could be due to the absence of microbial consortia. The metal is easily leached out. The bioavailability of Pb in the leachate contributed for 67 per cent. This was in agreement with Gerritse *et al.* (1982) who reported that micro-organisms are being increasingly seen as important components of models designed to predict metal mobility since they can significantly immobilize and mineralize soluble metals in the environment and the immobilized metal are difficult to remobilize. The heavy metals are retained on and in the biomass by the process of adsorption and precipitation and are released slowly as the biomass gradually diminishes on further microbial cycling.

Table 61.1 : Biosorption and Subsequent Bioavailability of Lead and Copper Through Soil Column Inoculated with Rhizosphere Microbial Isolates–Leachate and Soil Analysis

| Particulars | Lead (mg kg^{-1}) | | | | Copper (mg kg^{-1}) | | | |
| | Days | | | | Days | | | |
	7th	8th	9th	10th	7th	8th	9th	10th
Leachate					Leachate			
C_0	0.001	0.002	0.001	0.001	0.004	0.005	0.004	0.004
C_1	0.625	0.710	0.735	0.830	3.464	7.892	11.668	17.738
C_2	2.620	4.185	11.384	15.946	5.982	8.674	14.525	19.412
Column soil					Column soil			
C_0		0.002				0.034		
C_1		36.423				3.468		
C_2		11.625				0.964		

C_0 - Control (without metal and microbial consortia);

C_1 - Metal + Microbial Consortia;

C_2 - Metal Alone.

Biosorption and Mobility of Copper

Copper at 50 mg l^{-1} concentration with rhizosphere microbial consortia inoculation in the column C_1 recorded 3.46, 7.89, 11.67 and 17.74 mg l^{-1} of Cu on the 7th, 8th, 9th and 10th day leachate samples. While the column (C_2) that received only 50 mg l^{-1} of metal concentration without rhizosphere microbial consortia recorded higher concentration of Cu in the leachate collected from 7th, 8th, 9th and 10th day viz. 5.98, 8.67, 14.52 and 19.41 mgl^{-1} of Cu. The residual soil from the column after 10th day, on analysis, showed that C_1 recorded 3.47 mg l^{-1} of Cu and C_2 with 0.96 mg l^{-1} of Cu. Control column (C_0) registered very low Cu which is negligible.

Biosorption potential was observed to be lesser (6.94 per cent) in case of the soil packed in the column that received the microbial consortia. The corresponding leachate samples recorded 88 per cent (Table 61.1) of Cu. Though the microbes are involved, Cu was estimated at a higher rate in the leachate. The organisms could not accumulate much of Cu because of the highly mobile nature of the metal. Similar trend was observed in the column C_2 where no microbial consortia was added. The leachate sample accumulated 97 per cent of the total applied Cu (50 mg kg^{-1}). Their corresponding soil showed a sorptive potential of 1.93 per cent only. Because of the high mobility of Cu, the microbes were not able to sorb the metal in their cells. In general, however, the presence of potential toxic elements in sludge-treated soil was confined to the cultivation zone with very little movement below. Detailed analyses of soil profile samples from the long-term Woburn field experiment by

McGrath and Lane (1989) showed that approximately 1 per cent of the metals applied had moved 3.5 cm below the plough layer (0-23 cm) or less, but there was no evidence of accumulation of metals in deeper horizons down to 46 cm.

Conclusions

Mobility of Pb was reduced in the rhizosphere microbial consortia inoculated soil column. A maximum of 72.8 per cent was sorbed in the soil inoculated with microbial consortia. This higher biosorption subsequently reduced the mobility of Pb which was evident from the lower concentration of Pb in the leachate. In the column that received no microbial consortia, a maximum of 31.8 per cent of Pb was recorded as evidenced by the presence of 31.8 per cent of Pb in the leachate.

With respect to copper, the biosorption potential was only 6.95 per cent at 50 mg kg^{-1} concentration in the soil inoculated with microbial consortia and the corresponding leachate has 35.48 per cent of Cu. In column where no microbial consortia was inoculated, the mobility of Cu was maximum as evidenced by the presence of 38.8 per cent in the leachate. Hence, it is evident that Copper is not biosorbed as that of Pb by the microbia consortia.

Acknowledgement

The financial assistance rendered by the Jawahar Lal Nehru Memorial Fund through student followship to carry out this study is greatly acknowledged.

References

Alloway, B.J. and Jackson, A.P. 1991. The behaviour of heavy metals in sewage-sludge amended soils. Sci. Total Environ., 100 : 151-176.

Gerritse, R.G., Vriesema, R. Dalenberg, J.W. and Roos, H.P. de 1982. Effect of sewage sludge on trace metal mobility in soils. J. Environ. Qual., 11 : 359-364.

Gunasekaran, P. 1999. Microbial bioremediation of heavy metal toxicity - A molecular approach. In : Proceedings of Second International Conference on contaminants in the soil environment in the Australasia-Pacific Region, New Delhi. pp. 94-95.

McBride, M.B. 1995. Toxic metal accumulation from agricultural use of sludge : Are USEPA regulations protective? J. Environ. Qual. 24 : 5-18.

McGrath, S.P. and Zanc, P.W. 1989. An explanation for the apparent losses of metals in a long-term field experiment with sewage sludge. Environemntal Pollution. 60 : 235-256.

ENVIRONMENTAL BIOTECHNOLOGY

Edited By :   **Professor (Dr.) Arvind Kumar**

Published By :   **DAYA PUBLISHING HOUSE**

62

PHYTOEXTRACTION OF LEAD AND COPPER BY FINGER MILLET AND SORGHUM

• *A. Bharani, S. Mahimai Raja and D. Augustine Selvaseelan*

Department of Environmental Sciences, Tamil Nadu Agricultrual
University Coimbatore

Abstract

The influence of rhizosphere microbial isolates on phytoextractability of Pb and Cu by finger millet and sorghum was studied through a short-term seedling technique. The phytoextractability of Pb by the seedlings of finger millet and sorghum under the influence of microbial consortia was 30.8 and 32.8 per cent respectively. The seedling of both the minor millets recorded a higher Pb and Cu extraction potential without microbial consortia. In sorghum, the phytoavailability of Cu in the seedling biomass was 15.3 per cent under the influence of rhizosphere microbial consortia. The seedling biomass was able to phytoextract 54.18 per cent of the total applied Pb concentration at 25 mg kg^{-1} and its corresponding soil bioavailable Pb registered 40.8 per cent under the influence of microbial consortia.

Key Words : Phytoextraction, Lead, Copper, Finger millet, Sorghum.

Introduction

The effects of heavy metals on soil microorganisms and microbial processes are complex-and the published data concerning the impacts of sludge applications on soil fertility are frequently contradictory (Smith, 1991). This is explained principally because of the differences in reported experimental conditions particularly in relation to time-scales and the form and rate of applied metals. A further complication is that most studies generally makes assessments based on measurements of total metal concentrations in soil. However, the bioavailability of metals and the toxicity to soil micro-organisms is strongly influenced-by differences in soil physico-chemical properties such as pH and texture. For example, metal toxicity may be reduced in soils with high clay and organic matter contents (Chander and Brookes, 1991) due to the lower availability of metals to soil micro-organisms (Tyler, 1981, Baath, 1989). Another possibility is that the soil microbial population is highly adaptive to changing soil conditions and can develop resistance to toxic heavy metals.

Walton and Todd (1990) examined the possibility wherein the vegetation may be used to actively promote microbial restoration of chemically contaminated soils. They tested it by using rhizosphere and non-vegetated soils collected from a trichloroethylene (TCE) contaminated field site. The results showed that microbial activity is greater in rhizosphere soils and that TCE degradations occur faster in the rhizosphere than in the odaphosphere. Hence, an attempt was made to study the activity of microbes under the influence of heavy metals to minor millets.

Materials and Methods

To assess the influence of rhizosphere isolates on Pb and Cu availability to minor millets (sorghum and Finger millet), a short-term seedling technique was carried out. About 500 g of red soil was taken in Aluminium trays of 1 kg capacity. Microbial consortia consisted of mixed inoculum of *Bacillus* sp. (@ 10^7 cells per gram of soil) and *Aspergillus niger + Penicillium* sp. (@ 10^4 cells per gram of soil) were applied. Twenty-five seeds of sorghum (var. Co S 9) and finger millet (var Co 13) were sown. The following test crop Sorghum (var Co S9) and finger millet (var Co 13) were used for the study.

Treatments replicated twice were as follows

T_1 - Control (50 mg kg^{-1} of the metal + soil)

T_2 - 25 mg kg^{-1} of the metal (microbial consortia + soil)

T_3 - 50 mg kg^{-1} of the metal (microbial consortia + soil)

Two sets were maintained, one each for Pb and Cu. The seeds were sown and watered daily. The duration of the study was 17 days. After the seventeenth day, the seedlings were uprooted and preserved for the analysis. The soils were also analysed for total Pb and Cu. Microbial consortia consisted of mixed inoculum of *Bacillus* sp. (@ 10^7 cells per gram of soil) and *Aspergillus niger + Penicillium* sp. (@ 10^4 cells per gram of soil) were applied.

Results and Discussion

Phytoextraction of Lead

The 25 mg kg^{-1} Pb concentration in the treatment T_2 with microbial consortia recorded 8.32 mg kg^{-1} in the seedlings of finger millet and 10.68 mg kg^{-1} in sorghum biomass. The corresponding residual soil samples recorded higher concentraion of Pb than the seedlings *viz.*, 12.41 mg kg^{-1} under finger millet and 13.58 mg kg^{-1} under sorghum.

With regard to the treatment T_3 (50 mg kg^{-1} of Pb with rhizosphere microbial consortia), the finger millet seedlings accumulated 15.44 mg kg^{-1} of Pb and in Sorghum the seedlings accumulated 16.38 mg kg^{-1} of Pb. The soil samples of the treatment T_3 was found to accumulate higher concentration of Pb than the seedlings in both the test crops.

The control treatment (T_1) which received only 50 mg kg^{-1} of Pb without the microbial consortia, showed that the seedlings and soil samples recorded higher Pb content when compared to T_2 and T_3 under both the test crops.

Phytoextraction by finger millet seedling biomass was 33.28 per cent at 25 mg kg^{-1} of Pb concentration, which included the rhizosphere microbial consortia. The corresponding soil bioavailability of Pb was 49.62 per cent. When the concentration of Pb was increased to 50 mg kg^1, the bioavailability or biosorption potential of the rhizosphere microbial consortia in the soil declined (40.6 per cent). Also, the pH content in the seedling biomass decreased (30.8 per cent). In the control, the seedling biomass showed a marginal increase in the bioavailability of Pb (50 mg kg^{-1}). This difference is because of the activity of the microbial consortia in the treatment T_2 and T_3. The microbial consortia was able to biosorb the applied Pb and reduce the availability to the seedlings.

With regard to sorghum, the seedling biomass was able to phytoextract 54.18 per cent of the applied Pb concentration at 25 mg kg⁻¹ and its corresponding soil biovailable Pb registered 40.8 per cent under the influence of microbial consortia. When the concentration of Pb increased (50 mg kg⁻¹) the phytoextractable Pb in the seedling biomass registered 32.8 per cent under the influence of microbial consortia whereas, the seedling biomass of sorghum in the control (50 mg kg⁻¹ Pb alone) has 35.8 per cent. The bioavailable Pb of the corresponding soils recorded 45.2 (with microbial consortia) and 29.2 per cent (without microbial consortia). Boyle and Shann (1995) investigated biodegradation in rhizosphere soil collected from field grown plants, grouped for analysis as monocots or dicots. Microbial activity was highest in monocot rhizosphere soils followed by dicot rhizosphere soils and finally non-rhizosphere soils.

Phytoextraction of Copper

The 25 mg kg⁻¹ treatment which received the rhizosphere microbial consortia recorded 9.10 mg kg⁻¹ and 10.20 mg kg⁻¹ in the seedling biomass of finger millet and sorghum. The corresponding soil sample recorded a higher concentration of Cu (9.71 and 11.86 mg kg⁻¹). As regards the treatment T₃ (50 mg kg⁻¹ of Cu with rhizosphere microbial consortia), the finger millet seedlings accumulated 23.53 mg kg⁻¹ of Cu, while sorghum biomass recorded 26.63 mg kg⁻¹ of Cu respectively. But the corresponding soil samples exhibited a higher Cu concentration of 15.57 and 17.46 mg kg⁻¹ in finger millet and sorghum respectively.

The control treatment (T₁) where no microbial consortia was added, recorded a higher concentration of Cu than T₂ but concentration was lesser than T₃.

As regards Copper, when compared to control (where no microbial consortia was inoculated to soil with 50 mg kg⁻¹ concentration) phytoextractable Cu by the seedlings of finger millet registered 25.5 per cent and the corresonding soil bioavailable Cu had 22.5 per cent. When the microbial consortia was inoculated to soil with 50 mg kg⁻¹ of Cu, the phytoextractable Cu by the seedling biomass was 47 per cent and that retained by the soil was 31.1 per cent of Cu.

In sorghum, the seedling biomass of the control where no microbial consortia was inoculated that received 50 mg kg⁻¹ of copper, registered 27.8 per cent of phytoextractable Cu, its corresponding soil retained 21.1 per cent of Cu. The phytoextractable Cu of seedling biomass was 53.3 per cent under the influence of microbial consortia at 50 mg kg⁻¹ concentration of Cu. The corresponding soil bioavailable Cu retained was 34.9 per cent.

Conclusions

Phytoextractability of Lead by the seedling biomass of finger millet was 30.8 per cent under the influence of microbial

Table 62.1 : Phytoextraction of Lead and Copper by Finger Millet and Sorghum Under the Influence of Rhizosphere Microbes

Treamen	Lead (mg kg⁻¹)		Cooper (mg kg⁻¹)	
	Finger millet	Sorghum	Finger millet	Sorghum
Seedlings				
T₁	17.2	17.94	12.74	13.93
T₂	8.31	10.68	9.10	10.20
T₃	15.44	16.38	23.53	26.63
Soil				
T₁	16.19	14.59	11.24	10.81
T₂	12.41	13.55	9.71	11.86
T₃	20.31	22.57	15.56	17.46

T₁ - 50 mg 1⁻¹ (without microbial consortia);
T₂ - 25 mg 1⁻¹ (with microbial consortia);
T₃ - 50 mg 1⁻¹ (with microbial consortia).

consortia at 50 mg kg^{-1} of Pb concentration. The corresponding soil retained 40.6 per cent of Pb applied. It is evident that the seedlings are able to extract only 30.8 per cent from the soil because of the presence of microbial consortia in the soil.

With respect to sorghum, where rhizosphere microbial consortia was inoculated, the seedling biomass was able to phytoextract a maximum of 32.7 per cent at 50 mg kg^{-1} concentration of Pb but the corresponding soil had 45.2 per cent of bioavailable Pb. Hence, sorghum is able to phytoextract more than finger millet.

The seedling biomass of finger millet was able to phytoextract a maximum of 47.1 per cent of Cu under the influence of rhizosphere microbial consortia at 50 mg kg^{-1} concentration. But the corresponding soil retained 31 per cent of Cu. This indicated that the microbial isolates inoculated were able to biosorb maximum of added Cu thus reducing the phytoavailability of Cu to finger millet.

In sorghum, the phytoavailability of Cu in the seedling biomass was 15.3 per cent under the influence of rhizosphere microbial consortia. The corresponding soil retained 34 per cent of Copper.

Acknowledgement

The financial assistance rendered by the Jawahar Lal Nehru Memorial Fund through student fellowship is greatly acknowledged.

References

Baath, E. 1989. Effects of heavy metals in soil on microbial processes and populations (a review). Water, Air and Soil Pollution. 47 : 335-379.

Chander, K. and Brookes, P.C. 1991. Effects of heavy metals from past applications of sewage sludge on microbial biomass and organic matter accumulation in a sandy loam and a silty loam UK soil. Soil Biol. Biochem. 23 : 927-932.

Smith, S.R. 1991. Effect of sewage sludge application in soil microbial processes and soil fertility. Advances in Soil Science. 16 : 191-212.

Tylor, G. 1981. Heavy metals in soil biology and biochemistry. In : Soil Biochemistry, Paul, E.A. and J.N. Ladd (Eds.). Volume 5. Marcel Dekker, Inc., New York, pp. 371-414.

Walton, T.B. and Anderson, A. Todd. 1990. Microbial degradation of Trichlorocthylene in the rhizosphere : Potential application to biological remediation on waste sites. Appl. Environ. Microbiol. 56 : 1012-1016.

ENVIRONMENTAL BIOTECHNOLOGY

Edited By :   Professor (Dr.) Arvind Kumar

Published By : DAYA PUBLISHING HOUSE

63

IMPACT OF SEWAGE BIOSOLID ON SUSTAINING SOIL HEALTH AND PHYTOAVAILABILITY OF HEAVY METALS AS INFLUENCED BY AMELIORANTS

• *A. Bharani, S. Mahimai Raja and D. Augustine Selvaseelan*

Department of Environmental Sciences,
Tamil Nadu Agricultural University, Coimbatore

Abstract

The biosolid application, with the addition of ameliorants, to red and black soils registered a decreasing pH as the stages advanced. EC recorded 17.7 per cent increase and organic carbon content increased by 64.5 per cent under the influence of gypsum + composted coirpith. Gypsum and composted coirpith in combination reduced the phytoavailable Pb and Cu by 58.5 and 13.3 per cent in black soil. Phytoavailable Cr was not influenced by the addition of any of the ameliorant tested. Rock phosphate + composted coirpith was able to decrease phytoavailable Ni by 8.9 per cent.

Keywords : Phytovailable, heavy metals, ameliorants, impact soil health

Introduction

For most of the heavy metals of concern, the experimental evidence in the field suggests that there is relatively little movement of sludge applied below the surface soil, ever over periods of several decades (Alloway and Jackson, 1991). In general, however, unless soils are coarse textured or very acidic, the mobile fraction of most sludge applied metals is small, at least as long as organic matter from the biosolid persists (McBride, 1995). The mobility of heavy metals is governed by a sereis of chemical and biological mechanisms. It is therefore essential to study the behaviour of heavy metals in soil to characterize the mechanism of soil pollution and arrive at a suitable land disposal system for sewage biosolid. An understanding of distribution of heavy metals between the solid and solution phase in various soils that differ in their physico-chemical characteristics is desirable in establishing loading limits for each soil to ensure the contamination to the chain is regulated. The present investigation has been carried out to evaluate the impact of sewage biosolid applicaton on the bioavailability of heavy metals as influenced by bio and chemical ameliorants in two different soil types (red and black).

Materials and Methods

Application of sewage biosolid on land might lead to contamination of the land. Hence, to understand the long-term effect of sewage biosolid application, an incubation experiment was conducted. Sewage biosolid collected were dried and seived. Two different soil types were chosen for the study *viz.*, black soil and red soil having the water holding capacity of 35 and 25% respectively. About 500 g of the soil (previously dried and sieved) were weighed into plastic containers. Sewage biosolid at the rate of 50 t ha^{-1} of soil was added. The following were the treatments with three replications.

T_1 - Absolute Control (Soil alone without sewage biosolid and ameliorant)

T_2 - Control (With biosolid and without ameliorant)

T_3 - Biosolid + Potash alum (@ 500 kg ha^{-1})

T_4 - Biosolid + Gypsum (@ 5 t ha^{-1})

T_5 - Biosolid + Composted Coirp i th (@ 2 t ha^{-1})

T_6 - Biosolid + Rockphosphate (@ 1t ha^{-1})

T_7 - Biosolid + Gypsum + Composed Coirpith

T_8 - Biosolid + Composted Coirpith + Rockphosphate.

The soil and sludge along with ameliorants were mixed thoroughly moistened at water holding capacity. Samples were drawn at ever fortnight representing four stages.

After each sampling the weight of the plastic containers were recorded. Samples drawn were analysed for total (aqua regia extract) and phytoavailable (DTPA extractable) heavy metals *viz.*, Pb, Cu, Cr, Ni and Mn.

Results and Discussion

Phytoavailable Pb (Table 63.1).

Sewage biosolid application to farm land over a period of time would alter the soil physical and chemical status. Availability of the nutrients to the soil may be enhanced by the addition of sewage sludge. Likewise the heavy metal content also would be brought to the available pool, as the phytoavailable fraction, thus being released into the soil. This may differ in different soil types. Among the treatments, potash alum added treatment had increased the phytoavailable Pb in red soil by 17.3 per cent and the rock phosphate aded treatment increased the phytoavailable Pb by 69.4 per cent. While the other treatments have reduced the phytoavailable Pb. The reason might be due to complexation and precipitation of Pb. Gypsum added treatment (T_4) showed lower phytoavailable Pb. The calcium sulphate might have precipitated the Pb content of the sewage biosolid amended soil and made it immobile. As regards the black soil, gypsum along with composted coirpith recorded 12.8 per cent of the phytoavailable Pb than control. This was in agreement with Davis (1984) where he reported that Pb, Hg and Cr have very low phytoavailability in sludge treated soil and are likely to have detrimental effects on soil quality.

Phytoavailable Copper (Table 63.2)

The maximum of DTPA extractable Cu was 3.742 and 3.143 mg kg^{-1} in red and black soil respectively. The next higher value was recorded at T_4 (sewage biosolids + gypsum) and T_6 (sewage biosoild + rockphosphate) were 2.602 mg kg^{-1} and 2.402 mg kg^{-1} respectively. Combination of rock phosphate and CCP has shown an increase of phytoavailabe Cu in both red and black soils (79.8 per cent and 83.4 per cent) over control. The gypsum added treatment in combination with CCP has reduced the phytoavailable Cu by 13.3 per cent in black soil and 19.2 per cent in red soil. Calcium sulphate could have combined with copper and precipitated it restricting the entry of Cu into the labile pool. The rock phosphate added increased the phytoavailable Cu whereas gypsum reduced the Cu that is phytoavailable. Addition of ameliorants, like gypsum would redue the Cu in sewage biosolid amended soils. The reduction in Cu availability was attributed to the fact that the proportion bound to low molecular weight species in soil solution decreased whereas that bound to high molecular weight species in unavailable form increased in soils containing 200 and 400 mg Cu kg^{-1} (Berrow *et al.*, 1990).

Table 63.1 : Phytoavailable Lead (mg kg^{-1}) from Sewage Biosolids Applied Soils as Influenced by Aeliorants

Treatments	Red Soil					Black Soil				
	S_1	S_2	S_3	S_4	Mean	S_1	S_2	S_3	S_4	Mean
T_1	0.70	1.02	1.08	1.24	1.01	0.59	1.00	1.67	1.93	1.29
T_2	0.97	1.42	1.62	1.73	1.44	0.80	1.32	2.27	2.45	1.71
T_3	0.88	1.67	1.90	2.31	1.69	0.94	1.16	2.18	2.41	1.67
T_4	0.50	1.24	1.57	1.70	1.24	1.00	1.47	2.13	2.21	1.71
T_5	0.47	1.45	1.72	1.87	1.38	1.11	1.18	2.41	2.53	1.81
T_6	0.67	1.23	2.13	2.55	1.64	0.77	1.17	1.56	1.73	1.31
T_7	0.63	1.33	1.59	2.16	1.43	1.32	1.71	2.25	2.45	1.93
T_8	0.66	1.41	1.93	2.39	1.59	1.23	1.67	1.58	2.36	1.71
Mean	0.68	1.35	1.69	1.99	1.43	0.97	1.34	2.01	2.26	1.64

	SEd	CD(0.05)					SEd	CD(0.05)	
Stages	0.06	0.12					0.07	0.11	
Treatments	0.08	0.17					0.09	0.12	
Stages x Treatments	0.17	0.34					0.19	0.38	

Table 63.2 : Phytoavailable Copper (mg kg^{-1}) from Sewage Biosolids Applied Soils as Influenced by Ameliorants

Treatments	Red Soil					Black Soil				
	S_1	S_2	S_3	S_4	Mean	S_1	S_2	S_3	S_4	Mean
T_1	0.03	0.86	0.60	0.38	0.47	0.05	0.14	0.34	0.45	0.25
T_2	0.16	0.14	0.25	0.27	0.21	0.49	0.63	2.33	1.97	1.35
T_3	2.61	3.17	1.09	1.52	2.10	1.83	2.01	1.66	2.03	1.88
T_4	2.06	2.15	2.39	2.58	2.29	1.52	2.46	3.24	3.42	2.66
T_5	2.46	2.78	4.00	4.15	3.35	0.26	0.43	0.49	0.76	0.78
T_6	2.21	2.56	3.02	3.25	2.76	1.79	2.35	2.99	2.48	2.40
T_7	1.20	1.72	1.78	2.01	1.68	0.99	1.52	1.48	1.94	1.48
T_8	2.38	2.64	4.68	5.27	3.74	2.77	3.42	3.11	3.27	3.14
Mean	1.64	2.00	2.22	2.43	2.08	1.21	1.62	1.96	2.04	1.71

	SEd	CD(0.05)					SEd	CD(0.05)	
Stages	0.04	0.09					0.10	0.20	
Treatments	0.06	0.13					0.14	0.28	
Stages x Treatments	0.13	0.25					0.29	0.57	

Phytoavailable Chromium (Table 63.3)

The treatment, T_8 that received RP + CCP registered the highest phytoavailable Cr (1.189 mg kg^{-1}) in black soil but in red soil T_4 (Gypsum) recorded the highest (0.852 mg kg^{-1}). In both the soils the treatments T_4 and T_8 were statistically significant and had similar effect registering higher DTPA extractable Cr compared to the other treatments. Potash alum added treatment (T_3) in black soil recorded the least (0.860 mg kg⁻¹) but in the red soil, the composted coirpith registered the lowest DTPA extractable Cr of 0.703 mg kg⁻¹. Addition of ameliorants, composted coirpith and rock phosphate is the cause for the higher phytoavailable Cr in black soils (T_8) registering 38.4 per cent increase of Cr compared to control which could be attributed to the Cr content of rock phosphate. But addition of CCP has increased just 10.5 per cent of phytoavailable Cr than the treatment that received only rock phosphate with sewage biosolid (T_4). The trend was different in red soil, where rock phosphate alone observed 14.86 per cent increase over the control, but when composted coirpith was added with rock phosphate, the phytoavailable Cr was 6.67 per cent high over the control. Hence, the phytoavailability of Cr was reduced due to addition of CCP with rock phosphate. The reason might be that the phytoavailable Cr is bound within the organic matter complex and made unavailable to the soil. This was inline with Senesi *et al.* (1989) who reported that as the organic loading of the soil increased, the mental humic acid adsorption-desorption equilibria shift preferentially binding Cu, Ni, Zn and Cr which form stronger complexes with readily available sites on humic acid compared with more labile metals such as Mn, U, Ti and Mo which are desorbed and replaced.

Table 63.3 : Phytoavailable Chromium (mg kg⁻¹) from Sewage Biosolids Applied Soils as Influenced by Ameliorants

Treatments	Red Soil					Black Soil				
	S_1	S_2	S_3	S_4	Mean	S_1	S_2	S_3	S_4	Mean
T_1	0.49	0.51	0.54	0.52	0.52	0.47	0.52	0.61	0.57	0.54
T_2	0.53	0.55	0.94	0.91	0.74	0.59	0.74	0.99	1.13	0.86
T_3	0.39	0.47	1.03	1.15	0.76	0.59	0.71	0.99	1.15	0.86
T_4	0.37	0.47	1.22	1.34	0.85	0.82	1.28	1.29	1.24	1.16
T_5	0.41	0.39	0.93	1.08	0.70	0.95	1.16	1.06	1.12	1.07
T_6	0.52	0.61	0.94	1.05	0.78	0.62	0.71	1.13	1.32	0.95
T_7	0.49	0.57	0.96	1.11	0.79	0.64	0.73	1.15	1.22	0.94
T_8	0.53	0.61	0.99	1.05	0.79	0.61	1.14	1.47	1.54	1.19
Mean	0.47	0.52	0.95	1.03	0.74	0.66	0.87	1.09	1.16	0.95

	SEd	CD(0.05)				SEd	CD(0.05)
Stages	0.02	0.04				0.02	0.05
Treatments	0.03	0.06				0.03	0.07
Stages x Treatments	0.06	0.12				0.07	0.13

Phytoavailable Nickel (Table 63.4)

As the period of incubation advanced, the phytoavailable Ni showed an increasing trend in red soil. With reference to treatments, T_6 where rock phosphate was added, registered highest DTPA extractable Ni (0.641 mg kg⁻¹) followed by T_5 (composted coirpith) recording 0.597 mg kg⁻¹ of phytoavailable Ni in red soil. The treatments, T_7 and T_8, which received ameliorants in combination, recorded low DTPA extractable Ni in red soil.

In black soil, T_8 recorded maximum phytoavailable Ni (0.766 mg kg⁻¹) followed by T_7 (0.695 mg kg⁻¹) and T_4 (0.616 mg kg⁻¹) than control.

Red soil on analysis for phytoavailable Ni revealed that the ameliorants in combination increased the Ni content than when it is applied alone. The rock phosphate added treatment showed an increase of 14.3 per cent over control but with composted coirpith by 5.4 per cent increase over control was recorded. But combination of two ameliorants, either rock phosphate + CCP (T_7) or gypsum + CCP (T_8), decreased the phytoavailable Ni by 8.92 per cent and 14.28 per cent respectively.

In black soil, rock phosphate when applied alone recorded only 13.6% of Ni which was low compared to other treatments that is in agreement with Karapanagiotis *et al.* (1991) who reported that Ni was a weakly bounded metal in the sludge applied soil.

Table 63.4 : Phytoavailable Nickel (mg kg⁻¹) from Sewage Biosolids Applied Soils as Influenced by Ameliorants

Treatments	Red Soil					Black Soil				
	S_1	S_2	S_3	S_4	Mean	S_1	S_2	S_3	S_4	Mean
T_1	0.29	0.38	0.40	0.43	0.37	0.28	0.37	0.40	0.45	0.38
T_2	0.43	0.47	0.64	0.69	0.56	0.42	0.55	0.37	0.42	0.44
T_3	0.37	0.42	0.67	0.71	0.54	0.39	0.50	0.55	0.72	0.54
T_4	0.35	0.45	0.65	0.74	0.55	0.42	0.59	0.66	0.79	0.62
T_5	0.40	0.49	0.71	0.79	0.59	0.48	0.59	0.52	0.59	0.55
T_6	0.31	0.39	0.92	0.93	0.64	0.49	0.51	0.48	0.52	0.50
T_7	0.31	0.46	0.59	0.67	0.51	0.58	0.68	0.75	0.78	0.69
T_8	0.36	0.48	0.53	0.57	0.48	0.63	0.71	0.81	0.92	0.77
Mean	0.35	0.44	0.64	0.69	0.53	0.46	0.56	0.57	0.65	0.56

	SEd	CD(0.05)					SEd	CD(0.05)
Stages	0.01	0.02					0.02	0.03
Treatments	0.02	0.04					0.03	0.05
Stages x Treatments	0.04	0.07					0.05	0.10

Phytoavailble Manganese (Table 63.5)

In red soil, T_8 registered 15.97 mg kg⁻¹ of phytoavailable Mn followed by T_7 (14.76 mg kg⁻¹). In black soil, T_3 (potash alum) recorded the highest DTPA extractable Mn (13.76 mg kg⁻¹) followed by T_4 (12.46 mg kg⁻¹) wherein gypsum was added along with swage biosolid.

Table 63.5 : Phytoavailable Manganese (mg kg⁻¹) from Sewage Biosolids Applied Soils as Influenced by ameliorants

Treatments	Red Soil					Black Soil				
	S_1	S_2	S_3	S_4	Mean	S_1	S_2	S_3	S_4	Mean
T_1	4.91	5.62	6.39	5.99	5.73	4.97	5.09	6.28	6.59	5.73
T_2	8.52	11.12	13.42	14.24	11.82	10.41	12.19	9.58	8.73	10.23
T_3	10.31	13.64	13.81	14.71	13.12	11.35	13.41	14.85	15.51	13.76
T_4	11.64	12.92	13.62	15.05	13.30	7.53	9.50	16.45	16.36	12.46
T_5	13.02	14.08	13.94	14.14	13.79	6.67	10.17	9.99	11.34	9.54
T_6	9.35	11.32	12.16	14.84	11.92	8.51	13.03	10.49	12.61	11.16
T_7	10.67	12.58	17.59	14.22	14.77	10.75	15.03	8.54	9.62	10.98
T_8	11.42	13.63	18.99	19.82	15.97	8.75	11.79	7.44	8.59	9.14
Mean	9.98	11.86	13.74	14.63	12.55	8.62	11.28	10.45	11.15	10.38

	SEd	CD(0.05)					SEd	CD(0.05)
Stages	0.17	0.35					0.26	0.51
Treatments	0.25	0.49					0.36	0.72
Stages x Treatments	0.49	0.98					0.72	1.44

The treatment where composted coirpith was added recorded the least (9.54 mg kg^{-1}). Addition of gypsum and CCP (alone) reduced the phytoavailable Mn. Addition of rock phosphate and composted coirpith has increased the phytoavailable Mn to 35.11 per cent in red soil but it showed a decrease of 10.65 per cent in black soil.

Conslusions

The phytoavailable Pb in sewage biosolid amended soil with gypsum and composted coirpith decreased by 58.77 per cent in black soil proving to be an effective amendment combination for reducing the bioavailability of Pb. Phytoavailable Cu registered a decrease by 13.3 and 19.2 per cent in black and red soils respectively when amended with gypsum and composted coirpith. The bio and chemical ameliorants added did not reduce the bioavailable Cr with sewage biosolid at 50 t ha^{-1} dose application in both red and black soils. The bioavailable Ni in the red soil recorded a decrease of 8.92 and 14.28 per cent upon the addition of rock phosphate composted coirpith and gypsum + composted coirpith respectively. It could be concluded that addition of gypsum + CCP and rock phosphate + CCP to the sewage biosolid incorporated soils could reduce the bioavailable heavy metals by the precipitation and complexation reactions and make them unavailable to the crops.

Acknowledgement

The financial assistance rendered by the Jawahar Lal Nehru Memorial Fund through student fellowship is greatly acknowledged.

References

Alloway, B.J. and Jackson, A.P. 1991. The behavior of heavy metals in sewage sludge amended soils. Sci. Total Environ. 100 : 151-176.

Berrow, M., Morrison, A.R., Park, J.S. and Sharp, B.L. 1990. The long-term partitioning of copper in water extracted from polluted soils using high performance/size reduction exclusion liquid chromatography. In : Proceedings Fourth International Conference–Environmental Contamination, Barcelona. CEP Consultants Ltd, Edinburgh, pp. 85-87.

Davis, R.D. 1984. Crop uptake of metals (cadmium, lead, mercury, copper, nickel. zinc and chromium) from sludge-treated soil and its implications for soil fertility and for the human diet. In : Processing and Use of Sewage Sludge L' Hermite P. and H. Ott (eds)., D. Reidel Publishing Company, Dordrecht, pp. 349-357.

Karapanagiotis, N.K., Stcrritt, R.M. and Lester, J.N. 1991. Heavy metal complexation in sludge-amended soil : The role of organic matter in metal retention. Environmental Technol., 12 : 1107-1116.

McBride, M.B. 1995. Toxic metal accumulation from agricultural use of sludge : Are USEPA regulations protective? J. Environ. Qual., 24 : 5-18.

Senesi, N., Sposito, G. Holtzclaw, K.M. and Bradford, G.R. 1989. Chemical properties of metal-humic, acid fractions of sewage sludge-amended aridisol. J.Environ. Qual., 18 : 186-194.

ENVIRONMENTAL BIOTECHNOLOGY

Edited By :   **Professor (Dr.) Arvind Kumar**

Published By : **DAYA PUBLISHING HOUSE**

64

ROLE OF EPIDEMIOLOGICAL PARAMETERS ON DEVELOMPMENT OF COLLAR ROT OF APPLE

• *Bhupesh Gupta, L.N. Bhardwaj, S.V. Bhardwaj* and Anil Handa*
Department of Mycology and Plant Pathology,
Dr. Y.S. Parmar University of Horticulture and Forestry, Nauni, Solan (H.P.)
** Department of Biotechnology,*
Dr. Y.S. Parmar University of Horticulture and Forestry, Nauni, Solan (H.P.)

Abstract

Experiments on epidemiological parameters of collar rot were conducted under pot culture conditions. A combination of 25°C temperature, 90 per cent of moisture, 5.0 soil pH and clay soil was found to be most conducive for the development and spread of the disease.

Key Words : Apple, collar rot, epidemiological parameters.

Introduction

Apple (*Malus domestica* Borkh.) is grown throughout the world wherever the agro-climatic conditions are suitable for its cultivation. Apple ranks first among fruit crops of Himachal Pradesh both in terms of quality and quantity. In order to increase apple production, many new cultivars of apple have been introduced for commercial cultivation in the recent past as a result of which incidence of diseases has also been increased considerably. The increased incidence of diseases has led to low productivity of apple in the state. In this context, soil-borne diseases of apple especially white root-rot (*Dematophora necatrix*) and collar rot (*Phytophthora cactorum*) has contributed significantly because, there is a direct loss to the plant due to its death as a result of pathogen attack. Collar rot caused by *P. cactorum* (Leb. and Cohn) Schroet is the second most important soil-borne disease of apple after root rot in Himachal Pradesh, and is globally prevalent wherever apple is grown.

Materials and Methods

In order to study the effect of temperature on disease development, excised twig method (Borecki and Millikan, 1969) was used. The petriplates containing excised twigs inoculated with *Phytophthora cactorum* were incubated in BOD incubator at 5, 10, 15, 20, 25 and 30°C. Each treatment was replicated thrice and observations regarding incubation period and lesion colour were recorded periodically. Effect of different levels of soil moisture (50, 60, 70, 80 and 90%) on development and spread of collar rot of apple was studied by using soil moisture meter. The effect of different levels of soil pH (5.0, 5.5, 6.0, 6.5 and 7.0) on the development of collar rot of apple was also studied. The pH of soil was adjusted by adding 2.82, 5.64, 8.64 and 11.28 g of lime to 15 kg clay soil separately, having pH 5.0 to get the desired pH *i.e.* 5.5, 6.0, 6.5 and 7.0, respectively. The effect of different soil types *i.e.* sandy loam, clay (heavy), forest loam and sandy loam (gravelly) on disease development was studied. The soils of different types were sterilized in autoclave at 15 p.s.i. for an hour for two consecutive days. The sterilized soil was filled in the pots and three grafted plants were planted in each pot. One year old grafted plants of apple were planted in earthen pots (6″ diameter) with graft union of the potted plants placed at or near soil line. Three seedlings replicated thrice were kept for each treatment. Clay (heavy) soil with pH nearing 5.0, sterilized at 20 lbs p.s.i. for one hour for two consecutive days was used to conduct the experiment. After transplanting, the plants were kept in polyhouse where the temperature varied between 20-25°C during the period of experimentation. The standard cultural practices were used for raising plant. The plants were inoculated in the first week of July with the active culture of *P. cactorum* grown on maize-sand medium. Observations on disease incidence and disease severity were recorded 90 days after pathogen inoculation.

The severity was recorded by using the following scale :

Grade	Per cent area of collar region showing infection
0	Apparently free from infection
1	1-10
2	10-20
3	20-50
4	50-75
5	>75

The Plant Disease Index was calculated according to McKinney (1923)

$$\text{Plant Disease Index} = \frac{\text{Sum of all ratings}}{\text{Total no. of ratings} \times \text{maximum disease grade}} \times 100$$

Results and Discussion

It is clear from data (Table 64.1) that the temperature has a definite role to play in disease development as the symptoms of collar rot appeared on the susceptible root stock *M. prunifolia* Shaishie at all the temperatures, however, the incubation period and lesion colour varied with the temperature. A temperature of 25°C was found optimum for the development of collar rot as the pathogen produced the disease symptoms within 3 days after inculation, while at 5°C, the disease appeared after 19 days. However, at 20°C, the disease appeared on 6th day after pathogen inoculation, whereas at 30°C, it appeared 12 days after pathogen inoculation and the colour of lesion was light brown. It can be concluded from aforesaid experiment that 25°C is the optimum temperature for the development and spread of the collar rot of apple. These findings are in line with Buddenhagen (1955), Braun and Krober (1958), Andreeva, (1977), Gupta and Singh (1979) and Rana and Gupta (1979).

A perusal of the data presented in Table 64.2 indicates that with increase in soil moisture levels from 50 to 90 per cent both incidence as well as severity of collar rot also increases from 22.22 to 100.00 per cent and 6.67 to 97.78 per cent, respectively. However, maximum incidence (100.00%) and severity (97.78%) was observed at 90 per cent soil moisture level while, minimum incidence (22.22%) and severity (6.67%) was recorded at 50 per cent moisutre level. The aforesaid study concludes that 50 per cent soil moisture is requried for disease initiation and 90 per cent for the development and spread of collar rot. Welsh (1942) also reported increased susceptibility of apple bark to *P. cactorum* with increasing soil moisture from 61 to 96 per cent under pot culture. Mc Intosh (1963) also observed high incidence of collar rot of apple in irrigated orchards. Browne and Mircetich (1988) also reported high incidence of root and bark rot (*P. cactorum, P. cambivora* and *P. cryptogea*) in apple seedlings due to waterlogging and flooding of soils.

Table 64.1 : Effect of different temperature regimes on the development of collar rot of apple

Temperature (°C)	Incubation period (days)	Lesion colour
5	19	Light brown
10	17	Light to slight dark brown
15	12	Dark brown
20	6	Dark brown
25	3	Dark brown
30	12	Light brown

Table 64.2 : Effect of soil moisture levels on the incidence and severity of collar rot of apple

Soil moisture (%)	Disease incidence (%)	Disease severity (%)
50	22.22 (23.51)	6.67 (12.13)
60	44.44 (41.69)	20.00 (26.36)
70	77.78 (66.69)	42.22 (40.48)
80	88.89 (78.25)	86.67 (69.02)
90	100.00 (90.00)	97.78 (85.01)
$CD_{0.05}$	30.11	13.44

Figures in parentheses are arcsin transformed values

It is evident from the data (Table 64.3) that lower soil pH range *i.e.* between 5.0 to 5.5 was found more favourable for the development and spread of collar rot pathogen, while higher pH ranges *i.e.*> 6.0 were unfavourable for its development. However, maximum incidence of collar rot (100.00%) was observed at pH 5.0 followed by pH 5.5 (88.89%), though both the treatments were statistically at par. Least disease incidence (9.33%) was recorded at pH 7.0. The data with regard to disease severity also revealed that pH of 5.0 was optimum for the development and spread of collar rot. This was followed by pH 5.5 (66.67%). Both these treatments were found statistically at par with each other. However, minimum disease severity was recorded at pH 7.0 (2.22%). The study concludes that a soil pH of 5.0 to 5.5 is conducive for the development and spread of collar rot in apple. Jain (1961) in his survey report also recorded the prevalence of collar rot in apple orchards planted on newly cleared soils with pH ranging from 5.4 to 7.0. Lower soil pH ranging from 4.5 to 6.5 were also found favourable for the development and spread of red stele of strawberry caused by *P. fragariae* (Hickman and English, 1951, Montgomerie and Kennedy, 1973, Maas, 1976, Bolay, 1982, Gupta, 2001).

A perusal of data presented in Table 64.4 shows that clay soil had high incidence of disease (88.89%) followed by forest loam soil (33.33%) and sandy loam soil (6.67%). While least incidence of collar rot (5.86%) was observed in sandy loam (gravelly). The data with regard to disease severity also indicated high mortality of plants in clay soil (66.67%) and least (2.22%) in sandy loam (gravelly). Statistically, no significant differences were observed in sandy loam and forest loam as the disease severity was 4.44 and 6.67 per cent, respectively 90 days after pathogen inoculation. From the aforesaid findings, it is evident that clay soils are highly conducive for the development and spread of collar rot in apple. The results of present investigations are in conformity with Smith (1950), Woodhead (1957), McIntosh (1963), Braun and Wilcke (1963) and Utkhede (1999).

Table 64.3 : Effect of soil pH regimes on the incidence and severity of collar rot of apple

Soil PH	Disease incidence (%)	Disease severity (%)
5.0	100.00 (90.00)	80.00 (63.64)
5.5	88.89 (78.25)	66.67 (55.15)
6.0	33.33 (30.00)	15.55 (18.89)
6.5	11.11 (11.75)	4.44 (7.14)
7.0	9.33 (10.67)	2.22 (4.98)
$CD_{0.05}$	32.53	18.69

Figures in parentheses are arcsin transformed values

Table 64.4 : Effect of soil types on the incidence and severity of collar rot of apple

Soil type	Disease incidence (%)	Disease severity (%)
Sandy loam	6.67 (8.86)	4.44 (7.14)
*Clay (heavy soil)	88.89 (78.25)	66.67 (46.67)
Forest loam	33.33 (30.00)	15.56 (18.40)
Sandy loam (gravelly)	5.33 (7.86)	2.22 (4.98)
$CD_{0.05}$	32.53	18.05

Figures in parentheses are arcsin transformed values
*Clay = Containing 16 to 20 % clay content

References

Andreeva, N.F. 1977. Phytophthorosis of strawberry. Zashchita Rastenii 4:45.

Bolay, A. 1982. Red root disease of strawberry in Switzerland. Revue Suisse de Viticulture d' arboriculture et d' Horticulture 14 : 228-231.

Borecki, Z. and Millikan, D.F. 1969. A rapid method for determining pathogenicity of *Phytophthora cactorum*. Phytopathology 59 : 247-248.

Braun, H. and Krober, H. 1958. Studies of collar rot of apple caused by *Phytophthora cactorum* (Leb. and Cohn) Schroet. Phytopathol. Z. 35 : 35-94).

Braun, H. and Wilcke, D.E. 1963. Untersuchungen uber Bodenverdichtungen und ihre Beziehungen zum Auftreten der Kragenfaule (*Phtophthora cactorum*). Phytophathol. Z. 46 : 71-86.

Browne, G.T. and Mircetich, S.M. 1988. Effects of flood duration on the development of Phytophthora root and crown rots of apple. Phytopathology 78 : 846-851.

Buddenhagen, I.W. 1955. Various aspects of *Phytophthora cactorum* collar rot of apple trees in Netherlands. T.Pl. Ziekten. 61 : 122-129.

Gupta, M. 2001. Studies on Phytophthora diseases of strawberry. Ph.D. Thesis, Dr. Y.S. Parmar UHF, Nauni, Solan (H.P.) 83 p.

Gupta, V.K. and Singh, K. 1979. Factors affecting the development of collar rot (*Phytophthora cactorum*) of apple. Gartenbauwissens Chaft 44 : 29-32.

Hickman, C.J. and English, M.P. 1951. Factors influencing the development of red core in strawberries. Trans. Br. Mycol. Soc. 34 : 223-236.

Jain, S.S. 1961. Root and collar rot disease in Himachal Pradesh. Himachal Hort. 2 : 19-23.

Maas, J.L. 1976. Red stele disease of strawberry : soil, pH, pathogen and cultivar interactions. Hort. Science 11 : 258-260.

McIntosh, D.L. 1963. The collar rot problem in our orchards. Proc. Washington State Hort. Assoc. 59 : 131-136.

McKinney, H.H. 1923. Influence of soil temperature and moisture on infection of wheat seedling by *Helminthosporium sativum*. J. Agri. Res. 26 : 215-217.

Montgomerie, I.G. and Kennedy, D.M. 1973. Scot. Hort. Res. Instt. 19th Annual Report for the year 1972 : 87p.

Rana, K.S. and Gupta, V.K. 1979. Collar rot of apple, its development and control. Indian Phytopath. 32 : 162 (Abstr).

Smith, H.C. 1950. Crown rot of apple and gooseberrty. Orchardist N.Z. 23 : 13-14.

Utkhede, R.S. 1999. Influence of drip, microjet and sprinkler irrgation systems on the severity of corwn and root rot of M 26 apple rootstock trees in clay soil. Australian Plant Pathology 28 : 254-259.

Welsh, M.F. 1942. Studies on crown rot of apple trees. Can. J. Res. 20 : 457-490.

Woodhead, G.E. 1957. Collar rot and root rot of Cox's Orange Pippin and other apple varieties. Orchardist N.Z. 30 : 16-17.